Waves, Particles and Fields

Introducing Quantum Field Theory

Waves, Particles and Fields

Introducing Quantum Field Theory

A. C. Fischer-Cripps

CRC Press is an imprint of the
Taylor & Francis Group, an informa business

CRC Press
Taylor & Francis Group
6000 Broken Sound Parkway NW, Suite 300
Boca Raton, FL 33487-2742

Printed on acid-free paper

International Standard Book Number-13: 978-0-367-19876-3 (Paperback)
International Standard Book Number-13: 978-0-367-19878-7 (Hardback)

Visit the Taylor & Francis Web site at
http://www.taylorandfrancis.com

and the CRC Press Web site at
http://www.crcpress.com

Contents

Acknowledgements

I am by no means an expert in particle physics or quantum field theory, but wanted to know more than what I learned as an undergraduate in applied physics. There are various sources of information and learning about quantum field theory and Feynman diagrams on the internet, some kindly made available for public viewing by authors who have evidently put a great deal of time into their preparation. To this end, I am grateful in particular to the lecture notes of David Tong and Mark Thomson and to a very clear series by Freeman Dyson dated 1951. I must pay particular thanks to two sources of video lectures, one hosted by PBS Space Time on YouTube – a very nice introductory level series – and an outstanding series by Professor Leonard Susskind of Stanford University hosted at https://theoreticalminimum.com/. Professor Susskind is a man of infinite patience, and his lectures are essential viewing for the first-time student of particle physics. How I longed to be in the audience and ask a question or two. My greatest admiration is for Richard Feynman's book, *QED, the Strange Theory of Light and Matter*, an incredible work that brings quantum electrodynamics within reach of the layperson in what must rate as a first-class demonstration of excellence of scientific communication let alone advanced theoretical physics.

Thank you to Rebecca Davies and Kirsten Barr at CRC Press, and also Conor Fagan at Deanta Global Publishing, for their excellent support.

–Tony Fischer-Cripps

1 Mathematics

1.1 INTRODUCTION

Physics is largely written in the language of mathematics, which generally makes it inaccessible to those who don't have the opportunity to learn the language. In this book, it is assumed that a middle-level high-school level of mathematics is known to the reader. You should be familiar with the basic methods of differentiation and integration. In Part 1 of this book, we venture beyond that into what might be called first-year college-level mathematics. In Part 2, we go a little further. As we go on, the apparent complexity arises more from the numerous conventions of notation rather than new mathematical theorems. Familiarity with the notation will make things easier to understand.

But, in order to begin, we need to know the language. The aim is to go deep enough to be intellectually satisfying, but not so deep as to be impenetrable.

1.2 COMPLEX NUMBERS

1.2.1 Complex Numbers

Real numbers are the set of all the numbers, positive and negative, integers and fractions, from negative infinity to positive infinity, including zero, including rational and irrational numbers. These are the numbers we are usually familiar with.

Imaginary numbers are a duplicate set of all these numbers but are kept separated from the set of real numbers. There is nothing "imaginary" about imaginary numbers, this is just a label that distinguishes them from the set of "ordinary" real numbers.

Complex numbers consist of a vector combination of real and imaginary numbers. A complex number can be thought of as a two-dimensional number which contains a real component (say along the x axis), and an imaginary component (say along the y axis). This is shown in Figure 1.1.

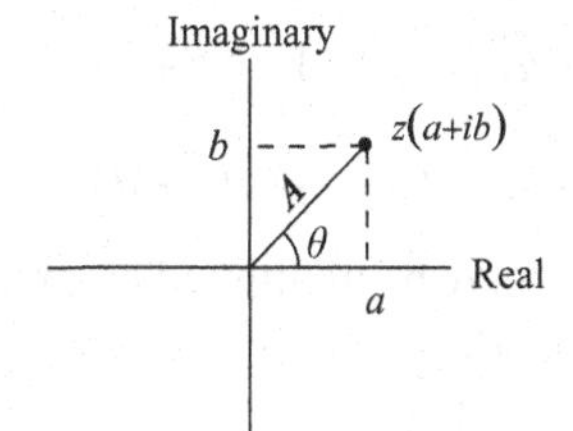

FIGURE 1.1 Real and imaginary coordinate axes and a complex number z.

A complex number has the form:

$$z = a + ib \tag{1.1}$$

where a is the real part and b is the imaginary part of the number. The symbol i is used to identify the imaginary part of a complex number.

A complex number is represented by a single point $z(a, b)$ on the complex plane. The magnitude, or amplitude, A, of a complex number is given by:

$$|z| = \sqrt{a^2 + b^2} \tag{1.2}$$

and is the distance from the origin to the point z. The symbol i has mathematical significance. It is equal to the square root of –1. Of course, the square root of a negative number is not defined, and it is in this sense that we use the term "imaginary" because we can imagine that if there *were* such a number, then this number, when squared, would equal –1. Most often, it is the quantity $i^2 = -1$ that is used in our calculations rather than i itself.

If $a + ib$ represents a complex number, then $a - ib$ is called the complex conjugate of the number. Therefore, it can be seen that:

$$\begin{aligned}(a+ib)(a-ib) &= a^2 - iab + iab + b^2 \\ &= a^2 + b^2 \\ &= |z|^2\end{aligned} \tag{1.3}$$

That is, the product of a complex number and its conjugate is the square of the magnitude of the complex number.

It is readily seen that a complex number can be expressed in trigonometric terms:

$$\begin{aligned} a &= |z|\cos\theta \\ b &= |z|\sin\theta \\ z &= (|z|\cos\theta) + i(|z|\sin\theta) \\ &= |z|(\cos\theta + i\sin\theta)\end{aligned}$$

The symbol i is present to remind us that we cannot just add the quantity $\cos\theta$ to $\sin\theta$ directly. The two components of a complex number must be kept separated.

Complex numbers may be added and subtracted by adding and subtracting the real and imaginary parts separately:

$$\begin{aligned}(a+ib)+(c+id) &= (a+c)+i(b+d) \\ (a+ib)-(c+id) &= (a-c)+i(b-d)\end{aligned} \tag{1.4}$$

Complex numbers may be multiplied by multiplying the terms as in a normal multiplication of factors as shown (where $i^2 = -1$):

$$\begin{aligned}(a+ib)(c+id) &= ac + iad + ibc + i^2bd \\ &= (ac - bd) + i(ad + bc)\end{aligned} \tag{1.5}$$

or, in trigonometric terms:

$$
\begin{aligned}
z_1 &= |z_1|(\cos\theta_1 + i\sin\theta_1) \\
z_2 &= |z_2|(\cos\theta_2 + i\sin\theta_2) \\
z_1 z_2 &= |z_1||z_2|(\cos\theta_1 + i\sin\theta_1)(\cos\theta_2 + i\sin\theta_2) \\
&= |z_1||z_2|\left(\cos\theta_1\cos\theta_2 - \sin\theta_1\sin\theta_2 + i(\cos\theta_1\sin\theta_2 + \sin\theta_1\cos\theta_2)\right) \\
z_1 z_2 &= |z_1||z_2|\left(\cos(\theta_1+\theta_2) + i\sin(\theta_1+\theta_2)\right)
\end{aligned}
\tag{1.6}
$$

by making use of:

$$
\begin{aligned}
\sin(a \pm b) &= \sin a\cos b \pm \cos a\sin b \\
\cos(a \pm b) &= \cos a\cos b \mp \sin a\sin b
\end{aligned}
\tag{1.7}
$$

Complex numbers may be divided by making use of the complex conjugate to transfer the imaginary part from the denominator to the numerator. This makes it easier to deal with the complex number, especially when it is desired to perform algebraic operations on them:

$$
\begin{aligned}
\frac{(a+ib)}{(c+id)} &= \frac{(a+ib)}{(c+id)}\frac{(c-id)}{(c-id)} \\
&= \frac{(a+ib)(c-id)}{(c^2+d^2)}
\end{aligned}
\tag{1.8}
$$

or, in trigonometric terms:

$$
\frac{z_1}{z_2} = \frac{|z_1|}{|z_2|}\left(\cos(\theta_1-\theta_2) + i\sin(\theta_1-\theta_2)\right) \tag{1.9}
$$

Complex numbers may be raised to a power. In trigonometric terms, this is known as de Moivre's theorem:

$$
(\cos\theta + i\sin\theta)^n = \cos n\theta + i\sin n\theta \tag{1.10}
$$

Euler's formula expresses a complex number in terms of an exponential function:

$$
e^{(a+ib)x} = e^{ax}(\cos bx + i\sin bx) \tag{1.11}
$$

The addition of two particular exponential functions can be written:

$$
C_1 e^{(a+ib)x} + C_2 e^{(a-ib)x} = e^{ax}(A\cos bx + iB\sin bx) \tag{1.12}
$$

where:

$$
\begin{aligned}
A &= C_1 + C_2 \\
B &= C_1 - C_2
\end{aligned}
\tag{1.13}
$$

which we shall see has an application in quantum mechanics.

1.2.2 Complex Quantities

A common example of a complex number is the resistance and reactance in a series AC circuit consisting of an inductor and a resistor.

The resistance of the circuit R is given by Ohm's law (where we ignore the resistance of the inductor):

$$R = \frac{V_R}{I} \tag{1.14}$$

The inductive reactance of the capacitor, X_L, is represented by:

$$X_L = \frac{V_L}{I} = \omega L \tag{1.15}$$

where V and I are the peak values of voltage and current respectively. The inductive reactance is a measure of the inductor's resistance to AC, where it can be seen that, unlike the case of a resistor, the reactance increases as the frequency of the AC increases.

Even though X_L and R have the same units (Ohms), we cannot just add them together directly because the current flowing through the inductor is out of phase with the voltage across it by 90°. The maximum current in the inductor occurs when the rate of change of voltage across it is a maximum – and this occurs when the voltage across the inductor passes through zero. On the other hand, the voltage across the resistor is in phase with the current flowing through it. In a series circuit, the current has to be the same in all components in the circuit, and so in this case, the voltage across the resistor is not in phase with the voltage across the inductor.

As far as the resistance to current flow is concerned, the total impedance Z of the circuit, as seen by the AC voltage generator, is therefore a vector combination of the inductive reactance Z and the resistance R.

That is, Z is a complex number with the real part being R and the imaginary part being X_L. This can be seen graphically in Figure 1.2.

$$Z = R + iX_C \tag{1.16}$$

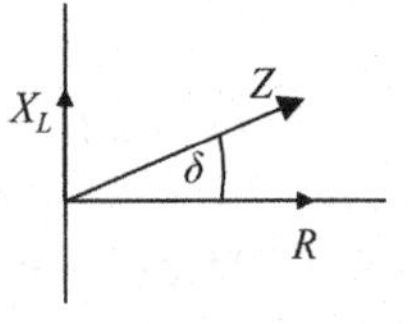

FIGURE 1.2 Complex impedance Z comprising inductive reactance X_L and resistance R.

The magnitude of the impedance Z is found from:

$$|Z| = \sqrt{R^2 + X_L{}^2} \tag{1.17}$$

The angle between the Z vector and the R axis is called the phase angle and is found from:

$$\tan \varphi = \left[\frac{X_L}{R}\right] = \frac{\omega L}{R} \tag{1.18}$$

1.2.3 Complex Functions

A function is a formula or an equation in which one number, y, depends in a particular way upon the value of the other number, x. When y is a function of x, we say that y is the dependent variable and x is the independent variable. The value of y at a particular value of x is written:

$$y = f(x) \tag{1.19}$$

The symbol f can be thought of as an operator which acts upon x to give y.

Examples of functions are:

$$\begin{aligned} y &= 5 \\ y &= mx + b \\ y &= ax^2 + bx + c \end{aligned} \tag{1.20}$$

The last example above is a quadratic function. When y is plotted as a function of x, the shape of the graph is a parabola. A quadratic equation arises when we wish to find where the function crosses the x axis, that is, values of x when $y = 0$. These values of x are called the roots of the equation.

$$ax^2 + bx + c = 0 \tag{1.21}$$

A quadratic equation can have no roots (the graph of the function lies completely above or below the x axis), one root (touches the x axis) or two roots but no more than two, since it can cross the x axis ($y = 0$) no more than twice.

The roots of a quadratic equation can be obtained from the quadratic formula:

$$x = \frac{-b \pm \sqrt{b^2 - 4ac}}{2a} \tag{1.22}$$

The quantity $b^2 - 4ac$ has special significance and is called the discriminant. When the discriminant > 0, the function crosses x axis at two points, thus, two roots. When the discriminant $= 0$, the function touches the x axis at a single point. When the discriminant < 0 the function lies completely above or below the x axis.

For the case of the discriminant being less than zero, the quadratic equation has no roots. However, this does not mean to say that the equation cannot be solved. The function may have complex roots.

For example, consider the function:

$$f(z) = z^2 - 2z + 2 \tag{1.23}$$

and its associated quadratic equation:

$$z^2 - 2z + 2 = 0 \tag{1.24}$$

This function is concave upwards and lies completely above the z axis. It does not cross the vertical, or $f(z)$, axis for any real values of z.

However, using the quadratic formula, the roots of the equation can be expressed in terms of the imaginary number i:

$$\begin{aligned} z &= \frac{2 \pm \sqrt{4-4(2)}}{2} \\ &= \frac{2 \pm 2\sqrt{-1}}{2} \\ &= 1 \pm i \end{aligned} \tag{1.25}$$

What this is telling us is that the original function $f(z)$ might represent some quantity whose value depends on z, but can only have a zero value when z is expressed as a complex number at $z = 1 + i$ and also at $z = 1 - i$. The magnitude of z for these zero values of $f(z)$ in this example would be:

$$|z| = \sqrt{1^2 + 1^2} = \sqrt{2} \tag{1.26}$$

We might refer to $f(z)$ as a complex function.

1.3 SCALARS AND VECTORS

1.3.1 Scalars

In many situations, a physical quantity has a direction associated with it. That is, the physical quantity is described not only by its value but also by what direction is associated with that value. Say, for example, the concept of velocity. This quantity describes the rate of change of position with respect to time in a certain direction. We are familiar with the concept of velocity being associated with a vector. A vector is an example of a general class of mathematical objects called a tensor, in this case, a tensor of rank 1. The rank indicates the number of directions associated with that quantity. On the other hand, a quantity such as the temperature at a point in a room has no direction associated with it. It is a tensor of rank 0 and is called a scalar. In the case of stresses acting throughout a solid, there is the possibility of the material supporting a stress which is the force acting over an area parallel to a surface (i.e. shear stress), as well as the force acting perpendicular to the surface (i.e. tension or compression). There are two directions involved, the direction of the force and the orientation over which the force acts. Stress is a tensor of the second rank.

Scalars are described by a single quantity called the magnitude. For example, the term speed is a scalar and is the magnitude of the velocity vector. Other familiar scalars are energy, mass and time.

1.3.2 Vectors

A physical quantity which depends on a direction is called a vector. Familiar examples are velocity, force and momentum. A vector consists of two parts, a magnitude and a direction.

As shown in Figure 1.3(a), if a vector **a** extends from point P_0 to point P_1, then we write this as: $\overrightarrow{P_0P_1}$ or simply **a**.

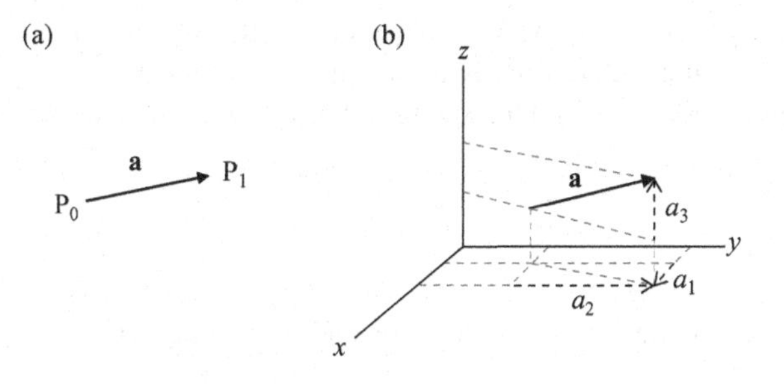

FIGURE 1.3 (a) Vector **a** joining points P_0 to P_1. (b) Components of the vector a_1, a_2, a_3.

The magnitude of the vector is written: $\left|\overrightarrow{P_0P_1}\right|$ or $|\mathbf{a}|$.
The components of the vector with respect to the x, y, z coordinate axes are written:

$$\begin{aligned}\overrightarrow{P_0P_1} &= \mathbf{a} = \langle a_1, a_2, a_3 \rangle \\ &= \langle x_1 - x_o, y_1 - y_o, z_1 - z_o \rangle\end{aligned} \tag{1.27}$$

These are shown in Figure 1.3(b). The magnitude of a vector is in effect the distance formula:

$$|\mathbf{a}| = \sqrt{a_1^2 + a_2^2 + a_3^2}$$

The special vectors:

$$\begin{aligned}\mathbf{i} &= \langle 1,0,0 \rangle \\ \mathbf{j} &= \langle 0,1,0 \rangle \\ \mathbf{k} &= \langle 0,0,1 \rangle\end{aligned} \tag{1.28}$$

are unit vectors in the direction of the x, y and z coordinate axes.
Thus

$$\mathbf{a} = a_1\mathbf{i} + a_2\mathbf{j} + a_3\mathbf{k} \tag{1.29}$$

This expression is a vector sum, where it is understood that the resultant is a vector formed by the addition of the components separately. So, if we have two vectors to be added, then:

$$\begin{aligned}\mathbf{a} &= a_1\mathbf{i} + a_2\mathbf{j} + a_3\mathbf{k} \\ \mathbf{b} &= b_1\mathbf{i} + b_2\mathbf{j} + b_3\mathbf{k} \\ \mathbf{a} + \mathbf{b} &= (a_1 + b_1)\mathbf{i} + (a_2 + b_2)\mathbf{j} + (a_3 + b_3)\mathbf{k}\end{aligned} \tag{1.30}$$

Much the same thing is done for the subtraction of one vector from another.

We might ask about the product of two vectors. To answer this question, let's go back a moment to two dimensions. Does it make sense to write:

$$\mathbf{ab} = (a_1\mathbf{i} + a_2\mathbf{j})(b_1\mathbf{i} + b_2\mathbf{j}) \tag{1.31}$$

The answer is no, this is not defined. However, we shall see that when two vectors are to be multiplied, there are two possible ways to perform the calculation.

What about a vector squared? Surely, we are used to something like the square of the velocity; it is used all the time – for example, the kinetic energy is ½mv^2. A commonly used equation of motion is $v^2 = u^2 + 2ax$. But what do we actually mean by the square of a vector? The square of the vector is the magnitude of the vector all squared, and is a scalar.

$$|\mathbf{a}|^2 = a^2 = a_1^2 + a_2^2 \tag{1.32}$$

Much like a complex number, the magnitude is given by the complex conjugate, and so we may write:

$$\begin{aligned} |\mathbf{a}|^2 &= (a_1\mathbf{i} + a_2\mathbf{j})(a_1\mathbf{i} - a_2\mathbf{j}) \\ &= a_1^2 + -a_1a_2 + a_2a_1 + a_2^2 \\ &= a_1^2 + a_2^2 \end{aligned} \tag{1.33}$$

Or, in the three-dimensional case:

$$|\mathbf{a}|^2 = a^2 = a_1^2 + a_2^2 + a_3^2 \tag{1.34}$$

We shall see that this special kind of multiplication is called the dot product of two vectors.

A unit vector **u** in the direction of **a** is given by:

$$\mathbf{u} = \frac{\mathbf{a}}{|\mathbf{a}|} = \left\langle \frac{a_1}{|\mathbf{a}|}, \frac{a_2}{|\mathbf{a}|}, \frac{a_3}{|\mathbf{a}|} \right\rangle \tag{1.35}$$

As seen in Figure 1.4, a vector drawn from the origin to a point $P(x, y, z)$ is called the position vector **r** of that point.

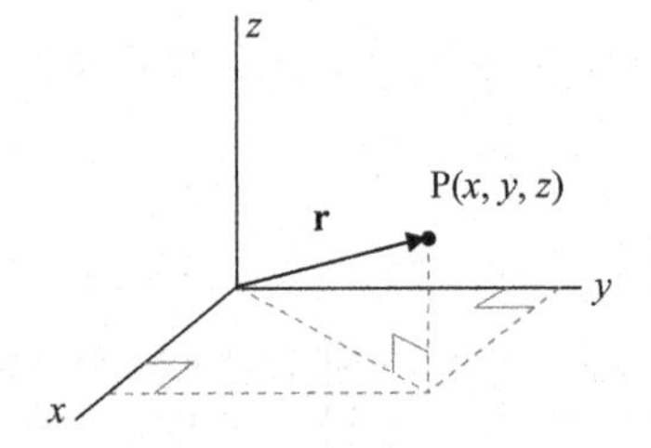

FIGURE 1.4 Position vector **r** on the coordinate axes.

Thus, the vector **r** and its magnitude are:

$$\begin{aligned} \mathbf{r} &= x\mathbf{i} + y\mathbf{j} + z\mathbf{k} \\ |\mathbf{r}| &= \sqrt{x^2 + y^2 + z^2} \end{aligned} \tag{1.36}$$

A vector function or vector field describes a vector in terms of components whose magnitudes are functions of x, y and z. That is, for each point P(x, y, z) there is a corresponding vector **F**:

$$\begin{aligned} \mathbf{F} &= f(x,y,z)\mathbf{i} + g(x,y,z)\mathbf{j} + h(x,y,z)\mathbf{k} \\ &= F_1\mathbf{i} + F_2\mathbf{j} + F_3\mathbf{k} \end{aligned} \tag{1.37}$$

1.3.3 Dot and Cross Product

As mentioned previously, there are two ways to multiply vectors.

If two vectors are:

$$\mathbf{a} = \langle a_1, a_2, a_3 \rangle$$
$$\mathbf{b} = \langle b_1, b_2, b_3 \rangle \tag{1.38}$$

then the dot product of them is defined as:

$$\mathbf{a} \cdot \mathbf{b} = a_1 b_1 + a_2 b_2 + a_3 b_3$$
$$= |\mathbf{a}||\mathbf{b}| \cos\theta \tag{1.39}$$

But the quantity $|\mathbf{a}|\cos\theta$ is the component of **a** in the direction of **b**. This is shown in Figure 1.5. Thus, the dot product is a scalar and represents the magnitude of the component of **a** in the direction of **b** multiplied by the magnitude of **b** or the magnitude of the component of **b** in the direction of **a** multiplied by the magnitude of **a**.

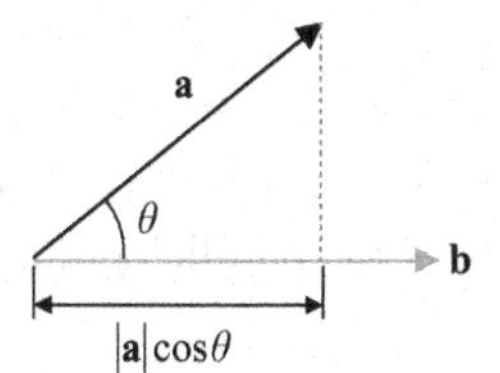

FIGURE 1.5 Dot product of two vectors.

To find the component of **a** in **b**'s direction, we form the dot product of **a** with a unit vector in the direction of **b**.

$$|\mathbf{a}|\cos\theta = \mathbf{a} \cdot \frac{\mathbf{b}}{|\mathbf{b}|} = \mathbf{a} \cdot \mathbf{u} \tag{1.40}$$

The magnitude of a vector is given by the square root of the dot product of it and itself:

$$|\mathbf{a}| = \sqrt{\mathbf{a} \cdot \mathbf{a}} = \sqrt{a_1^2 + a_2^2 + a_3^2} \tag{1.41}$$

In matrix form, the dot product of two vectors can be expressed:

$$\begin{bmatrix} a_1 & a_2 & a_3 \end{bmatrix} \begin{bmatrix} b_1 \\ b_2 \\ b_3 \end{bmatrix} = \begin{bmatrix} a_1 b_1 + a_2 b_2 + a_3 b_3 \end{bmatrix} \tag{1.42}$$

The result is a single element matrix, a scalar.

Consider now two vectors **a** and **b** that lie in a plane as shown in Figure 1.6(a).

The vector cross product of the two vectors is defined as:

$$\mathbf{a} \times \mathbf{b} = (a_2 b_3 - a_3 b_2)\mathbf{i} - (a_1 b_3 - a_3 b_1)\mathbf{j} + (a_1 b_2 - a_2 b_1)\mathbf{k} \tag{1.43}$$

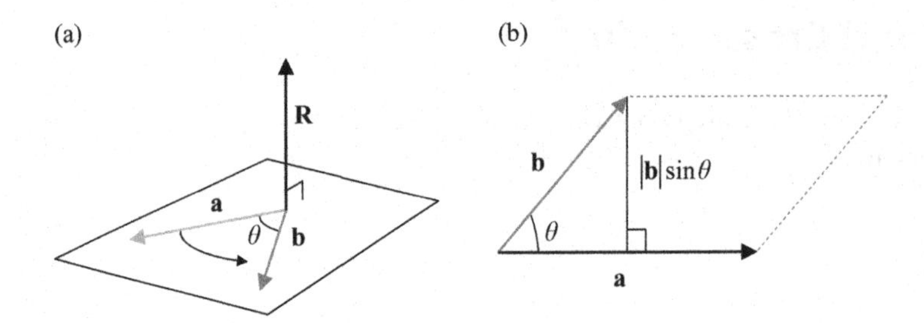

FIGURE 1.6 (a) Cross product of two vectors. (b) Magnitude of the cross product.

We shall see later that this is the result of a determinant of a matrix, whereby:

$$\mathbf{a}\times\mathbf{b}=\begin{vmatrix}\mathbf{i} & \mathbf{j} & \mathbf{k}\\ a_1 & a_2 & a_3\\ b_1 & b_2 & b_3\end{vmatrix}=\left(a_2b_3-a_3b_2\right)\mathbf{i}-\left(a_1b_3-a_3b_1\right)\mathbf{j}+\left(a_1b_2-a_2b_1\right)\mathbf{k} \tag{1.44}$$

The resultant vector **R** is perpendicular to the plane of the two vectors **a** and **b**. The magnitude of the resultant vector is:

$$|\mathbf{R}| = |\mathbf{a}||\mathbf{b}|\sin\theta \tag{1.45}$$

But the quantity $|\mathbf{b}|\sin\theta$ is the height of a parallelogram formed by the two vectors. This is shown in Figure 1.6(b). The magnitude of the resultant **R** is therefore the area of the parallelogram formed by the two vectors **a** and **b**.

1.3.4 Vector Differentiation

The position vector of a point in space is given by:

$$\mathbf{r} = x\mathbf{i} + y\mathbf{j} + z\mathbf{k} \tag{1.46}$$

A curve C in space may be represented by a series of points, each with its own position vector.

For any two points on the curve, the length of the chord joining P_0 and P_1 is $\Delta\mathbf{r}$. It is often of interest to determine the rate of change of the vector **r** with respect to the arc length s along the curve:

$$\begin{aligned}\frac{d\mathbf{r}}{ds} &= \lim_{\Delta t\to 0}\frac{\mathbf{r}\left(s_0+\Delta s\right)-\mathbf{r}\left(s_0\right)}{\Delta s}\\ &= \lim_{\Delta s\to 0}\frac{\Delta\mathbf{r}}{\Delta s}\end{aligned} \tag{1.47}$$

As $\Delta s \to 0$, $|\Delta\mathbf{r}|$ becomes equal to Δs and so: $d\mathbf{r}/ds = 1$

That is, the vector $d\mathbf{r}/ds$ is a unit vector with a direction tangent to the curve at any point P.

The position vector may also be expressed in terms of time (t) such that:

$$\mathbf{r} = x(t)\mathbf{i} + y(t)\mathbf{j} + z(t)\mathbf{k} \tag{1.48}$$

Differentiating with respect to t gives:

$$\begin{aligned}\frac{d\mathbf{r}}{dt} &= \lim_{\Delta t\to 0}\frac{\mathbf{r}(t_0+\Delta t)-\mathbf{r}(t_0)}{\Delta t}\\ &= \lim_{\Delta t\to 0}\frac{\Delta\mathbf{r}}{\Delta t}\end{aligned} \tag{1.49}$$

As $\Delta t \to 0$, evaluated at $t = t_0$ is a vector whose direction is tangent to the curve at P and represents the instantaneous velocity of a body which may be at point P.

In general, a vector function is said to be differentiable if the limit:

$$\lim_{\Delta t\to 0}\frac{\mathbf{r}(t_0+\Delta t)-\mathbf{r}(t_0)}{\Delta t} \tag{1.50}$$

exists. Whereupon:

$$\frac{d\mathbf{r}}{dt}=\frac{dx}{dt}\mathbf{i}+\frac{dy}{dt}\mathbf{j}+\frac{dz}{dt}\mathbf{k} \tag{1.51}$$

And the magnitude of the resulting vector is given by:

$$\left|\frac{d\mathbf{r}}{dt}\right|=\sqrt{\frac{d\mathbf{r}}{dt}\cdot\frac{d\mathbf{r}}{dt}}=\sqrt{\left(\frac{dx}{dt}\right)^2+\left(\frac{dy}{dt}\right)^2+\left(\frac{dz}{dt}\right)^2} \tag{1.52}$$

That is, the derivative of a vector gives a new vector whose components have magnitudes which are the derivatives of the original vector's components.

1.4 DIFFERENTIAL EQUATIONS

1.4.1 Differential Equations

A differential equation is an equation that contains derivatives. Some examples of differential equations are:

$$\frac{dy}{dx}=x+8;\ \frac{d^2y}{dx^2}+3\frac{dy}{dx}+5y=0;\ \left(\frac{d^2y}{dx^2}\right)^2+\left(\frac{dy}{dx}\right)^3+4y=x^2;\ \frac{\partial^2 z}{\partial x^2}+\frac{\partial z}{\partial y}=x^2+y \tag{1.53}$$

The first three equations above are *ordinary* differential equations because they involve one independent variable x. The last example is a *partial* differential equation because the equation has two independent variables, x and y. That is, part of the equation involves derivatives with respect to one variable x, while another part involves a derivative with respect to another variable y.

When we find a solution to an equation, say $y = f(x)$, we find values of x that provide the required quantity y. When we deal with *differential* equations, the solution to such an equation is a function rather than a specific quantity x. That is, we find a *function* whose derivative, or second or third derivative, satisfies the differential equation.

1.4.2 Solutions to Differential Equations

The general solution to a differential equation is a function of the form $y = f(x)$. The function $f(x)$ will usually have a number of arbitrary constants. A particular solution to the differential equation is found from the general solution for particular values of the constants. The constants are usually determined from the stated initial conditions referred to as the boundary conditions.

There are various methods available to find the solution to differential equations. The most common is a technique called the separation of variables.

Consider the differential equation:

$$\frac{dy}{dx} = x+5 \tag{1.54}$$

We are looking for a function, $y = f(x)$, whose derivative is $x + 5$. The first step is to separate the variables x and y so that they on opposite sides of the = sign.

$$dy = (x+5)dx \tag{1.55}$$

Then, we take the integral of both sides:

$$\begin{aligned} \int dy &= \int x+5dx \\ y &= \frac{x^2}{2}+5x+C \end{aligned} \tag{1.56}$$

We have one arbitrary constant C. The function $y = f(x)$ above is the general solution to the differential equation. The value of C can be obtained if we know the boundary, or initial conditions. For example, if it is known that $y = 1$ when $x = 0$, then C must be = 1. We call:

$$y = \frac{x^2}{2}+5x+1 \tag{1.57}$$

a particular solution to the differential equation. A different boundary condition leads to a different particular solution (i.e. a different value of C).

In general, there may be many general solutions to a differential equation each differing by a *multiplicative* constant.

The collection of functions $f(x)$ that are general solutions to the differential equations are called eigenfunctions. For example, a general solution to a differential equation might be:

$$\psi_n(x) = A_n \cos k_n x \quad \text{where } n = 1,3,5,7\ldots \tag{1.58}$$

That is, $\psi_n(x)$ are the eigenfunctions, and they differ from each other depending on the value of n.

Differential equations may be complex equations whose solutions may be complex functions (that is, containing complex numbers). We shall see later that the Schrödinger equation is a complex partial differential equation.

1.4.3 Differential Operators

We will find in our study of quantum mechanics that the equations involved are often differential equations. A short-hand way of writing differential equations is to write the equation using what are called differential operators.

For example, consider the familiar equation often encountered in kinematics that relates the distance travelled s for a body to its initial velocity u, the acceleration a and the time taken t. We will assume here that u and a are constants, and so s is a function of t alone:

$$s = ut + \frac{1}{2}at \tag{1.59}$$

Now to get velocity v as a function of t, we take the derivative with respect to t:

$$\frac{ds}{dt} = u + at \tag{1.60}$$

Instead of writing ds/dt, we can define d/dt as a differential operator D, so that:

$$D = \frac{d}{dt} \tag{1.61}$$

When the differential operator D acts on the displacement variable s, then, in this case, we get the velocity v.

$$Ds = v \tag{1.62}$$

As is often the case, a quantity like s may be a function of more than one variable. Differential operators are then expressed as partial derivatives.

Operators obey the laws of algebra for the operations of addition and multiplication. For example, consider three differential operators A, B and C, then:

$$\begin{gathered} A + B = B + A \\ (A + B) + C = A + (B + C) \\ (AB)C = A(BC) \\ A(B + C) = AB + AC \end{gathered} \tag{1.63}$$

and, if the operators have constant coefficients:

$$AB = BA \tag{1.64}$$

Also, for m and n being positive integers:

$$D^m D^n = D^{m+n}$$

A consequence of this last condition is that:

$$DD = D^2 \tag{1.65}$$

That is, an alternate way of writing the second derivative which we will come to use later.

Operators are sometimes written using a hat symbol to identify them as operators rather than variables, e.g. $\hat{D}$.

1.5 PARTIAL DERIVATIVES

In many physical situations, a dependent variable is a function of two independent variables. This type of relationship is expressed as a surface in a three-dimensional coordinate system. To determine the rate of change of z as x and y vary, we hold one variable constant and differentiate with respect to the other. This is called partial differentiation. For example, if $z = f(x, y)$, then the partial derivative of z with respect to x is written:

$$\frac{\partial z}{\partial x} = \lim_{x \to x_0} \frac{f(x, y_0) - f(x_0, y_0)}{x - x_0} \tag{1.66}$$

If we wish to know the rate of change of z with respect to x at $x = x_0$, then we hold y constant at y_0 and take:

$$\frac{\partial z}{\partial x} \tag{1.67}$$

This gives the slope of the tangent to the surface in the x direction at (x_0, y_0).

If we wish to know the rate of change of z with respect to y at $y = y_0$, then we hold x constant at x_0 and take:

$$\frac{\partial z}{\partial y} \tag{1.68}$$

This gives the slope of the tangent to the surface in the y direction at (x_0, y_0).

Higher order derivatives are written as:

$$\frac{\partial^2 z}{\partial x \partial y} = \frac{\partial}{\partial x}\left(\frac{\partial z}{\partial y}\right); \frac{\partial^2 z}{\partial y \partial x} = \frac{\partial}{\partial y}\left(\frac{\partial z}{\partial x}\right); \frac{\partial^2 z}{\partial x^2} = \frac{\partial}{\partial x}\left(\frac{\partial z}{\partial x}\right); \frac{\partial^2 z}{\partial y^2} = \frac{\partial}{\partial y}\left(\frac{\partial z}{\partial y}\right) \tag{1.69}$$

The first two equations in the above are equal.

1.6 MATRICES

A matrix is a rectangular array of numbers with m rows and n columns. When $m = n$, the matrix is square. Numbers in the matrix are called elements and are written a_{jk} where the subscripts refer to the row and column, respectively.

$$\mathbf{A} = \begin{bmatrix} a_{11} & a_{12} & a_{13} \\ a_{21} & a_{22} & a_{23} \\ a_{31} & a_{32} & a_{33} \end{bmatrix} \tag{1.70}$$

Addition and subtraction of matrices:

$$\mathbf{A} + \mathbf{B} = (a_{jk} + b_{jk}) \tag{1.71}$$

Multiplication by a number:

$$k\mathbf{A} = (ka_{jk}) \tag{1.72}$$

Multiplication of two matrices: if **A** is an $m \times p$ matrix and **B** is a $p \times n$ matrix, then the product **C** of **A** and **B** is given by:

$$c_{jk} = \sum_{i=1}^{n} a_{ji} b_{ik} \tag{1.73}$$

The number of columns in **A** must be the same as the number of rows in **B**. The result matrix **C** has dimensions $m \times n$. That is, C has rows equal to the number of rows in **A** and columns equal to the number of columns in **B**.

For example, for the product of a 1×3 matrix **A** with a 3×1 matrix **B**, we have:

$$\mathbf{AB} = \begin{bmatrix} a_1 & a_2 & a_3 \end{bmatrix} \begin{bmatrix} b_1 \\ b_2 \\ b_3 \end{bmatrix} = \begin{bmatrix} a_1 b_1 + a_2 b_2 + a_3 b_3 \end{bmatrix} \tag{1.74}$$

Or:

$$\begin{bmatrix} a_{11} & a_{12} & a_{13} \\ a_{21} & a_{22} & a_{23} \end{bmatrix} \begin{bmatrix} b_{11} & b_{12} \\ b_{21} & b_{22} \\ b_{31} & b_{32} \end{bmatrix} = \begin{bmatrix} a_{11}b_{11} + a_{12}b_{21} + a_{13}b_{31} & a_{11}b_{12} + a_{12}b_{22} + a_{13}b_{32} \\ a_{21}b_{11} + a_{22}b_{21} + a_{23}b_{31} & a_{21}b_{12} + a_{22}b_{22} + a_{23}b_{32} \end{bmatrix} \tag{1.75}$$

The transpose of a matrix is when the rows and columns are interchanged:

$$\mathbf{A} = (a_{jk}), \ \mathbf{A}^{\mathrm{T}} = (a_{kj}) \tag{1.76}$$

The diagonal elements of a square matrix are called the principal or main diagonal, and the sum of these is called the trace of the matrix. A square matrix in which the main diagonal elements are all 1 and all other elements are 0 is called a unit matrix **I**.

$$\mathbf{I} = \begin{bmatrix} 1 & 0 & 0 \\ 0 & 1 & 0 \\ 0 & 0 & 1 \end{bmatrix} \tag{1.77}$$

For a square matrix **A**, if there exists a matrix **B** such that **AB** = **I**, then **B** is the inverse of **A** and is written $\mathbf{A}^{-1}$. Not every square matrix has an inverse. If a matrix has an inverse, then there is only one inverse. If the matrix is a singular matrix, then the inverse does not exist.

If two square matrices **A** and **B** are multiplied so that the product **AB** = **BA**, then the two matrices are said to commute. If the product is such that **AB** = –**BA** then the two matrices are said to anti-commute.

A matrix **A** is orthogonal if $\mathbf{A}^{\mathrm{T}}\mathbf{A} = \mathbf{I}$. Two column matrices, or vectors, are orthogonal if $\mathbf{A}^{\mathrm{T}}\mathbf{B} = 0$.

Just as we saw that there was the conjugate of a complex number, we can also have the conjugate of a matrix. That is, if the elements of the matrix are complex numbers, then the conjugate of the matrix comprises the complex conjugates of its elements.

For example, if **A** is the matrix of complex numbers:

$$\mathbf{A} = \begin{bmatrix} a_{11} + ib_{11} & a_{12} + ib_{12} & a_{13} + ib_{13} \\ a_{21} + ib_{21} & a_{22} + ib_{22} & a_{23} + ib_{23} \\ a_{31} + ib_{31} & a_{32} + ib_{32} & a_{33} + ib_{33} \end{bmatrix} \tag{1.78}$$

Then the complex conjugate of **A** is written:

$$\mathbf{A}^* = \begin{bmatrix} a_{11}-ib_{11} & a_{12}-ib_{12} & a_{13}-ib_{13} \\ a_{21}-ib_{21} & a_{22}-ib_{22} & a_{23}-ib_{23} \\ a_{31}-ib_{31} & a_{32}-ib_{32} & a_{33}-ib_{33} \end{bmatrix} \tag{1.79}$$

The conjugate of the product of two matrices is the product of their conjugate matrices written in the same order.

$$(\mathbf{AB})^* = \mathbf{A}^* \cdot \mathbf{B}^* \tag{1.80}$$

The Hermitian conjugate of a matrix $\mathbf{A}^+$ is found from the complex conjugate of the transpose of the matrix:

$$\mathbf{A}^+ = \left(\mathbf{A}^{\mathrm{T}}\right)^* = \left(\mathbf{A}^*\right)^{\mathrm{T}} \tag{1.81}$$

A linear system of equations is an ordered set of coefficients and unknowns.

$$\begin{aligned} a_{11}x_1 + a_{12}x_2 + \ldots a_{1n}x_n &= y_1 \\ a_{21}x_1 + a_{22}x_2 + \ldots a_{2n}x_n &= y_2 \\ \\ a_{m1}x_1 + a_{m2}x_2 + \ldots a_{mn}x_n &= y_n \end{aligned} \tag{1.82}$$

The linear system can be written in matrix form:

$$\begin{bmatrix} a_{11} & a_{12} & & a_{1n} \\ a_{21} & a_{22} & & a_{2n} \\ \\ a_{m1} & a_{m2} & & a_{mn} \end{bmatrix} \begin{bmatrix} x_1 \\ x_2 \\ \\ x_n \end{bmatrix} = \begin{bmatrix} y_1 \\ y_2 \\ \\ y_n \end{bmatrix} \tag{1.83}$$

$$\mathbf{AX} = \mathbf{Y}$$

The determinant of a square matrix is denoted:

$$\det \mathbf{A} = \Delta\mathbf{A} = |\mathbf{A}| \tag{1.84}$$

The determinant of a second order matrix is given by:

$$\begin{vmatrix} a_{11} & a_{12} \\ a_{21} & a_{22} \end{vmatrix} = a_{11}a_{22} - a_{12}a_{21} \tag{1.85}$$

The determinant of a third order matrix is given by:

$$\begin{aligned} \begin{vmatrix} a_{11} & a_{12} & a_{13} \\ a_{21} & a_{22} & a_{23} \\ a_{31} & a_{32} & a_{33} \end{vmatrix} &= a_{11}\begin{vmatrix} a_{22} & a_{23} \\ a_{32} & a_{33} \end{vmatrix} - a_{12}\begin{vmatrix} a_{21} & a_{23} \\ a_{31} & a_{33} \end{vmatrix} + a_{13}\begin{vmatrix} a_{21} & a_{22} \\ a_{31} & a_{32} \end{vmatrix} \\ &= a_{11}\left(a_{22}a_{33} - a_{23}a_{32}\right) - a_{12}\left(a_{21}a_{33} - a_{23}a_{31}\right) + a_{13}\left(a_{21}a_{32} - a_{22}a_{31}\right) \end{aligned} \tag{1.86}$$

Determinants are only defined for square matrices.

The determinant components of the determinant are called minors of the determinant. For example, using the above third order square matrix, we have the minors M_{11}, M_{12} and M_{13}:

$$M_{11} = \begin{vmatrix} a_{22} & a_{23} \\ a_{32} & a_{33} \end{vmatrix}; \; M_{12} = \begin{vmatrix} a_{21} & a_{23} \\ a_{31} & a_{33} \end{vmatrix}; \; M_{13} = \begin{vmatrix} a_{21} & a_{22} \\ a_{31} & a_{32} \end{vmatrix} \tag{1.87}$$

When the sign is included, such that:

$$(-1)^{i+j} M_{ij} \tag{1.88}$$

then the minors are called the cofactors α_{ij} of the determinant:

$$\begin{aligned} \alpha_{11} &= (-1)^2 M_{11} = M_{11} = \begin{vmatrix} a_{22} & a_{23} \\ a_{32} & a_{33} \end{vmatrix} \\ \alpha_{12} &= (-1)^3 M_{12} = -M_{12} = -\begin{vmatrix} a_{21} & a_{23} \\ a_{31} & a_{33} \end{vmatrix} \\ \alpha_{13} &= (-1)^4 M_{13} = M_{13} = \begin{vmatrix} a_{21} & a_{22} \\ a_{31} & a_{32} \end{vmatrix} \end{aligned} \tag{1.89}$$

A matrix of cofactors of a square matrix is called the adjoint of the matrix. For the above example, the adjoint of **A** is:

$$\text{adj}\,\mathbf{A} = \bar{\mathbf{A}} = \begin{bmatrix} \alpha_{11} & \alpha_{21} & \alpha_{31} \\ \alpha_{12} & \alpha_{22} & \alpha_{32} \\ \alpha_{13} & \alpha_{23} & \alpha_{33} \end{bmatrix} \tag{1.90}$$

Note the ordering of the elements of the cofactor matrix is that of the transpose of **A**.

Waves

2

2.1 INTRODUCTION

The concept of a wave is absolutely essential for the understanding of quantum mechanics. The periodic motion of a particle undergoing a harmonic oscillation is the basis for just about everything in particle physics. The wave–particle duality of existence arises because at the macro-scale, we perceive these things to be very different, but on the micro-scale, they merge into an unfamiliar reality that sometimes can only be described in mathematical terms.

2.2 PERIODIC MOTION

Consider the motion of a mass attached to a spring as shown in Figure 2.1.

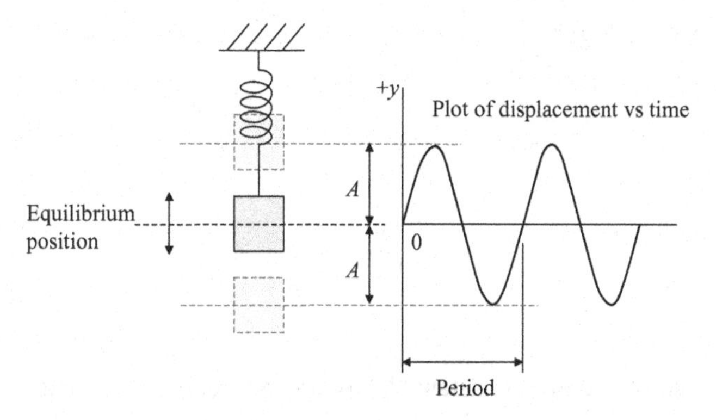

FIGURE 2.1 Simple harmonic motion of a mass suspended by a spring.

Periodic motion of the mass consists of displacements y around an equilibrium position. When y is plotted against time, we find the relationship is sinusoidal. The motion of the mass would be the same as if it were attached to the rim of a rotating wheel and viewed edge on. That is, the angle θ (called the phase) would be the angular displacement of the wheel.

The displacement y of the mass as a function of θ is:

$$y = A\sin\theta \tag{2.1}$$

where A is the amplitude of the motion. In units of angular measure, the angular displacement θ is given by the angular velocity ω times the time.

$$\theta = \omega t \tag{2.2}$$

Thus, the displacement y from the equilibrium position as a function of time is expressed:

$$y = A\sin\omega t \tag{2.3}$$

The general expression for position (i.e. displacement from the equilibrium position) for the mass is thus:

$$y = A\sin(\omega t + \varphi) \tag{2.4}$$

where ϕ is the *initial* phase, or phase advance, that is added to the angle θ to account for the case when the time $t=0$ does not correspond to $y=0$.

The velocity of the mass is given by dy/dt:

$$v = \frac{dy}{dt} = A\omega\cos(\omega t + \varphi) \tag{2.5}$$

The acceleration of the mass is given by dv/dt:

$$a = \frac{dv}{dt} = -A\omega^2\sin(\omega t + \varphi) \tag{2.6}$$

2.3 SIMPLE HARMONIC MOTION

Consider the force applied to a body moving with an oscillatory motion. Now:

$$F = ma = -mA\omega^2\sin(\omega t + \varphi) \tag{2.7}$$

But:

$$y = A\sin(\omega t + \varphi) \tag{2.8}$$

And so:

$$F = -m\omega^2 y \tag{2.9}$$

However, the product $m\omega^2$ is actually a constant "k" for a fixed frequency ω. Thus:

$$F = -\mathrm{k}y \tag{2.10}$$

The minus sign indicates that the force acting on the mass is in a direction opposite to the displacement and acts so as to bring the mass back to the equilibrium position. The magnitude of this restoring force is a function of the displacement of the mass from the equilibrium position. k is often called the spring or force constant.

The combination of periodic motion and a restoring force whose magnitude depends on the displacement from the equilibrium position is called simple harmonic motion.

Now:

$$\omega = \sqrt{\frac{\mathrm{k}}{m}} = 2\pi f = \frac{2\pi}{T} \tag{2.11}$$

where T is the period of the motion.

Since:

$$T = \frac{2\pi}{\omega} \tag{2.12}$$

then:

$$T = 2\pi\sqrt{\frac{m}{\mathrm{k}}} \tag{2.13}$$

Note that the period of oscillation only depends on the mass and the spring (or force) constant and not on the amplitude.

In general, at any instant, the mass has both kinetic and potential energy.

The potential energy E_{PE} is the sum of the force times the displacement in the y direction, but the force is a function of the displacement, and so we must integrate between 0 and some value of $y(t)$:

$$\begin{aligned} E_{PE} &= \int F dy = \int_0^{y(t)} -\mathrm{k}y dy \\ &= \frac{1}{2}\mathrm{k}y(t)^2 \\ &= \frac{1}{2}\mathrm{k}A^2 \sin^2(\omega t + \varphi) \\ &= \frac{1}{2}m\omega^2 A^2 \sin^2(\omega t + \varphi) \end{aligned} \tag{2.14}$$

In terms of position, the integral is simply ½ky^2.

The kinetic energy E_{KE} depends upon the velocity v, which in turn is a function of the angular displacement ωt:

$$E_{KE} = \frac{1}{2}mv^2 = \frac{1}{2}m\omega^2 A^2 \cos^2(\omega t + \varphi) \tag{2.15}$$

The total energy E_T possessed by the mass m at a time t is thus:

$$E_T = \frac{1}{2}m\omega^2 A^2 \sin^2(\omega t + \varphi) + \frac{1}{2}m\omega^2 A^2 \cos^2(\omega t + \varphi) \tag{2.16}$$

But:

$$\sin^2(\omega t + \varphi) + \cos^2(\omega t + \varphi) = 1 \tag{2.17}$$

Therefore:

$$E_T = \frac{1}{2}m\omega^2 A^2 \tag{2.18}$$

As the $\sin^2$ term increases, the $\cos^2$ term decreases. In other words, for a constant m, ω and A, the total energy is a constant.

Note that the frequency of oscillation depends on the square root of the spring constant and the mass. These are usually fixed for a given mechanical system, and so the energy stored within the system is, in practice, proportional to the amplitude.*

In simple harmonic motion, energy is continually being transferred between potential and kinetic energies.

2.4 WAVE FUNCTION

Consider a transverse wave in a string. A wave is a disturbance which travels from place to place, say in the x direction. The particles in the string that carry the wave move up and down with simple harmonic motion in the y direction. The particles themselves do not move in the x direction; it is the disturbance that travels in the x direction.

Figure 2.2 shows a snapshot of the wave at some time t where the shape of the wave indicates the displacement of the particles in the y direction along the direction of travel of the wave in the x direction. The plot has been drawn for three different times to show that the shape of the wave is travelling from left to right, while the particles within the wave travel up and down.

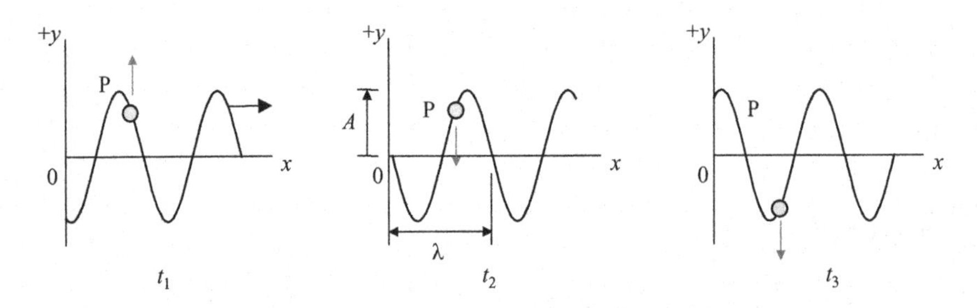

FIGURE 2.2 Snapshot view of a wave at time T showing displacement of particles in the y direction as a function of x, the direction of travel of the wave.

We note the following:

- The shape of the wave is a repeating pattern.
- λ is called the wavelength and is the length of one complete cycle.
- The wave travels along the x axis with a velocity v.
- The time for a particle to complete one complete cycle of vertical movement is the period T.

* There is a subtlety here that requires some discussion. We might wonder why the energy content of a harmonic oscillator depends upon the frequency. Say we have two oscillators, each with the same mass but having a different spring constant. They are set into motion with the same amplitude. They are therefore vibrating with different frequencies. The one with the larger spring constant, the "stiffer spring", has a higher frequency of vibration compared to the one with the lower spring constant, the "softer" spring. Each system is converting a certain amount of energy from kinetic to potential energy. Is it not true that the one with the higher frequency of oscillation is just making the transfer from kinetic to potential energy more rapidly than the other? Surely the rate at which this happens should not affect the energy content? Why then is the total energy different in each system? It is because *for the same amplitude*, we had to put in more energy for the case of the system with the stiffer spring in the first place. That is, $F = ky$ and so for a larger value of k, the force applied through the distance $y = A$ for the stiffer spring had to be more compared to the case of the softer spring to get things going. That is, more work is done by the external force. Thus, the system with the softer spring might indeed be oscillating with the same amplitude, but it is alternately converting a lower amount of energy at each oscillation. The rate of oscillation doesn't matter.

One complete wavelength passes a given point in the x direction in a time T, the period. Thus, since the velocity is d/t, where d is the distance in the x direction, the velocity of the wave is:

$$v = \frac{\lambda}{T} \tag{2.19}$$

Now, the frequency f of the oscillations, in cycles per second, is the reciprocal of the period:

$$f = \frac{1}{T}$$

And so:

$$v = f\lambda \tag{2.20}$$

That is, the velocity of the wave, the disturbance, is expressed as the frequency of the vertical oscillations times the wavelength.

We wish to calculate the displacement y of any point P located at some distance x on the string as a function of time t.

Let's consider the motion of the point located at $x=0$. If the points on the string are moving up and down with simple harmonic motion, then:

$$y = A\sin(\omega t + \varphi) \tag{2.21}$$

If $y=0$ at $t=0$, then $\phi=0$.

The disturbance, or wave, travels from left to right with velocity $v=x/t$. Thus, the disturbance travels from 0 to a point x in time x/v.

Now, let us consider the motion of a point P located at position x. As shown in Figure 2.3, the displacement of point P located at x at time t is the same as that of point P′ located at $x=0$ at time $(t - x/v)$.

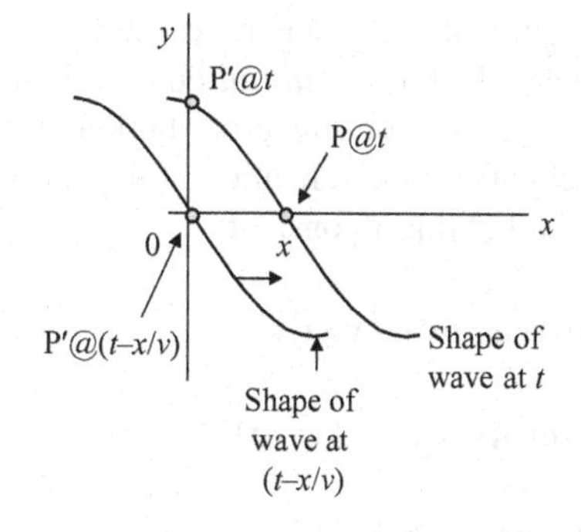

FIGURE 2.3 Displacement of a point P located at x at time t compared with the displacement of the same point at $t - x/v$.

Thus, to get the displacement of the point P at (x, t) we use the same formula for the motion of point P′ located at $x=0$ but put in the time $t=(t-x/v)$:

$$y = A\sin(\omega t + \varphi) = A\sin\left(\omega\left(t - \frac{x}{v}\right) + \varphi\right) \tag{2.22}$$

Now, it is convenient to let:

$$k = \frac{2\pi}{\lambda} \tag{2.23}$$

Where k is called the wave number (different from the spring constant k used previously).

Since:

$$v = f\lambda \tag{2.24}$$

Then:

$$\begin{aligned} v &= f\frac{2\pi}{k} \\ &= \frac{\omega}{k} \end{aligned} \tag{2.25}$$

Thus:

$$y = A\sin\left(\omega\left(t - x\frac{k}{\omega}\right) + \varphi\right) \tag{2.26}$$

or:

$$y(x,t) = A\sin(\omega t - kx + \varphi) \tag{2.27}$$

This equation gives displacement of the point P from its equilibrium position as a function of x and t. It is called the wave function.

When viewed as a plot against time, the coordinate x gives us a place from which to view the wave. Or, we could think of x as the particle whose movement in the y direction we are plotting. When viewed as a snapshot at a particular time, the plot shows all the positions of all the particles at some instant. Thus, we can hold x or t constant in the above equation and plot the y coordinate as a function of the quantity that is left free to vary.

A wave that can be represented by either a sine or cosine function is called a sinusoidal wave or a harmonic wave.

The argument to the sine function $(\omega t - kx + \phi)$ is often called the phase of the wave – not to be confused with ϕ, which is the *initial* phase. The initial phase ϕ is an offset, or a constant which may exist at $t = 0$ and so makes a constant contribution to the phase independent of x and t.

The velocity and acceleration of the particles in the string are found by differentiating y with respect to time while holding x constant:

$$v_y = \omega A\cos(\omega t - kx + \varphi) \tag{2.28}$$

$$a_y = -\omega^2 A\sin(\omega t - kx + \varphi) \tag{2.29}$$

These equations give the velocity and acceleration of the particles within the medium of the wave, not the velocity of the wave itself.

Although we have described a transverse wave on a string, the same principles apply to any wave motion.

2.5 WAVE EQUATION

In the above equations, we differentiated y with respect to t while holding x a constant. Let us now differentiate with respect to x while holding t a constant. Setting $\phi = 0$ for convenience, we obtain:

$$y = A\sin(\omega t - kx)$$

$$\frac{\partial y}{\partial x} = -kA\cos(\omega t - kx) \tag{2.30}$$

$$\frac{\partial^2 y}{\partial x^2} = -k^2 A\sin(\omega t - kx)$$

but:

$$\omega = vk \tag{2.31}$$

where v is the velocity of the *wave*. And so:

$$\frac{\partial^2 y}{\partial x^2} = -\frac{\omega^2}{v^2} A\sin(\omega t - kx) \tag{2.32}$$

or:

$$\frac{\partial^2 y}{\partial x^2} = \frac{1}{v^2}\frac{\partial^2 y}{\partial t^2} \tag{2.33}$$

This is called the (one-dimensional) wave equation and gives information about all aspects of the wave by tying together the motion of the particles within the medium and the wave (the disturbance which travels through the medium).

Note that on the left-hand side of this equation, we have the positional information of the particle in the y direction as a function of x, while on the right-hand side, we have the time dependence of the position in the y direction as a function of t.

The equation is a second order partial differential equation. The solution to this equation is a function. That is, the solution to the wave equation is the original wave function:

$$y = A\sin(\omega t - kx) \tag{2.34}$$

This wave function satisfies the wave equation. That is, when we take the second partial derivative of y with respect to x, this equals to $1/v^2$ times the second derivative of y with respect to t.

The collection of equations which satisfy the wave equation are the eigenfunctions, sometimes referred to as eigenvectors. They may differ according to the value of A. Two different solutions of the wave equation (having different values of the constant A) may be added to yield a third solution – this is the principle of superposition.

We could have just as easily equated the first derivatives of y with respect to x and t, and we would have obtained a first order differential equation involving v instead of v^2. However, in many cases, we deal with the superposition of two waves travelling in opposite directions to form standing waves in which case we need two constants to completely specify the resultant wave. In such a situation, with two arbitrary constants, the second order differential equation is therefore the most general expression of a wave.

It has been shown that the wave function is written:

$$y(x,t) = A\sin(\omega t - kx + \varphi) \tag{2.35}$$

A very important assumption in the above is that we were dealing with the transverse motion of a particle in a two-dimensional situation. That is, the particle was moving up and down along the y axis, and the wave, or shape of the disturbance, moved towards the right along the x axis as shown in Figure 2.2 (a). As the wave moves to the right, the particle at $x=0$ moves upwards.

However, in some physical situations, the wave itself may not represent the motion of a particle in the xy plane. For example, in quantum physics, the matter wave associated with a particle, such as an electron, has a connection with the probability of finding that particle at some place x at a time t.

In cases such as those, it is convenient to reverse the signs in the phase terms such that:

$$y(x,t) = A\sin(kx - \omega t + \varphi) \tag{2.36}$$

Note, both representations of the wave are correct, but not equivalent – their phases differ by a factor of π.

2.6 COMPLEX REPRESENTATION OF A WAVE

Representation of waves using sine and cosine functions can become unwieldy. An often more convenient method is to use complex numbers.

Imagine a point that is represented by the complex number $z = a + ib$ on the complex plane. z is rotating about the origin. As shown in Figure 2.4, viewed from side-on, z would be moving up and down with simple harmonic motion.

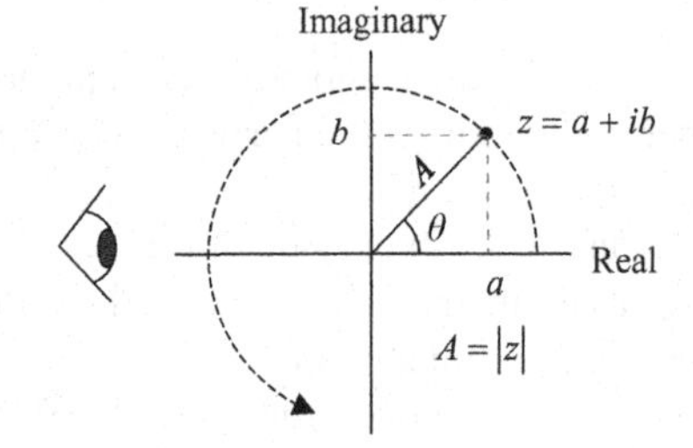

FIGURE 2.4 Complex number representation of simple harmonic motion.

Now, in polar coordinates, we have:

$$\begin{aligned} a &= A\cos\theta \\ b &= A\sin\theta \end{aligned} \tag{2.37}$$

Let:

$$z = a + ib \tag{2.38}$$

And:

$$A = \sqrt{a^2 + b^2} \tag{2.39}$$

Therefore:

$$\begin{aligned} z &= A\cos\theta + iA\sin\theta \\ &= A(\cos\theta + i\sin\theta) \end{aligned} \tag{2.40}$$

Using Euler's formula, where:

$$e^{i\theta} = (\cos\theta + i\sin\theta) \tag{2.41}$$

We can write:

$$z = Ae^{i\theta} \tag{2.42}$$

The amplitude A of the motion is given by the magnitude of z.

$$A = |z| = \sqrt{z^* z} \tag{2.43}$$

where z^* is the complex conjugate: $z^* = a - ib$.

At first sight, we might wonder why complex numbers involving real and imaginary quantities would have any connection at all with wave motion. Previously, we were concerned with the motion of the particle in the y direction and in fact, often made the point that the particles themselves in the medium carrying the wave did not move in the x direction at all, but only up and down in the y direction. So, we might ask, why bother with complex numbers?

The reason is that in many situations, the mathematics associated with the exponential expression of a wave, even though it might contain what appear to be inconvenient quantities such as e and i, are easier to deal with than the trigonometric form of the wave function. The cyclic nature of the wave is carried by the simple harmonic motion of the particles at any point. This in turn is embodied in a smoothly varying value of the b (or the a) coordinate in the complex plane as if a complex number z were being rotated around the origin in this plane and viewed edge-on.

You will find that it is never necessary to evaluate "i" on its own. In equations involving i, it is usually paired with another i and cancels out or is squared, and the product $i = -1$ used.

It can be seen from the above, that the cosine function is associated with the real part of the motion, and the sine function is associated with the imaginary part. Either cosine or sine can be used to represent a wave. It doesn't matter which is used.

Thus, a travelling wave can be represented by the real (or the complex) part of:

$$z = A(\cos\theta + i\sin\theta) \tag{2.44}$$

Or, in exponential form:

$$z = Ae^{i\theta} \tag{2.45}$$

Note that the trigonometric functions and the exponential give the function z its periodic character. These provide the "phase", which is a value between −1 and +1. This is multiplied with the amplitude A to give the value of z as a function of θ.

Note that z is a complex number. If the wave it represents has a constant amplitude, then of course the amplitude of z is a constant, but the angle, or phase, of z varies in time since $\theta = \omega t$.

We saw before that in the xy plane, $y(x, t)$ gives the y axis position of a particle in the medium of the wave at x and t. How can we relate this to the exponential expression for a wave?

In the complex case, we note that $\theta = \omega t$, and so we might write, for a particular value of x, say $x = 0$:

$$y(t) = Ae^{i\omega t} \tag{2.46}$$

At some time t, say $t = 0$, we wish to know the value of y at some point x, which is the same as that of the point located at $x = 0$ but with the time $t = (t - x/v)$:

$$y(x) = Ae^{i\omega(-x/v)} \tag{2.47}$$

but $v = \frac{\omega}{k}$, and so:

$$\begin{aligned} y(x) &= Ae^{i\omega(-kx/\omega)} \\ &= Ae^{-ikx} \end{aligned} \tag{2.48}$$

Generally speaking, the wave function in exponential terms is therefore written:

$$y(x,t) = Ae^{i(\omega t - kx)} \tag{2.49}$$

As is often the case, the sign of the phase term is reversed, and we write:

$$y(x,t) = Ae^{i(kx - \omega t)} \tag{2.50}$$

As before, we can hold either x or t constant and plot the wave as a function of time or distance along the x axis.

2.7 ENERGY CARRIED BY A WAVE

A wave transfers energy between two locations. Consider the example of a transverse wave on a stretched string. In the case of a stretched string, the potential energy is stored as the elastic strain energy. That is, the string is initially straight and stretched between two supports which are fixed. A wave motion is imparted, say at one end of the string. The resulting displacement of that portion of the string means that the string is deformed away from its initial straight profile. Elastic strain energy is thus stored within the stretched bonds holding the particles of the string together. Elastic strain energy (potential energy) is a maximum when the displacement of a particle is a maximum (at the amplitude) and a minimum as it passes through its zero or initial position. The kinetic energy of motion is a maximum at the zero position and a zero at the maximum displacement.

The external source performs work on the first particle in the string. The particle moves with simple harmonic motion. The energy of the particle is converted from potential to kinetic energy in a cyclic manner. The total energy of the particle is unchanged in time. However, because the particle is connected to its neighbouring particle, the first particle loses energy to the next particle at the same rate it receives energy from the external source. Although the total energy of the particle remains unchanged, energy from the source gets passed on from one particle to the next until it arrives at the target location. Thus, energy from the external source travels along the string with velocity v.

The total energy of each particle is the sum of the potential and kinetic energies of the particle:

$$\begin{aligned} E_T &= \frac{1}{2}m\omega^2 A^2 \sin^2(\omega t + \varphi) + \frac{1}{2}m\omega^2 A^2 \cos^2(\omega t + \varphi) \\ &= \frac{1}{2}m\omega^2 A^2 \end{aligned} \tag{2.51}$$

The total energy for all the oscillating particles in a segment of string one wavelength long is:

$$E_T = \frac{1}{2}\omega^2 A^2 (\rho\lambda) \tag{2.52}$$

Since:

$$m = \rho\lambda \tag{2.53}$$

where ρ is the mass per unit length of the string.

In one time period *T*, the energy contained in one wavelength segment of string will have moved on to the next wavelength segment. The power transmitted during is thus:

$$\begin{aligned} P &= \frac{E_{T\lambda}}{T} \\ &= \frac{1}{2}\omega^2 A^2 (\rho\lambda)\frac{1}{T} \\ &= \frac{1}{2}\omega^2 A^2 \rho v \end{aligned} \tag{2.54}$$

since:

$$v = \frac{1}{T}\lambda \tag{2.55}$$

Expressed in terms of frequency *f*, this is:

$$P = 2\rho\pi^2 v A^2 f^2 \tag{2.56}$$

For a stretched string, the tension T in the string is related to the velocity of the wave and mass per unit length as:

$$v = \sqrt{\frac{T}{\rho}} \tag{2.57}$$

And so, we have:

$$P = \frac{1}{2}\rho\sqrt{\frac{T}{\rho}}\omega^2 A^2 \tag{2.58}$$

The power transmitted by a wave on a stretched string of tension T and mass per unit length ρ oscillating with frequency ω and amplitude *A*, is therefore:

$$P = \frac{1}{2}\sqrt{T\rho}\ \omega^2 A^2 \tag{2.59}$$

The important observation to be made here is that the energy and power transmitted are proportional to the amplitude squared and also to the square of the angular frequency.

In the case of a stretched string, the energy is alternately transferred between potential and kinetic energy; the total energy of any particle within the string is a constant. The rate of propagation of energy is proportional to both ω^2 and A^2.

In the case of an electromagnetic wave, the electric and magnetic fields travel together, in step, and each sinusoidal time-varying field carries half the energy being transmitted. The instantaneous value of energy carried by the electric and magnetic fields at any point in space is not constant in time. In this case, much like in AC circuits in electricity, it is more useful to talk about the average energy, or average power being transmitted rather than the instantaneous values. When the power is integrated over a full cycle (0 to 2π) to obtain this average, the frequency term drops out, and we find that the average power transmitted is a function of the amplitude of the square of the E (or the B) field only.

2.8 SUPERPOSITION

Consider two waves on a stretched string that are travelling in opposite directions as shown in Figure 2.5. For simplicity, let the waves have the same amplitude, wavelength and frequency.

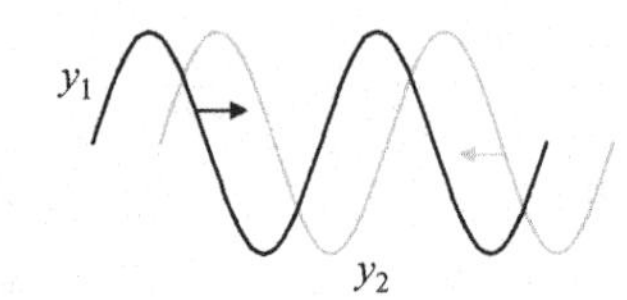

FIGURE 2.5 Two waves travelling in opposite directions on a stretched string.

The wave travelling to the right is represented by:

$$y_1 = A\sin(\omega t - kx) \tag{2.60}$$

and the wave travelling to the left by:

$$y_2 = A\sin(\omega t + kx) \tag{2.61}$$

The resultant wave, the superposition of the two, is thus:

$$y_1 = A\sin(\omega t - kx)$$
$$y_2 = A\sin(\omega t + kx)$$

$$y_1 + y_2 = A\left(2\sin\frac{(\omega t - kx)+(\omega t + kx)}{2}\cos\frac{(\omega t - kx)-(\omega t + kx)}{2}\right) \tag{2.62}$$

$$= 2A\cos kx \sin \omega t$$

Since:

$$k = \frac{2\pi}{\lambda} \tag{2.63}$$

And:

$$\sin A + \sin B = 2\sin\left(\frac{A+B}{2}\right)\cos\left(\frac{A-B}{2}\right) \tag{2.64}$$

Thus:

$$y_1 + y_2 = 2A\left[\cos\frac{2\pi x}{\lambda}\right]\sin\omega t \tag{2.65}$$

This equation shows that the amplitude of the superposed or resultant wave is not a constant but varies between 0 and $2A$ as x varies. The motion of the particles within the string is still simple harmonic motion but the *amplitude* varies according to the value of x.

The resulting displacement of the particles is always zero when:

$$\frac{2\pi x}{\lambda} = n\frac{\pi}{2} \text{ where } n = 1,3,5\ldots \tag{2.66}$$

These values of x are called nodes and occur when:

$$x = \frac{\lambda}{4}, \frac{3\lambda}{4}, \frac{5\lambda}{4}\ldots \tag{2.67}$$

The resultant displacements of the particles are a maximum when:

$$\frac{2\pi x}{\lambda} = m\pi \text{ where } m = 0,1,2\ldots \tag{2.68}$$

These positions are called antinodes and occur when:

$$x = 0, \frac{\lambda}{2}, \lambda, \frac{3\lambda}{2}\ldots \tag{2.69}$$

The positions of the nodes are shown in Figure 2.6.

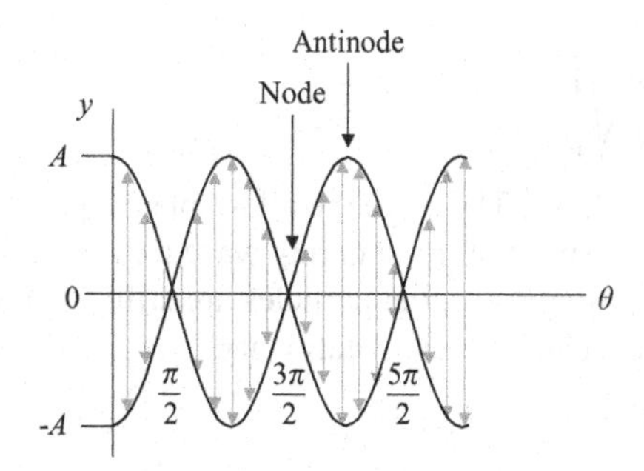

FIGURE 2.6 Position of nodes and antinodes of two superimposed waves in a stretched string.

All particles in the medium undergo simple harmonic motion. It just so happens that the particles at the nodes always have an amplitude of zero and hence are stationary. The other particles usually oscillate very rapidly, and the eye only sees the envelope of the waveform, hence the term standing wave.

2.9 STANDING WAVES

A very common example of standing waves appears in a stretched string. If the string is fixed at both ends, then there must be at least a node at each end because $y=0$ at the ends always.

For a standing wave to be produced, the length of the string must be equal to an integral number of half-wavelengths.

$$L = n\frac{\lambda}{2} \text{ where } n = 1,2,3,\ldots \tag{2.70}$$

The frequency of the mode of vibration when $n=1$ is called the fundamental frequency. The shape of this mode of vibration is shown in Figure 2.7(a).

Now:

$$v = f\lambda$$
$$f = \frac{v}{\lambda} \tag{2.71}$$

For a stretched string:

$$v = \sqrt{\frac{T}{\rho}} \tag{2.72}$$

where T is the tension in the string and ρ is the mass per unit length. And so:

$$f = \frac{1}{\lambda}\sqrt{\frac{T}{\rho}} \tag{2.73}$$

But:

$$L = \frac{n\lambda}{2} \tag{2.74}$$

Thus:

$$f = \frac{n}{2L}\sqrt{\frac{T}{\rho}} \tag{2.75}$$

Where $n = 1, 2, 3\ldots$ These are the allowable frequencies for standing waves (or normal modes) for a string length L. The shape of the standing wave for $n=2$ is shown in Figure 2.7(b).

In a standing wave, the particles are still moving with periodic motion in the vertical direction, but the disturbance, the wave, is stationary.*

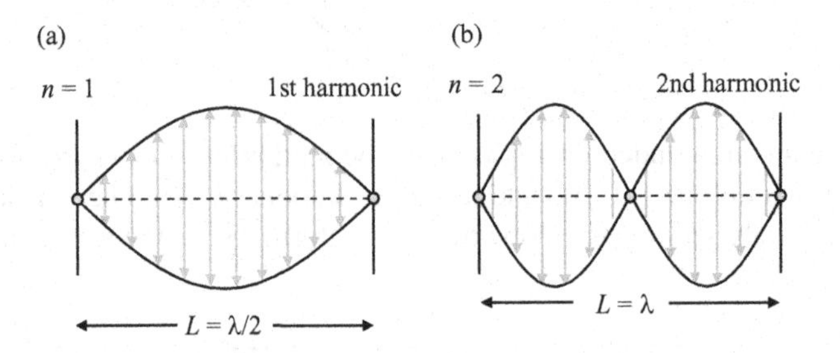

FIGURE 2.7 (a) Fundamental mode of vibration on a stretched string. (b) Second mode of vibration.

* The frequency of the motion depends on which mode of vibration is excited. In practice, frequency modes of vibration, that is, different harmonics, are excited depending on the way the string is set into motion. A violin bow drawn across a string produces a different set of harmonics compared to that same string being plucked. When there are multiple modes of vibration present, the particles themselves may no longer be oscillating in simple harmonic motion, but the motion is a superposition of a series of simple harmonic motions. Generally speaking, a "bright" sound may contain a high amplitude of high frequency components, whereas a mellow sound contains a high amplitude of lower frequency components, or harmonics.

2.10 BEATS

Consider two sound waves travelling in the same direction, with the same amplitude A, but with different frequencies ω_1 and ω_2. The wave functions are:

$$y_1 = A\sin(\omega_1 t - kx)$$
$$y_2 = A\sin(\omega_2 t - kx) \tag{2.76}$$

Figure 2.8 shows the displacement plotted as a function of time (keeping x fixed).

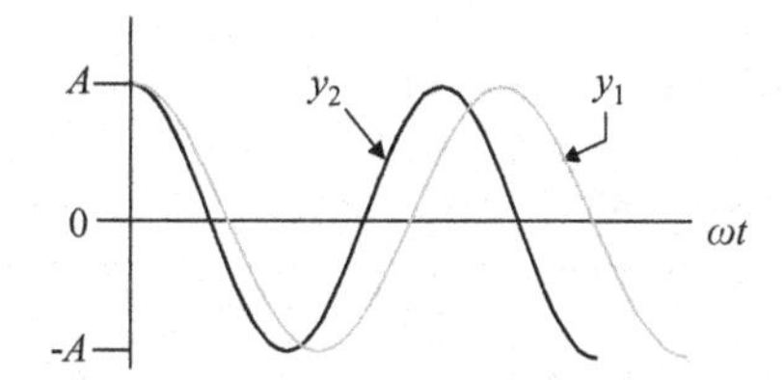

FIGURE 2.8 Two travelling waves with slightly different wavelengths.

The resultant superimposed wave y_R is:

$$\begin{aligned} y_R &= y_1 + y_2 \\ &= A\sin(\omega_1 t - kx) + A\sin(\omega_2 t - kx) \\ &= 2A\cos\left[\frac{\omega_1 - \omega_2}{2}t\right]\sin\left[\frac{\omega_1 + \omega_2}{2}t\right] \end{aligned} \tag{2.77}$$

Since $\sin A + \sin B = 2\sin\left(\frac{A+B}{2}\right)\cos\left(\frac{A-B}{2}\right)$

The cosine term in the equation for y_R represents the amplitude of the resultant wave; the sine term characterises the frequency. That is, the frequency of the resultant is the average of the two component frequencies. The amplitude term itself oscillates with a frequency of:

$$\frac{\omega_1 - \omega_2}{2} \tag{2.78}$$

However, the ear hears two pulses or beats in this one cycle since our ears don't distinguish between positive and negative values for y_R. The shape of the superposition is shown in Figure 2.9.

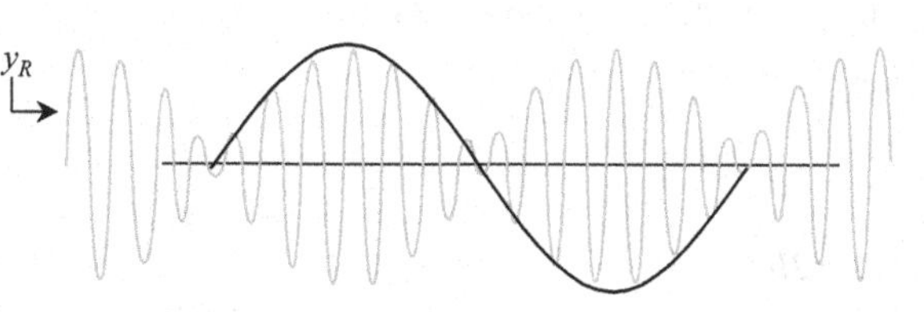

FIGURE 2.9 Superposition of two travelling waves with slightly different wavelengths.

The beat frequency (rad s^{-1}) actually heard is therefore:

$$2\frac{\omega_1 - \omega_2}{2} = \omega_1 - \omega_2 \tag{2.79}$$

2.11 SUPERPOSITION IN COMPLEX FORM

The solution to certain types of differential equations often takes the form of a superposition of two waves, one travelling in the $+x$ direction and the other in the $-x$ direction, each with different amplitudes C_1 and C_2. For the example here, we will consider the superposition of two waves at some point x (i.e. $y(x)$ only).

For a wave travelling to the right, we might write:

$$e^{ikx} = \cos kx + i \sin kx \tag{2.80}$$

And for a wave travelling to the left, we have:

$$\begin{aligned} e^{-ikx} &= \cos(-kx) + i \sin(-kx) \\ &= \cos kx - i \sin kx \end{aligned} \tag{2.81}$$

The superposition of these waves gives:

$$\begin{aligned} y(x) &= C_1 e^{ikx} + C_2 e^{-ikx} \\ &= C_1 (\cos kx + i \sin kx) + C_2 (\cos kx - i \sin kx) \end{aligned} \tag{2.82}$$

Let:

$$\begin{aligned} C_1 &= a + bi \\ C_2 &= a - bi \\ C_1 + C_2 &= 2a \\ C_1 - C_2 &= 2bi \end{aligned} \tag{2.83}$$

Thus, the superimposed wave is:

$$\begin{aligned} C_1 e^{ikx} + C_2 e^{-ikx} &= (a + bi)(\cos kx + i \sin kx) + (a - bi)(\cos kx - i \sin kx) \\ &= a \cos kx + ai \sin kx + bi \cos kx - b \sin kx + a \cos kx \\ &\quad - ai \sin kx - bi \cos kx - b \sin kx \\ &= 2a \cos kx - 2b \sin kx \\ C_1 e^{ikx} + C_2 e^{-ikx} &= A \cos kx - B \sin kx \end{aligned} \tag{2.84}$$

where:

$$\begin{aligned} A &= 2a \\ B &= 2b \end{aligned} \tag{2.85}$$

In this example, we have carefully chosen the values of C_1 and C_2 so that the resulting superimposed wave is represented by a real function. This need not necessarily be the case. It depends on the physical situation being modelled. For example, for superimposed waves on a stretched string, the waves are real. For matter waves in quantum physics, the waves are complex.

Electromagnetic Waves

3

3.1 ELECTROMAGNETISM

In 1830, Michael Faraday was experimenting with electricity and magnetism at the Royal Institution in London. A connection between the two had been earlier demonstrated by Hans C. Oersted in Denmark in 1820. It was Faraday's thought to represent the effect of one electric charge on another in terms of a field. The field could be represented by lines drawn coming radially outwards from a positive charge and pointing inwards towards a negative charge. The number of lines depended upon the strength of the field.

If the field is created by the presence of an isolated electric charge q, then the strength of the field at some radius r from the charge (in free space, or a vacuum), is given by:

$$E = \frac{1}{4\pi\varepsilon_0}\frac{q}{r^2} \tag{3.1}$$

where $\varepsilon_0 = 8.85 \times 10^{-12}\ \mathrm{Fm^{-1}}$ is called the permittivity of free space.

The total number of lines (not that we ever actually see or count them) is proportional to a quantity called the electric flux ϕ. The field strength E is the electric f lux per unit area through which the lines pass. For the case of an isolated charge, with the field lines directed radially outwards (or inwards for the case of a negative charge), the area A at some radius r is simply $A = \pi r^2$, and so the electric field is:

$$E = \frac{\varphi}{A} \tag{3.2}$$

and so the electric flux, for this case, is calculated from:

$$\varphi = \frac{q}{\varepsilon_0} \tag{3.3}$$

The direction of the field lines indicates the direction of the force which would be applied to a positive "test" charge if it were placed in the field. This is shown in Figure 3.1(a).

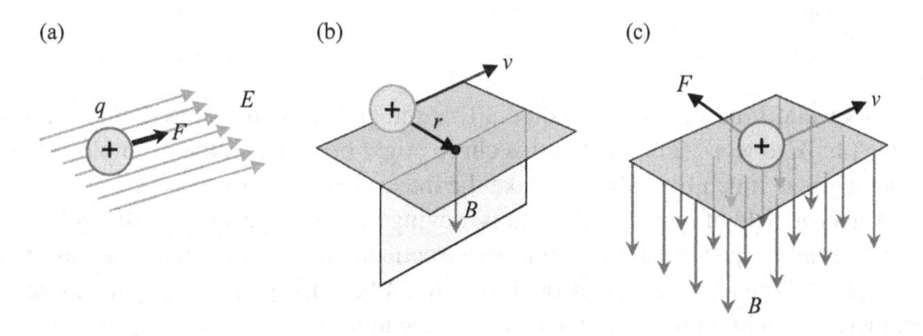

FIGURE 3.1 (a) Direction of force applied to a positive test charge in an electric field. (b) Direction of magnetic field lines for a moving charge. (c) Direction of force on a positive test charge moving in a magnetic field.

The magnitude of the force due to the electric field was shown by Coulomb to be:

$$F = qE \tag{3.4}$$

where q is the magnitude of the charge, and E is the magnitude of the field.

There are various ways of creating an electric field. As we shall see, it is not always necessary to have an actual electric charge present.

A magnetic field B can be created by the steady movement of an electric charge (i.e. an electric current). The charge q moves with a constant velocity v.

As shown in Figure 3.1(b), the magnitude of B at some distance r from the charge is found to be dependent on the magnitude of the charge q and the velocity of the charge v and inversely proportional to the perpendicular distance r squared.

$$B = \frac{\mu_0}{4\pi}\frac{qv}{r^2} \tag{3.5}$$

where $\mu_0 = 4\pi \times 10^{-7}$ Wb A^{-1} m^{-1} is called the permeability of free space.

If there is already a magnetic field present, as seen in Figure 3.1(c), when a test charge moves in this field, it experiences a force which is perpendicular to the direction of motion of the charge and the field.

The magnitude of the force is found to be dependent on the magnitude of the charge, the velocity and the strength of the field.

$$F = qvB \tag{3.6}$$

In a similar way to the electric field, Faraday envisioned that magnetism could be expressed in terms of magnetic field lines.

Expressed in terms of a magnetic flux, we have:

$$B = \frac{\Phi}{A} \tag{3.7}$$

where the magnetic flux can be calculated from:

$$\Phi = \mu_0 qv \tag{3.8}$$

Faraday discovered that, as well as being produced by a moving charge, a steady or constant magnetic field B could also be created by a changing electric flux. The magnitude of the B field was dependent on the rate of change of the electric flux ϕ. James C. Maxwell expressed this mathematically as*:

$$\oint \mathbf{B} \cdot d\mathbf{l} = \mu_0 \varepsilon_0 \frac{d\varphi}{dt} \tag{3.9}$$

This integral is a special kind of integral called a line integral around a closed curve. For our present purposes, we don't have to worry about what this curve might be, since in the example below, it will just be a straight-line segment much like when we take the integral along the x axis.

Consider an electric field of magnitude E that is moving through free space with a velocity v as shown in Figure 3.2(a). The field is just about to enter a cross-sectional area A. In a time interval dt, the field has travelled a distance vdt. The electric flux through the area A has changed. Initially it was zero, and at the end of time dt, there is a field on the right-hand edge of the area.

* This is also called Ampere's law.

By Ampere's law, a changing electric flux is one of the conditions that produce a magnetic field B. That is, because there is a change in electric flux over the area A, then we obtain a steady magnetic field B also passing through the area A but pointing at right angles to E and the direction of motion of E. This is shown in Figure 3.2(b).

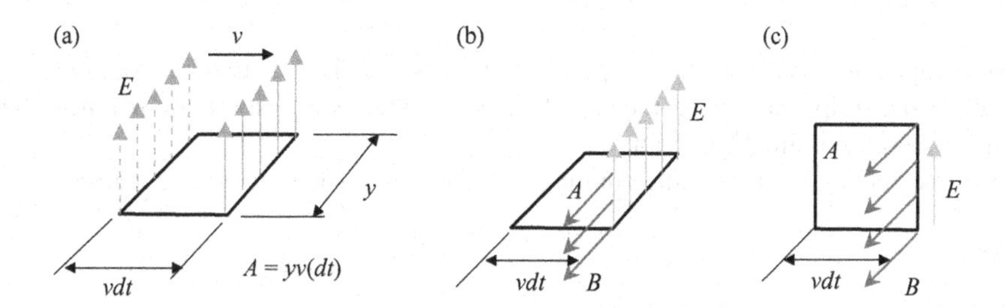

FIGURE 3.2 (a) Moving electric field entering area A. (b) Moving electric field leaving area A and associated magnetic field. (c) Change in magnetic field for area A.

The change in electric flux $d\phi$ is during time dt over the area dA:

$$\begin{aligned} d\varphi &= EdA \\ &= Eyvdt \end{aligned} \tag{3.10}$$

and so:

$$\frac{d\varphi}{dt} = Eyv \tag{3.11}$$

Thus, the resulting magnetic field can be obtained from:

$$\begin{aligned} \oint \mathbf{B} \cdot d\mathbf{l} &= \mu_0 \varepsilon_0 \frac{d\varphi}{dt} \\ &= \mu_0 \varepsilon_0 Eyv \end{aligned} \tag{3.12}$$

In this case, the line integral is the product of B and the length y (since y lies along the same direction as B, and after time dt, B now exists only on the right-hand edge of the area), and so:

$$\begin{aligned} \oint \mathbf{B} \cdot d\mathbf{l} &= By \\ &= \mu_0 \varepsilon_0 Eyv \\ B &= \mu_0 \varepsilon_0 Ev \\ E &= \frac{1}{\mu_0 \varepsilon_0 v} B \end{aligned} \tag{3.13}$$

In this example, the change in electric flux has been brought about by the passage of a field E across an area dA. We can also have an area A through which the field changes dE. Both represent a change in flux which in turn produces a magnetic field B. The strength of the B field depends upon the strength of the E field and the velocity (i.e. the rate of change of electric flux) of the E field.

Another way of looking at this is to use Faraday's law, where a changing magnetic field creates a changing electric field. Maxwell formulated Faraday's law as:

$$\oint E.dl = -\frac{d\Phi}{dt} \tag{3.14}$$

Now consider the upright area A which is perpendicular to the area in the previous example. This is shown in Figure 3.2(c). In the simple case here, the area A, in which there was initially no magnetic field, experiences a magnetic field B passing through it.

From this perspective, this is a change in magnetic flux $d\Phi$ through that area. Thus:

$$\begin{aligned} d\Phi &= BdA \\ &= Byvdt \\ \frac{d\Phi}{dt} &= Byv \end{aligned} \tag{3.15}$$

And so, Faraday's law gives:

$$\begin{aligned} Ey &= Byv \\ E &= vB \end{aligned} \tag{3.16}$$

The strength of the E field depends upon the strength of the B field and the velocity of the B field.

It was the great discovery of Maxwell to realise that if the velocity of the B field were to be the same as the velocity of the E field, the two fields would be self-sustaining and self-propagating. That is,

$$\begin{aligned} E &= \frac{1}{\mu_0\varepsilon_0 v}B = vB \\ \frac{E}{B} &= \frac{1}{\mu_0\varepsilon_0 v} = v \\ &= \frac{1}{\mu_0\varepsilon_0} = v^2 \end{aligned} \tag{3.17}$$

By measuring the strength of the electric field E and the accompanying magnetic field B, Maxwell was able to calculate this velocity, and to his great surprise, the value came out to be the velocity of light – a quantity with which he was familiar and so recognised immediately. Up until then, no one had any idea what light actually was. Maxwell showed that light consisted of a changing electric field accompanied by a changing magnetic field – an electromagnetic *wave*.

The velocity of light is usually given the symbol c, and so:

$$\begin{aligned} c^2 &= \frac{1}{\mu_0\varepsilon_0} \\ c &= 2.99792458 \times 10^8 \text{ m s}^{-1} \end{aligned} \tag{3.18}$$

Maxwell showed that:

$$\frac{E(x,t)}{B(x,t)} = c \tag{3.19}$$

That is, a self-sustaining electromagnetic wave can only travel at velocity c in free space and is independent of the frequency of the wave.

The most common type of travelling electromagnetic fields are those which vary continuously with time in a smooth periodic manner. In terms of the electric field, at a particular point in space, this might be represented as:

$$E(t) = E_0 \sin(\omega t) \tag{3.20}$$

where E_0 is the amplitude of the field.

A changing electric field travelling through space is accompanied by a changing magnetic field with a direction normal to the field and the velocity. This is the electromagnetic wave. Figure 3.3 shows a snapshot view of the wave travelling in the z direction.

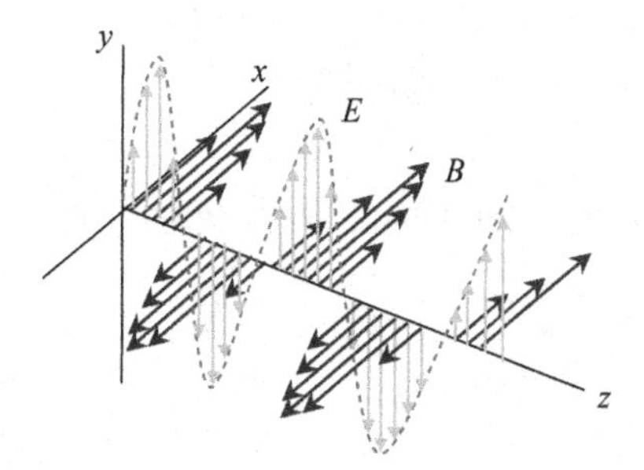

FIGURE 3.3 Time-varying electric and magnetic fields at some time t.

In this case, in terms of x and t:

$$\begin{aligned} E(x,t) &= E_0 \sin(kx - \omega t) \\ B(x,t) &= B_0 \sin(kx - \omega t) \end{aligned} \tag{3.21}$$

In exponential form, we can express the waves as:

$$\begin{aligned} E(x,t) &= E_0 e^{i(kx-\omega t)} \\ B(x,t) &= B_0 e^{i(kx-\omega t)} \end{aligned} \tag{3.22}$$

Note, the fields in Figure 3.3 have been drawn in accordance with the right-hand rule. Depending on how the axes are labelled, the E and B fields may have opposite signs. For example, if as shown in Figure 3.3 the positive x direction was into the page, the fields have the same sign. If the positive x direction is out of the page, then E and B have opposite signs and the positive z direction would be into the page.

These "field" waves have all the properties of mechanical waves, displaying diffraction, interference, reflection and refraction. The range of frequencies most commonly encountered is termed the electromagnetic spectrum.

Maxwell's prediction about the existence of electromagnetic waves, although begun as a mathematical treatment of Faraday's lines of force, was theoretical in nature. The existence of electromagnetic waves had only been demonstrated in so far as their velocity was predicted to be the same as that of light. The actual existence of oscillating, propagating electric and magnetic waves was not so easily accepted. In 1888, Heinrich Hertz verified their existence by showing that a spark generated between the electrodes of an induction coil would produce a spark in the gap of a nearby coil. Hertz deduced that electromagnetic waves travelled from the generator coil to the receiver coil. He measured the velocity of these waves, which were in the microwave region, to be that of the speed of light, and that this speed did not depend upon the frequency of the waves.

3.2 ENERGY IN ELECTROMAGNETIC WAVES

Consider the energy required to charge a capacitor which consists of two parallel plates as shown in Figure 3.4.

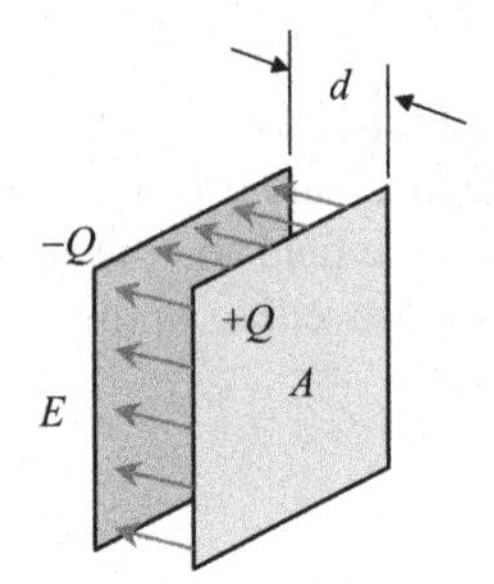

FIGURE 3.4 Electric field between two parallel plates.

When a voltage V is applied across the plates of a capacitor, energy U_E is stored within the field E between the plates of the capacitor.

$$U_E = \frac{1}{2}CV^2 \tag{3.23}$$

Now, the capacitance C of a parallel plate capacitor is:

$$C = \varepsilon_0 \frac{A}{d} \tag{3.24}$$

and the voltage across the plates is:

$$V = Ed \tag{3.25}$$

Thus:

$$\begin{aligned} U_E &= \frac{1}{2}\varepsilon_0 \frac{A}{d} E^2 d^2 \\ &= \frac{1}{2}\varepsilon_0 E^2 (Ad) \end{aligned} \tag{3.26}$$

The energy density u_E is the energy contained within the electric field per unit volume (J m^{-3} in SI units). Here, the volume occupied by the field E is the product Ad. Thus:

$$u_E = \frac{U_E}{Ad} = \frac{1}{2}\varepsilon E^2 \tag{3.27}$$

This formula says that the energy density of an electric field is proportional to the square of the magnitude of the E field.

Turning now to an inductor, the energy required to "charge" an inductor of length l, cross-sectional area A and N number of turns is:

$$U_B = \frac{1}{2}LI^2 \text{ where } L = \mu_0 A \frac{N^2}{l} \tag{3.28}$$

The energy is stored with the magnetic field B:

$$B = \mu_0 I \frac{N}{l} \tag{3.29}$$

And so, the energy stored is:

$$\begin{aligned} U_B &= \frac{1}{2}\left[\mu_0 A \frac{N^2}{l}\right]\left[\frac{B}{\mu_0}\frac{l}{N}\right]^2 \\ &= \frac{1}{2\mu_0} B^2 A l \\ u_B &= \frac{1}{2\mu_0} B^2 \end{aligned} \tag{3.30}$$

where μ_B is the energy density of a magnetic field.

If these steady state electric and magnetic fields appear together, the total energy density of the combined fields is thus:

$$\begin{aligned} u &= u_E + u_B \\ &= \varepsilon_0 E^2 \text{ or } \frac{1}{\mu_0} B^2 \end{aligned} \tag{3.31}$$

If, during time Δt, these steady state E and B fields pass a particular point in space, then the length of the "volume" containing the field is $l = c\Delta t$. If we consider a 1 m^2 area perpendicular to the direction of travel, then in terms of the energy density u, the intensity of the field (power per unit cross-sectional area A) is obtained from:

$$\begin{aligned} I &= \frac{P}{A} \\ &= \frac{u}{\Delta t} V \frac{1}{A} \\ &= \frac{u}{\Delta t} lA \frac{1}{A} \\ &= \frac{u}{\Delta t} c\Delta t \\ &= uc \end{aligned} \tag{3.32}$$

Since E and B are vectors, and it is the cross-sectional, or perpendicular, area through which the intensity is being calculated, the intensity I will be a vector also and will involve a cross-product.

Substituting $\mu = \varepsilon_0 E^2$, we have:

$$\begin{aligned} I &= \varepsilon_0 E^2 c \\ &= \varepsilon_0 EB \frac{1}{\varepsilon_0 \mu_0} \\ &= \frac{1}{\mu_0} EB \end{aligned} \tag{3.33}$$

In vector form, this is written:

$$\mathbf{S} = \frac{1}{\mu_0}\mathbf{E} \times \mathbf{B} \tag{3.34}$$

where **S** is called the Poynting vector and has units of W m^{-2}.

Now, for the case where $E=E(t)$ and $B=B(t)$ are functions of time, we have:

$$E = Bc \tag{3.35}$$

where here, E and B are the instantaneous values. Thus:

$$\begin{aligned} u_E &= \frac{1}{2}\varepsilon_0 B^2 c^2 \\ &= \frac{1}{2}\varepsilon_0 u_B 2\mu_0 c^2 \\ &= \varepsilon_0 \mu_0 c^2 u_B \end{aligned} \tag{3.36}$$

But:

$$c = \frac{1}{\sqrt{\varepsilon_0 \mu_0}} \tag{3.37}$$

therefore:

$$u_E = u_B \tag{3.38}$$

That is, the energy carried by the fields is divided equally between the electric and magnetic fields.

If the electric and magnetic fields both vary sinusoidally with time (an electromagnetic wave), and E_0 and B_0 are the amplitudes, then:

$$\begin{aligned} E(t) &= E_0 \sin(\omega t) \\ B(t) &= B_0 \sin(\omega t) \end{aligned} \tag{3.39}$$

and so, the intensity (W m^{-2} in SI units) as a function of time is:

$$I(t) = \frac{1}{\mu_0} E_0 B_0 \sin^2(\omega t) \tag{3.40}$$

In this case, the energy is being carried along by the wave, but the energy density at some particular point is changing sinusoidally with time as the wave passes through. If we were to take a snapshot of the wave at some time t, at some point in space, the energy density would be zero. A short time later, the energy density at that same point would be a maximum, the value being dependent on the amplitude of the wave. This does not mean that energy is being alternately created and destroyed at that point. It means that the energy is being transferred from that point to the next. The energy in the field at one point is being transferred over to the next point in space. At any point in space, the energy density, no matter what it might be – somewhere between zero and a maximum – is divided equally between the electric and magnetic fields. This is different from the case of a transverse wave on a stretched string or a swinging pendulum.

In the case of a pendulum, at points of zero velocity at the extremities of the motion, all the kinetic energy of the pendulum has been transferred to potential energy. The energy is not divided equally between kinetic and potential energy but continually converted from one to the other. Unlike the electromagnetic wave, the total energy at some point in the motion of the pendulum is a constant.

To find the average power and the average intensity of an electromagnetic wave, we integrate the $\sin^2$ function over 2π:

$$I_{av} = \frac{\int_0^{2\pi} I_i d\theta}{2\pi} = \frac{1}{\mu_0}\frac{E_0 B_0}{2\pi}\int_0^{2\pi} \sin^2\theta d\theta \tag{3.41}$$

$$= \frac{1}{2\mu_0} E_0 B_0$$

where E_0 and B_0 are the amplitudes of the fields. Note that the integral in the above evaluates to π.

Or, working in terms of the E field only:

$$I_{av} = \varepsilon_0 c\left[\frac{1}{2}E_0^2\right] \tag{3.42}$$

Note that here, the average intensity of an electromagnetic field is proportional to the amplitude of the E field squared and does not depend upon the frequency of the wave.

Just as we define the root mean square (rms) values of voltage and current, which, when multiplied give the average power dissipation in a resistive AC circuit, we can express the average value of the intensity of an electromagnetic wave in terms of the rms value of the E (or B) field. Letting:

$$E_{rms} = \frac{1}{\sqrt{2}} E_0 \tag{3.43}$$

where E_0 is the amplitude of the field, we have:

$$I_{av} = \varepsilon_0 c E_{rms}^2 \tag{3.44}$$

Kinetic Theory of Gases

4

4.1 INTRODUCTION

Consider N molecules of an ideal gas inside a container of volume V at an absolute temperature T. We recognise that in normal circumstances, there are a great many molecules, too small to be seen individually, moving with great velocity and undergoing collisions with the walls of the container and each other. Indeed, in a typical macroscopic system, the number of gas molecules is of the order of Avogadro's number, 6×10^{23} molecules – far too many to calculate the individual velocities and displacements. The average velocity of such molecules is of the order of 1000 m s^{-1}. The typical distance a gas molecule might move before colliding with another, or the container walls, is of the order of 100 μm – far greater than the size of the molecule itself. These figures set the scale of what we are about to study.

Unlike the motion of planets in the solar system, of which there are only a few easily seen bodies, the study of gases cannot be done on an individual basis. As such, any macroscopic properties that are required to be measured in the laboratory, like pressure and temperature, can only be described in terms of the average motion of a large number of individual molecules making up the gas.

The kinetic theory of gases uses the statistical average of individual motions to describe macroscopic quantities. In order to do this, several basic assumptions must be made to begin with:

- The molecules are negligibly small in comparison to the size of the container and the distance between them.
- The molecules collide elastically with each other and the walls of the container.
- The molecules do not otherwise interact (e.g. no attractive forces between them – such as there would be in a solid or liquid).
- The molecules have an initial random motion.

Because there are such large numbers involved, statistical treatments lead to great certainty in the predictions of macroscopic quantities, and this is the power of statistical mechanics. Unlike classical thermodynamics, which deals with experimentally derived equations of state to describe the relationship between macroscopic properties of gases, statistical mechanics seeks to derive these macroscopic relationships from microscopic properties which, in some circumstances, involve the quantum aspects of matter. Since the analysis of single molecules in these circumstances is not feasible, we gain access to these microscopic quantities by considering the statistics of the properties of large numbers of individual molecules.

4.2 PRESSURE AND TEMPERATURE

4.2.1 Pressure

Pressure is the result of the total average force acting on the walls of the container. Consider the collision of one molecule with the container wall as shown in Figure 4.1(a). During the collision, the velocity component v_y of the molecule is unchanged but v_x changes in direction but not in magnitude.

The change in velocity of a molecule during a time interval Δt is:

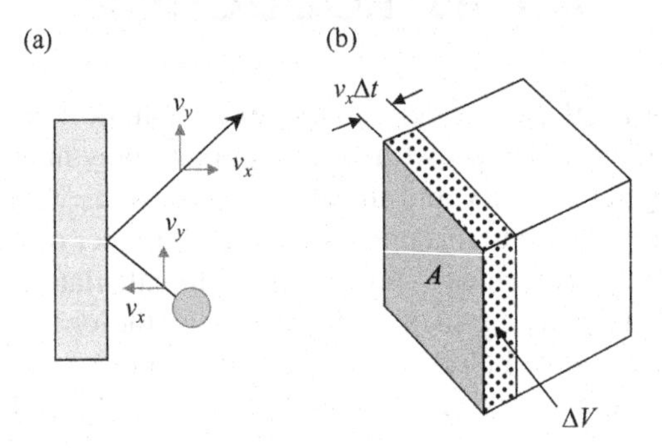

FIGURE 4.1 (a) Gas molecule bouncing off container wall showing velocity components. (b) Volume element that contributes to pressure on the wall.

$$v_x - -v_x = 2v_x$$
$$\Delta v = 2v_x \tag{4.1}$$

Thus, the force imparted to the wall by the molecule is:

$$F = \frac{m\Delta v}{\Delta t} = \frac{2mv_x}{\Delta t} \tag{4.2}$$

During a time Δt, molecules a distance less than or equal to $v_x\Delta t$ away from the wall will strike the wall. Thus, the number of collisions will be the number of molecules within the volume element $\Delta V = Av_x\Delta t$. This is shown in Figure 4.1(b).

If there are N molecules in the total volume V, then the number within the volume element ΔV is:

$$N\frac{\Delta V}{V} \tag{4.3}$$

The number of collisions is thus:

$$N = \frac{1}{2}N\frac{Av_x\Delta t}{V} \tag{4.4}$$

The total force on the wall at any instant during time Δt is thus:

$$F_{\text{total}} = \left[\frac{m2v_x}{\Delta t}\right]\left[\frac{1}{2}\frac{N}{V}Av_x\Delta t\right]$$
$$\frac{F}{A} = \frac{N}{V}mv_x^2 \tag{4.5}$$

The first bracketed term in the above is the force due to each collision. The second term is the total number of collisions.

But, so far, we have assumed that v_x is the same for each molecule. The molecules in the container actually have a range of speeds. The average value of v_x^2 components leads to the average force (and hence pressure) on the wall. The average value of the square of the velocities is:

$$\overline{v_x^2} = \frac{v_{x1}^2 + v_{x2}^2 + v_{x3}^2 + \ldots}{N} \tag{4.6}$$

And so, the average force per unit area is:

$$\frac{F_{\text{av}}}{A} = \frac{N}{V} m\overline{v_x^2} \tag{4.7}$$

It would be more convenient to have an expression which included the total velocity v rather than the x component v_x. The average squares of the velocity in each direction are equal:

$$\begin{aligned} \overline{v_x^2} &= \overline{v_y^2} = \overline{v_z^2} \\ \overline{v_x^2} &= \frac{1}{3}\overline{v^2} \end{aligned} \tag{4.8}$$

And so:

$$\begin{aligned} \overline{v^2} &= \overline{v_x^2} + \overline{v_y^2} + \overline{v_z^2} \\ \overline{v_x^2} &= \frac{1}{3}\overline{v^2} \end{aligned} \tag{4.9}$$

Thus, the pressure p becomes:

$$\begin{aligned} \frac{F_{\text{av}}}{A} &= \frac{N}{V} m\overline{v_x^2} \\ p &= \frac{1}{3}\frac{N}{V} m\overline{v^2} \\ &= \frac{2}{3}\frac{N}{V}\left(\frac{1}{2}m\overline{v^2}\right) \end{aligned} \tag{4.10}$$

Thus:

$$p = \frac{2}{3}\frac{N}{V}\left(\frac{1}{2}m\overline{v^2}\right) \tag{4.11}$$

The significance of the above is that a macroscopic quantity, pressure, has been expressed in terms of microscopic quantities: the mass and velocity of the individual molecules within the gas.

So, thus far, the only "statistics" that have been involved is the average value of the square of all the velocities as would be calculated (if this were possible) from each individual velocity for all the N molecules. We can see that pressure is a consequence of N, V and the average translational kinetic energy of a molecule, E. That is, $p=f(N,V,E)$.

4.2.2 Temperature

The macroscopic property temperature can also be related to the average kinetic energy of a single molecule.

Now, the pressure p is given by:

$$p = \frac{2}{3}\frac{N}{V}\left(\frac{1}{2}m\overline{v^2}\right) \tag{4.12}$$

Bringing V across to the left-hand side and making use of the general gas equation $pV = nRT$, we have:

$$\begin{aligned} pV &= \frac{2}{3}N\left(\frac{1}{2}m\overline{v^2}\right) \\ &= nRT \\ nRT &= \frac{2}{3}N\left(\frac{1}{2}m\overline{v^2}\right) \\ &= \frac{2}{3}nN_A\left(\frac{1}{2}m\overline{v^2}\right) \end{aligned} \tag{4.13}$$

since:

$$N = nN_A \tag{4.14}$$

Where N_A is Avogadro's Number. But R and N_A are both constants. The ratio of them is a new constant, Boltzmann's constant $k = 1.38 \times 10^{-23}$ J K^{-1}.

$$\frac{3}{2}kT = \frac{1}{2}m\overline{v^2} \tag{4.15}$$

or:

$$\frac{3}{2}kT = \frac{1}{2}mv_{\text{rms}}^2 \tag{4.16}$$

where:

$$v_{\text{rms}} = \sqrt{\overline{v^2}} \tag{4.17}$$

The square root of the average of all the velocity squared is called the root mean square velocity, or the rms velocity. The rms velocity does not equal the average velocity; it is a little less.

The quantity $3/2kT$ is therefore the average translational kinetic energy of a molecule. The average translational kinetic energy of a single molecule depends only on the temperature T.

The equation of state thus becomes:

$$\begin{aligned} pV &= \frac{2}{3}N\left(\frac{3}{2}kT\right) \\ &= NkT \end{aligned} \tag{4.18}$$

or:

$$
\begin{aligned}
T &= \frac{pV}{Nk} \\
&= f(p,V,N)
\end{aligned}
\tag{4.19}
$$

The equation of state describes the state of a system in terms of macroscopic quantities N, p, V and T. We have seen that for the case of translational kinetic energy, T is essentially a measurement of E, the energy in the system.

4.2.3 Degrees of Freedom

Molecules in a gas are capable of independent motion. Consider a diatomic gas molecule (one with two molecules, such as H_2):

- The molecule itself can travel as a whole from one place to another. This is a translational motion.
- The molecule can spin around on its axis. This is a rotational motion.
- The atoms within the molecule can vibrate backwards and forwards. This can happen when the molecule consists of more than one atom. This type of motion is more applicable to solids.

For a monatomic gas molecule, (e.g. He) only three translational velocity components are required to describe the motion before and after any collisions. These components are called degrees of freedom.

For a diatomic gas molecule, three translational and two rotational velocity components are required to describe the motion before and after any collisions. This represents five degrees of freedom.

One way to determine if a rotation actually counts as a degree of freedom is to look along the axis of rotation. If the outline of the shape of the molecule changes during a rotation, then it is counted. If one cannot see any difference in the orientation of the molecule, then it doesn't count.

4.2.4 Equipartition of Energy

The average translational kinetic energy of a gas molecule depends upon the temperature. This is true for a mixture of gases as well as just a single gas. For example, if there is a mixture of two monatomic gases present, one a heavy molecule and one a light molecule, then the average velocity of the heavy molecules adjusts downwards so as to keep the average kinetic energy ($1/2\ mv^2$) equal to that of the lighter ones. Although each molecule undergoes frequent changes of kinetic energy during collisions, the average kinetic energy is equally distributed amongst all the molecules present.

The average translational kinetic energy of a gas molecule is:

$$
\frac{3}{2}kT = \frac{1}{2}mv_{\text{rms}}^2 \tag{4.20}
$$

There is also kinetic energy of rotation to consider for gases with two or more atoms per molecule. For a diatomic molecule, there are two additional degrees of freedom (i.e. the two rotational ones). The average total kinetic energy will be equally distributed amongst the available degrees of freedom.

The average translational kinetic energy of molecules is evenly distributed amongst the gas molecules present. It is this component of the total kinetic energy that gives rise to pressure. The average translational kinetic energy depends only on the temperature T and not on the mass or velocity of the molecules. For a given set of molecules, the average velocities adjust for each species so that they all have the same average translational kinetic energy.

The total kinetic energy of the molecules (often called the internal energy) is equally distributed amongst the available degrees of freedom. This is why the specific heat of different gas species are all different. When a diatomic gas is heated, the input energy has to be evenly spread amongst five degrees of freedom. It is only the translational components that give rise to temperature. So, we have to put in more energy for a diatomic gas to get the same rise in temperature compared to a monatomic gas.

Molecules in solids do not have translational or rotational kinetic energy components in the same sense as gases (otherwise they would not be solid). Instead, in the simplest sense, internal energy is distributed through vibrations of atoms about their equilibrium positions. This vibration takes the form of simple harmonic motion, whereby energy is alternately stored (as strain potential energy) and then released (as kinetic energy). The emission of radiation from solids in general is a complicated affair, but for metals, it is more easily understood. Metals are characterised by a large number of conduction electrons which are free to move about within the solid. These electrons are in many ways similar to the molecules in a gas, and so their modes of vibration can encompass different energy levels. As a consequence, as the temperature rises, the frequency distribution of the radiant energy from the solid changes, with high frequencies gaining more amplitude – the solid begins to emit visible light and might become "red hot".

4.2.5 Internal Energy

Consider the heating of N molecules of a gas a constant volume V from T_1 to T_2 by an electric element as shown in Figure 4.2. The element is switched on for a certain time and then switched off. A certain amount of energy Q is fed into the system.

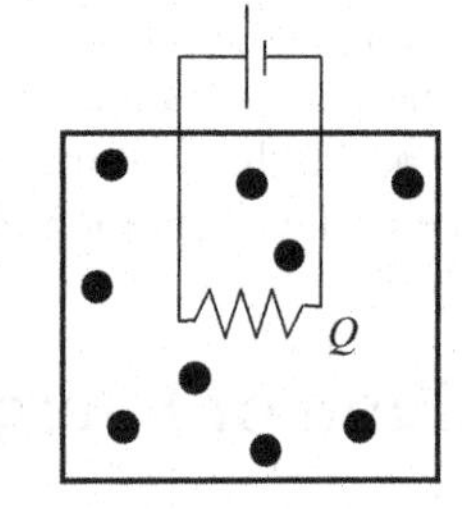

FIGURE 4.2 Heating of a constant volume of gas.

During the time the element is switched on, the atoms in the heating element are vibrating according to the flow of current and the resistance of the element. These vibrations are imparted to any molecules which strike the heating element, and so the molecules acquire additional kinetic energy. These molecules then go off and make collisions with other molecules, and the overall temperature of the gas increases with time.

When the element is switched off, and the system allowed to settle, the temperature and pressure reach new steady-state values. Thus, we have an initial state (p_1, V_1 and T_1) and a final state (p_2, V_2, T_2) where here, $V_1 = V_2$.

During the process (i.e. the transition from the initial state to the final state), the hot molecules near the element gradually give up some of their energy to surrounding cooler molecules until such time as the average kinetic energy of molecules in a representative volume anywhere in the container is the same. This is a condition of thermal equilibrium.

At thermal equilibrium, an equation of state can be used to relate macroscopic quantities. For example, the macroscopic quantity pressure can be expressed:

$$p = \frac{NkT}{V} = f(N,V,T) \tag{4.21}$$

Alternately, the macroscopic quantity temperature can be described by:

$$T = \frac{pV}{Nk} = f(N, p, V) \tag{4.22}$$

The quantity kT has physical significance: it is the kinetic energy of one degree of freedom of a molecule. For diatomic molecules, the internal kinetic energy is a combination of translational and rotational components – but due to the equipartition of energy, these components all depend equally on T. So, a measure of E (the total kinetic energy) is really a measure of T (we just multiply T by Boltzmann's constant k to get E). The actual relationship between E and T depends on the nature of the gas (i.e. the specific heat). In general terms, $E = E(T)$ (energy is some function of temperature).

4.3 STATISTICAL MECHANICS

4.3.1 Statistical Weight

At any instant, a thermodynamic system may have particular values of volume, pressure, temperature, etc., which we call its macrostate. There are usually very many different microstates (molecular velocities and individual kinetic energies) that correspond to one particular macrostate.

Consider a macrostate with a certain total energy E. One possible microstate corresponding to this macrostate might be if all the energy E was concentrated in one molecule and the rest had $E_1 = 0$. It is very unlikely that we would encounter this particular microstate, especially if we had a large number of molecules.

Let's examine the case of a volume V consisting of $N = 4$ molecules that can each have one of two distinct energy levels, 0 or E_1. There is only one degree of freedom. Molecules can move up or down from 0 to E_1 but not side to side or back to front. The model is a one-dimensional "gas". The value of the total energy we shall arbitrarily set (for the sake of example) at $E = 2E_1$. The possible microstates for the four molecules m_1 to m_4 are shown in Figure 4.3. Because molecules are identical, all these microstates represent the same macrostate $E = 2E_1$. The selection of $E = 2E_1$ as the total energy is equivalent to setting the temperature of the gas.

$$\begin{array}{l} E = 2E_1 \\ \Omega = 6 \end{array} \left\{ \begin{array}{cccc} m_1 & m_2 & m_3 & m_4 \\ E_1 & E_1 & 0 & 0 \\ E_1 & 0 & E_1 & 0 \\ E_1 & 0 & 0 & E_1 \\ 0 & E_1 & E_1 & 0 \\ 0 & E_1 & 0 & E_1 \\ 0 & 0 & E_1 & E_1 \end{array} \right.$$

FIGURE 4.3 Microstates for two energy levels 0 and E_1 for four gas molecules with total energy $2E_1$.

The number of possible microstates for a particular macrostate is given the symbol $\Omega(n)$ where n is the number of molecules with energy $E > 0$. From statistics:

$$\Omega(n) = \frac{N!}{n!(N-n)!} \tag{4.23}$$

Note, when N is large, Stirling's formula may be used:

$$\ln x! \approx x \ln(x) - x \tag{4.24}$$

Ω is called the statistical weight of the macrostate. In this example, $n=2$, $N=4$ and so $\Omega=6$.

What if the temperature of the system were lowered so that $E=1E_1$? Table 4.1 shows the values of Ω for different values of total energy (i.e. different macrostates) E from zero to $4E_1$. Note that the value of Ω is low for highly ordered macrostates ($E=0$ or $E=4E_1$) and high for macrostates where the energy is more evenly shared by each molecule. When a system is at equilibrium, it has a particular macrostate specified by N, V and E. This macrostate can be represented by a number of different possible microstates. For our two-level system with four molecules, for the macrostate $E=2E_1$ there are six possible microstates. The system will cycle through these six possible states (in random order) continually as collisions occur. Note, it is advantageous to treat systems in terms of their total energy rather than their temperature since this makes the method applicable to many other physical systems.

Let us now imagine that there exist many more available levels: E_1=$3E_1$, $4E_1$, $5E_1$, etc. For a total energy of $2E_1$ three of our molecules could have an energy 0 and the remaining one exist at the $2E_1$ level ($n=1$, $\Omega=4$), or we could have two molecules at 0 and two at the $1E_1$ level ($n=2$, $\Omega=6$). The total number of possible microstates is $6+4=10$. Note that there are more disordered states ($\Omega=6$) than ordered states ($\Omega=4$). The possibilities are shown in Figure 4.4(a).

Note, here we have a set of two groups of macrostates of different statistical weights corresponding to the same total energy $2E_1$.

For a total energy of $3E_1$, as shown in Figure 4.4(b), three of our molecules could have an energy zero and the remaining one exist at the $3E_1$ level ($n=1$, $\Omega=4$), or we could have one molecule at 0 and three at

TABLE 4.1 Statistical weight vs number of macrostates

n	$\Omega(n)$
0	1
1	4
2	6
3	4
4	1

(a)

$E=2E_1$	m_1	m_2	m_3	m_4
$\Omega=4$	$2E_1$	0	0	0
	0	$2E_1$	0	0
	0	0	$2E_1$	0
	0	0	0	$2E_1$
$\Omega=6$	$1E_1$	$1E_1$	0	0
	$1E_1$	0	$1E_1$	0
	$1E_1$	0	0	$1E_1$
	0	$1E_1$	$1E_1$	0
	0	0	$1E_1$	$1E_1$
	0	$1E_1$	0	$1E_1$

(b)

$E=3E_1$	m_1	m_2	m_3	m_4	
$\Omega=4$	$3E_1$	0	0	0	$n=1$
	0	$3E_1$	0	0	
	0	0	$3E_1$	0	
	0	0	0	$3E_1$	
$\Omega=4$	$1E_1$	$1E_1$	$1E_1$	0	$n=3$
	$1E_1$	$1E_1$	0	$1E_1$	
	$1E_1$	0	$1E_1$	$1E_1$	
	0	$1E_1$	$1E_1$	$1E_1$	
$\Omega=12$	$2E_1$	$1E_1$	0	0	
	$2E_1$	0	$1E_1$	0	
	$2E_1$	0	0	$1E_1$	
	0	$2E_1$	$1E_1$	0	
	0	0	$2E_1$	$1E_1$	
	0	$2E_1$	0	$1E_1$	
	$1E_1$	$2E_1$	0	0	
	$1E_1$	0	$2E_1$	0	
	$1E_1$	0	0	$2E_1$	
	0	$1E_1$	$2E_1$	0	
	0	0	$1E_1$	$2E_1$	
	0	$1E_1$	0	$2E_1$	

FIGURE 4.4 (a) Microstates for three energy levels 9, E_1, $2E_1$ for total energy $2E_1$. (b) Total energy $3E_1$.

the $1E_1$ level ($n=3$, $\Omega=4$). Or we could have two at $0E_0$ and one at $1E_1$ and one at $2E_1$. In Figure 4.4(b), $\Omega=12$ corresponds to two degenerate versions of n = 2, and so the total statistical weight is $\Omega=2\times 6=12$.

We have here three groups of macrostates corresponding to the same total energy $3E_1$. Each group might contain an additional external macroscopic variable.

4.3.2 Boltzmann Distribution

It is evident that as both N and total energy increase, the greatest statistical weight increases. For $E=2E_1$, the greatest statistical weight was 6, whereas for $E=3E_1$, the greatest statistical weight was 12. That is, the higher the temperature, the more disorder there is in the system.

Let's focus on the system at total energy $E=3E_1$. There are three macrostates representing this energy which have statistical weights of 4, 4 and 12. The total number of possible microstates is 20. The probability of each of these three macrostates occurring is 4/20, 4/20 and 12/20, respectively.

- For the first state (4/20), we have three molecules with $0E_1$ and one with $3E_1$.
- For the second state (4/20), we have one molecule with $0E_1$ and three with $1E_1$.
- For the third state (12/20), we have two molecules at energy $0E_1$, one molecule at $1E_1$ and one at $2E_1$.

To find the most likely number of molecules at any one of the available energy levels, we add up the numbers in each of the three states and weight the sum by the probability of that state. For example, for the case of $E=3E_1$, to determine the most likely number of molecules at the $0E_1$ level, we see that we have:

$$\begin{aligned} 3\times(4/20) &= 0.6 \\ 1\times(4/20) &= 0.2 \\ 2\times(12/20) &= 1.2 \\ 0.6+0.2+1.2 &= 2 \end{aligned} \tag{4.25}$$

The most likely number of molecules at $0E_1$ is therefore two.

Repeating for the other energy levels, we obtain the data shown in Table 4.2.

Plotting the most likely number of molecules $n(E_i)$ against the energy levels E_i, we obtain a graph as shown in Figure 4.5(a).

This number distribution is the most likely number of molecules at each energy level $n(E_i)$. It shows how the number of molecules is distributed over the energy levels.

When the number of molecules becomes very large, we find that the distribution is in the form of an exponential. This is shown in Figure 4.5(b).

$$n(E_i) = Ce^{-E_i/A} \tag{4.26}$$

TABLE 4.2 Number of molecules vs energy level

ENERGY LEVEL	*CALCULATION*	*NO. MOLECULES*
$0E_1$	$3\times(4/20) + 1\times(4/20) + 2\times 12/20) =$	2.0
$1E_1$	$0\times(4/20) + 3\times(4/20) + 1\times(12/20) =$	1.2
$2E_1$	$0\times(4/20)+0\times(4/20)+1\times(12/20) =$	0.6
$3E_1$	$1\times(4/20)+0\times(4/20)+0\times(12/20) =$	0.2

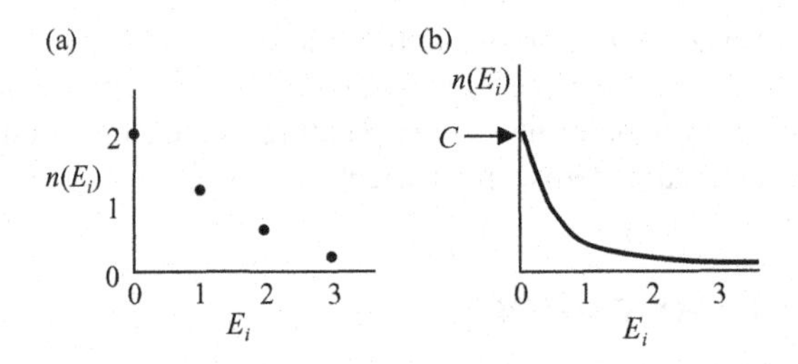

FIGURE 4.5 (a) Most likely number of molecules $n(E_i)$ against the energy levels E_i. (b) Exponential distribution of $n(E_i)$ vs E_i.

If a large proportion of the total energy of a system is stored in a higher energy level (say E_2), then we don't need as many molecules at that level to hold the energy, so $n(E_2)$ is very much less than $n(E_1)$.

A and C in the exponential function are constants. The constant C gives the value of $n(E_i)$ at $i=0$, that is: $E_0=0$ in our example system. The constant A sets the scale of the energies and is the average energy E over all the levels and is the product kT. The Boltzmann distribution is thus expressed:

$$n(E_i) = n(0)e^{-E_i/kT} \tag{4.27}$$

The Boltzmann distribution gives the number of molecules at a certain energy level in terms of the number of molecules that exist at the zero energy level. The zero energy level ($i=0$) is usually taken to be $E=0$. It is a number distribution.

The relative number of molecules in two different energy levels E_1 and E_2 can be found from the ratio:

$$\frac{n_1}{n_2} = \frac{e^{-E_1/kT}}{e^{-E_2/kT}} = e^{-\Delta E/kT} \tag{4.28}$$

If the energy interval ΔE is made vanishingly small, then the distribution is a continuous distribution. For continuous distributions, the chances of finding a molecule at a particular value of E_i is zero. In much the same way, it would be impossible to find a physics student of height exactly 1.6328 metres in a population of students. It is more meaningful to determine the number of molecules with an energy between two limits, E_i and E_i+dE.

In this sense, the gas molecules we are considering here are "classical" in the sense that we have not imposed any quantisation of energy levels. Energies can take on any values between 0 and infinity.

4.3.3 Velocity Distribution

We wish to know how to compute the fraction, or proportion, of molecules that have a velocity in the range v to $v + dv$. This will be given by the product of some function $f(v)$ evaluated at v times the width dv. The function and the velocity range of interest are shown in Figure 4.6.

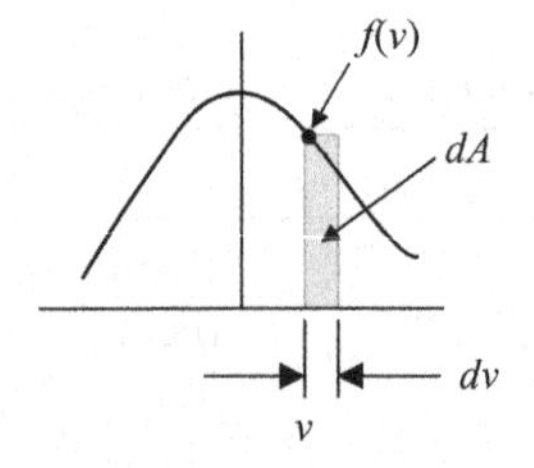

FIGURE 4.6 The fraction of molecules which have a velocity in the range $v + dv$.

The total area under the distribution curve represents all molecules and so has an area = 1. That is:

$$\int_{-\infty}^{\infty} f(v)\,dv = 1 \tag{4.29}$$

Can we now determine the form of $f(v)$?

Since the energies we are talking about are kinetic energies, it is not surprising to find that the velocity distribution of gas molecules follows a similar distribution to the Boltzmann distribution (since $E = 1/2mv^2$).

We can thus write:

$$f(v) = Ce^{-E/kT} \tag{4.30}$$

where C is a constant.

Or even better:

$$dA = f(v)\,dv = Ce^{-mv^2/2kT}\,dv \tag{4.31}$$

which gives the area dA under the curve, or the fraction of molecules with velocity in the range v to $v + dv$.

Since the sum over all velocities must equal 1, we have:

$$\int_{-\infty}^{\infty} f(v)\,dv = \int_{-\infty}^{\infty} Ce^{-mv^2/2kT}\,dv = 1 \tag{4.32}$$

Using a standard integral:

$$\int_{-\infty}^{\infty} e^{-x^2}\,dx = \sqrt{\pi} \tag{4.33}$$

we find:

$$C = \sqrt{\frac{m}{2\pi kT}} \tag{4.34}$$

and so, the one-dimensional Maxwell velocity distribution is:

$$f(v) = \left(\frac{m}{2\pi kT}\right)^{1/2} e^{-\frac{mv^2}{2kT}} \tag{4.35}$$

It should be stressed that this computation is for our sample "gas" in one dimension, i.e. one degree of freedom.

In Fig. 4.6, it is shown that the most probable velocity of a selected molecule is therefore 0 for one degree of freedom.

The complete velocity distribution for molecules of gas in a volume V has to include velocities for other degrees of freedom (side to side and back and forwards). In this case, we need to consider the velocity intervals v_x to $v_x + dv_x$, v_y to $v_y + dv_y$ and v_z to $v_z + dv_z$.

$$f(v_x, v_y, v_z)\,dv_x\,dv_y\,dv_z = A\left(e^{-mv_x^2/2kT}e^{-mv_y^2/2kT}e^{-mv_z^2/2kT}\right)dv_x dv_y dv_z \tag{4.36}$$

This can be more easily understood if we work with a three-dimensional velocity vector v. We draw a sphere of radius v which represents the vector sum of the velocities dv_x, dv_y, dv_z. This is drawn in Figure 4.7. We wish to compute the fraction of molecules with a velocity in the range $v+dv$. The sphere is a representation of "velocity space". The fraction we seek is the product of $f(v)$ times the volume enclosed by v and $v+dv$.

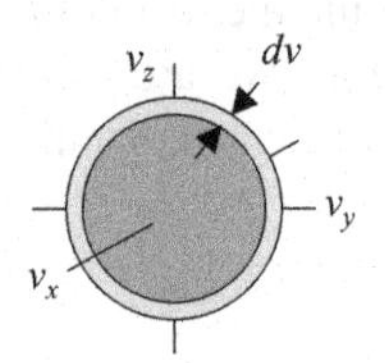

FIGURE 4.7 Fraction of molecules with velocity in the range $v + dv$ within a volume dv.

The volume of the sphere is:

$$V = \frac{4}{3}\pi v^3 \tag{4.37}$$

and so, an increment of volume expressed in terms of an increment of velocity is dV/dv:

$$dV = 4\pi v^2 dv \tag{4.38}$$

The fraction (volume) of molecules having velocity v in the range v to $v + dv$ is thus:

$$f(v)dV = Ae^{-mv^2/2kT}4\pi v^2 dv \tag{4.39}$$

Since the sum over all velocities must equal 1, we have:

$$\int_0^\infty f(v)dV = \int_0^\infty Ae^{-mv^2/2kT}4\pi v^2 dv = 1 \tag{4.40}$$

Using a standard integral, we find:

$$A = \left(\frac{m}{2\pi kT}\right)^{3/2} \tag{4.41}$$

Note that these integrals are taken from 0 to ∞. In spherical coordinates, we do not have negative "volumes".

The Maxwell velocity distribution for three degrees of freedom is thus:

$$f(v) = 4\pi\left(\frac{m}{2\pi kT}\right)^{3/2} v^2 e^{-\frac{mv^2}{2kT}} \tag{4.42}$$

Plotting $f(v)$ against v, we obtain the curve shown in Figure 4.8.

In the one-dimensional case, the maximum in the distribution occurred at $v=0$ because the one-dimensional distribution is based directly on the Boltzmann distribution – which is a probability density function. Here, the factor v^2 modifies the exponential, and so even though the probability density may be a maximum at 0, the most probable velocity is not.

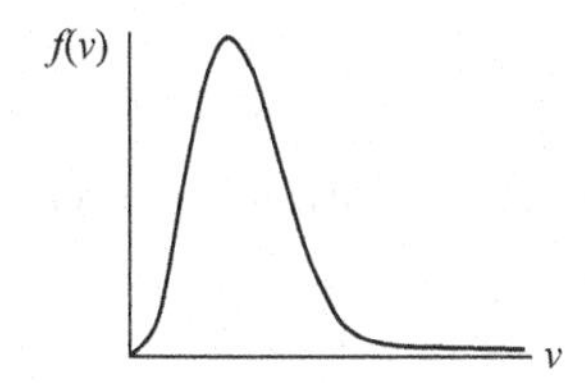

FIGURE 4.8 Maxwell velocity distribution.

4.3.4 Partition Function

In our model system of four gas molecules, we saw how those molecules could be arranged at different energy levels.

Over time, if the macroscopic quantities N, V and E are not changing, the system will cycle through every one of those possible microstates. But the probability of encountering a microstate within those of the greatest statistical weight is far greater (especially for large values of N) than the probability of finding a microstate not included in the macrostate of the greatest statistical weight – the latter being so small as to be negligible.

To find the most likely number of molecules at any one of the available energy levels, we add up the numbers in each of the microstates and weight the sum by the probability of that state. The Boltzmann distribution is the most likely number of molecules at each energy level $n(E_i)$.

$$n(E_i) = n(0)e^{-E_i/kT} \tag{4.43}$$

We should be clear about the difference between an energy level and an energy state. An energy level is just that, a particular value of E. For example, $E = 2E_1$. Let's say that there are multiple ways that molecules can arrange themselves at a particular energy level. (For example, in an atom, electrons can occupy the same energy level if they have different spins.) For a "two-fold" degeneracy, we say that there are two states in which molecules can occupy a particular energy level. This is depicted in Figure 4.9.

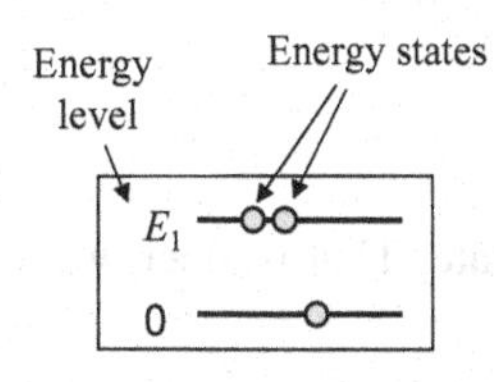

FIGURE 4.9 Energy states within an energy level.

In our model system, an individual molecule can have a range of energies over time. The total energy $E_T = E_0 + E_1 + \ldots$ may be a constant, but the energy level occupied by an individual molecule may change considerably with time.

For our one-dimensional model, the Boltzmann distribution shows how the most likely number of molecules at a certain energy level E_i at any instant can be calculated in terms of the number of molecules at the zero energy level.

If we normalise this to the total number of molecules, we can calculate the probability of encountering one molecule at energy level E_i when selected at random. The total number of molecules N is:

$$N = n_1 + n_2 + n_3 + \ldots \tag{4.44}$$

We may have n_1 molecules at E_1, n_2 molecules at E_2 and n_3 molecules at E_3, etc.

$$n(E_1) = n(0)e^{-E_1/kT} \tag{4.45}$$

$$n(E_2) = n(0)e^{-E_2/kT}$$

Thus, the total number of molecules can be found from the sum of the individual molecules:

$$N = n_0 + n_1 + n_2 + n_3 + \ldots = n(0)\sum_{i=0}^{\infty} e^{-E_i/kT} \tag{4.46}$$

The probability we seek is thus:

$$P(E_i) = \frac{n(E_i)}{N} = \frac{n(0)e^{-E_i/kT}}{n(0)\sum_{i=0}^{\infty} e^{-E_i/kT}} \tag{4.47}$$

Where the denominator here is the total number N. Eliminating $n(0)$, we obtain:

$$P(E_i) = \frac{e^{-E_i/kT}}{\sum_{i=0}^{\infty} e^{-E_i/kT}} \tag{4.48}$$

The normalising factor, the denominator in this expression, is called the partition function Z.

Note that the sum in the partition function is not just a count of the energy levels, but a weighted sum according to how these energy levels are likely to be occupied according to the prevailing temperature. The probability that a selected molecule has an energy E_i is thus:

$$P(E_i) = \frac{e^{-E_i/kT}}{Z} \tag{4.49}$$

But it is possible for a given energy level to have more than one state. That is, energy levels may have degeneracy $g(E_i)$.

Thus, to include degenerate states, we write:

$$N = n_0 + n_1 + n_2 + n_3 + \ldots = n(0)\sum_{i=0}^{\infty} g(E_i)e^{-E_i/kT} \tag{4.50}$$

The probability of selecting a molecule with energy E_i is thus:

$$P(E_i) = \frac{g(E_i)e^{-E_i/kT}}{\sum_{i=0}^{\infty} g(E_i)e^{-E_i/kT}} \tag{4.51}$$

Where the partition function is:

$$Z = \sum_{i=0}^{\infty} g(E_i)e^{-E_i/kT} \tag{4.52}$$

So, in general, the probability that a selected molecule has an energy E_i where degeneracy is accounted for is thus:

$$P(E_i) = \frac{g(E_i)e^{-E_i/kT}}{Z} \tag{4.53}$$

The partition function tells us, on average, how many energy states over which the molecules are going to be distributed at a given temperature T. It is a powerful function. It says something about the actual occupation of energy states, not just the existence of them.

The exponential form of the partition function comes from the number of ways in which molecules can arrange themselves amongst the available energy levels in different combinations.

4.3.5 Properties of the Partition Function

Consider a one-dimensional, two-level system. The energy levels are 0 and E with no degeneracy. We can express E in terms of the product kT for convenience. Here, E is essentially the energy gap between the two levels. At $\Delta E = kT = 150k$ (a modest energy gap), we can write the probability of finding a molecule at, say, the upper energy level $150k$ as:

$$P(150k) = \frac{e^{-150k/kT}}{Z} = \frac{e^{-150/T}}{e^{-0/T} + e^{-150/T}} \tag{4.54}$$

where:

$$Z = e^{-0/T} + e^{-150/T} \tag{4.55}$$

Let's calculate the probability $P(0)$ (a selected molecule being at the zero energy level) at $T = 200$K. In this example, there is no degeneracy.

$$\begin{aligned} Z &= e^{-0/200} + e^{-150/200} \\ &= 1.4724 \end{aligned} \tag{4.56}$$

The value of Z tells us, on average, how many energy levels of the two available are occupied when the temperature is 200 K.

$$\begin{aligned} P(0) &= \frac{e^{-0/200}}{1.4724} \\ &= \frac{1}{1.4724} \\ &= 0.679 \end{aligned} \tag{4.57}$$

The value of $P(0)$ tells us the number of molecules in the zero energy level expressed as a fraction of the total number of molecules.

If we repeat the calculation for different values of T (keeping $\Delta E = 150k$) we find $P(0)$ approaches 100% as T approaches 0 K and $P(0)$ approaches 50% as $T >> 150k$. This is shown in Figure 4.10(a).

At low temperature, the value of Z says that all the molecules tend to be distributed over one energy level. Z approaches 1. At low temperature, the value of $P(0)$ says that there is a high probability of selecting a molecule at the zero energy level. At high temperature, the value of Z says that all the molecules tend to be distributed evenly over both energy levels (Z approaches 2). At high temperature, the value of $P(0)$ says that there is about a 50% chance of selecting a molecule at the zero energy level.

There may be millions of molecules. For our simple two-level system, at high temperature, about half of them will be at the zero level state, and the other half in the upper energy state. Z tells us how many states are occupied; P tells us how many molecules will be in a particular state.

It is interesting to compute the partition function and probabilities for the case of a widely spaced energy interval: $\Delta E = 1000k$. This is plotted in Figure 4.10(b). In that figure, Z tells us, on average, how many states of the two available are occupied when the temperature is 1000 K. $P(0)$ tells us the number of molecules in the zero state expressed as a fraction of the total number of molecules.

We can see that at a widely spaced energy level, the temperature has to be raised much further (compared to the lower spaced energy level case) for molecules to enter the higher energy state.

The partition function tells us, on average, how many of the available energy levels are being used to accommodate the energy of the molecules. It doesn't tell us how many molecules may be in those levels. It just tells us how many of the available levels are being used. If we want to know the number of molecules that are in a particular energy level, we can determine this by calculating the probability (that is, we find the number of molecules at a particular level as a fraction of the total number of molecules present).

The value of Z depends on the temperature and the energy level spacing. If the temperature is high, the partition function increases because the molecules can be spread out over more states to accommodate the energy to be stored by the molecules. That is, the number of combinations of the ways in which molecules can be distributed increases.

In the above, we computed the partition function on the basis of no degeneracy – or single-state levels. For each energy level, there may be one or more states available for filling. When degeneracy is included, the two curves shown in Figure 4.10 for $P(0)$ and Z move apart from each other leaving a gap at the $T=0$ coordinate which depends upon the level of degeneracy.

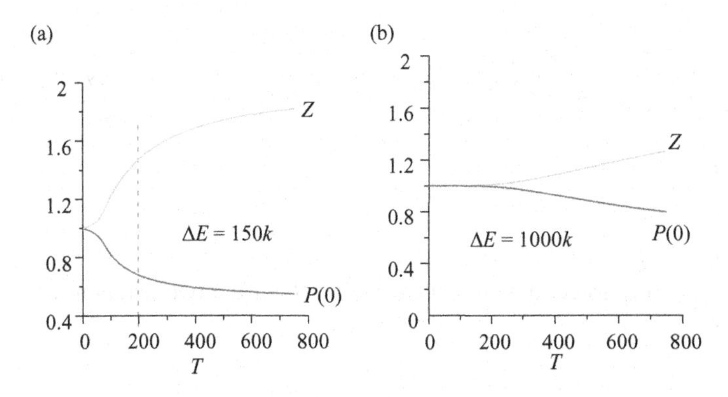

FIGURE 4.10 (a) Number of occupied energy levels Z, and the fractional number of molecules at the zero energy level $P(0)$ as a function of temperature for an energy gap of $150k$. (b) Same as (a) but with an energy gap of $1000k$.

The partition function tells us, on average, how many of the available states may be occupied. The value of Z depends on the temperature and the energy level spacing. If the temperature is high, then molecules are able to occupy higher levels, and so the partition function increases. If the spacing is high, then the energy levels may be beyond reach, and so the partition function decreases – unless the temperature is made higher to compensate.

Since we are dealing with the partition function with respect to energy levels (and not individual molecules) it is not beyond our capability to compute the partition function and probabilities for a modest range of levels. Let $T = 200\text{K}$ and energy spacing $\Delta E = 150k$, with 15 energy levels and no degeneracy. What then is the value of the partition function?

$$Z = \sum_{i=0}^{14} e^{-150i/200}$$
$$= 1.895 \tag{4.58}$$

If we were to plot Z vs the level, we observe that the sum for the partition function converges rather rapidly after about five terms when kT is of the same order as kE.

The partition function varies with the energy spacing and the temperature. Figure 4.11(a) shows the value of Z against T for different values of ΔE_i and shows how as the temperature rises, the partition function rises more steeply for lower values of ΔE_i.

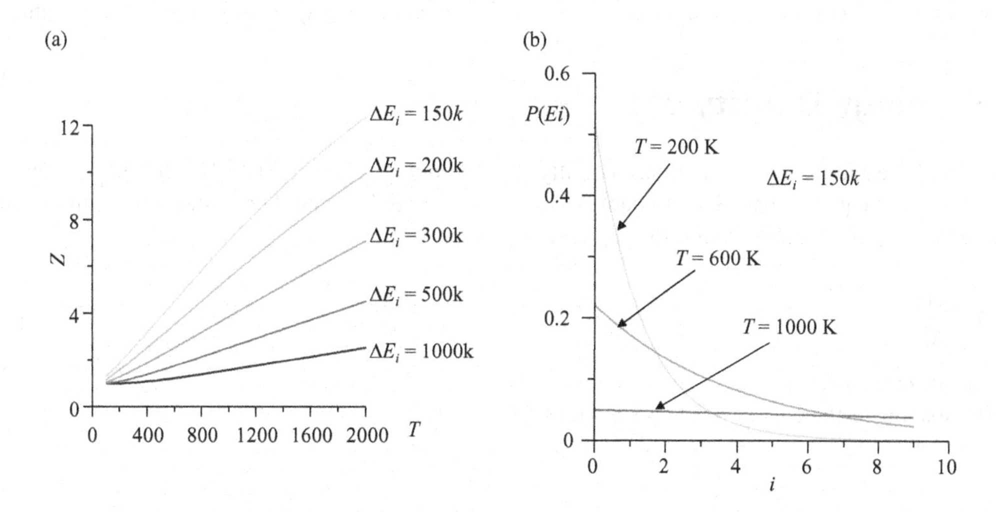

FIGURE 4.11 (a) Values for the partition function Z vs temperature for different energy spacings. (b) Probability of a molecule being at a particular energy level for an energy spacing of 150k for different temperatures.

As the spacing increases, a lower number of states are needed to accommodate the energy of the molecules, but as the temperature increases, more states are needed. At a small energy gap, at high temperatures, Z is high, indicating that nearly all of the 15 available states are required to hold the energy of the molecules.

The probability of a molecule being in a particular energy level is calculated from:

$$P(E_i) = \frac{e^{-E_i/kT}}{Z} \tag{4.59}$$

with no degeneracy, $g(E_i) = 1$

As the temperature increases, there are a decreasing number of molecules in the zero energy level, and the molecules become more spread out over the higher levels.

As the temperature approaches infinity, the molecules become evenly spread over all the 15 energy levels, including $E = 0$, and the probability of selecting a molecule at any energy is 1/15. These relationships are drawn in Figure 4.11(b).

The partition function shows how energy is distributed throughout a system. For a given energy spacing, as the temperature increases, molecules acquire more kinetic energy, and so a greater number of energy states are now accessible to them. The partition function increases. For a given temperature, as the spacing between the energy levels is reduced, more energy states become accessible for the molecules present, and so the partition function increases.

Generally speaking, the more states available and accessible to the molecules, the greater the partition function. As the partition function increases, the probability of being in the ground state decreases while the probability of being in higher energy states increases.

The partition function allows us to normalise the statistical weights of microstates to obtain probabilities of occupation of energy levels. It is a weighted sum of the total number of possible states for given values of T and ΔE. The partition function Z is a constant for a particular value of T and energy spacings.

We have used the partition function here as applied to discrete systems, although we originally derived the Boltzmann and Maxwell distributions in a classical sense (as continuous distributions). Dividing up a continuous (classical) situation into a number of infinitely small (quantum) steps is an extremely powerful method of analysis. It is important to appreciate that at the time this was developed by Maxwell and Boltzmann, the idea of quantisation of energy as applied to quantum physics was in the distant future.

4.3.6 Energy Density of States

In our one-dimensional model, there are equally spaced energy levels $E_i = 0E$, $1E$, $2E$, $3E$, $4E$, etc. The energy states are equally spaced an amount ΔE, and so $E_i = i\Delta E$. The number of states per unit of energy is called the density of states D, and for this case, would be:

$$D = \frac{1}{\Delta E} \tag{4.60}$$

The units of D are J^{-1}.

The spacing of the energy levels is shown in Figure 4.12(a).

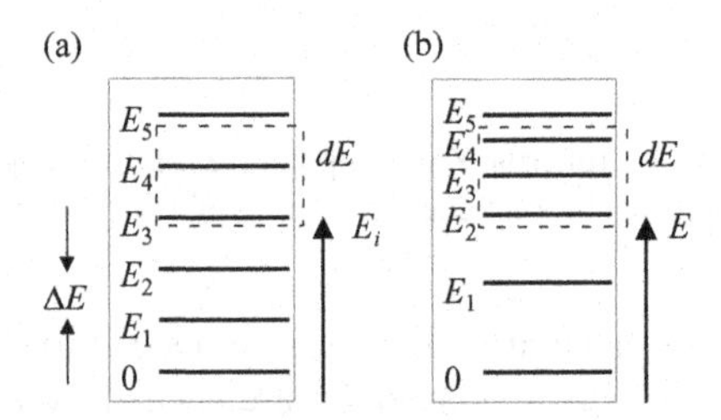

FIGURE 4.12 (a) Evenly spaced energy levels, fixed density of states. (b) Non-evenly spaced energy levels showing variable density of states.

At some energy E_i, the number of states that exist below this energy would be:

$$n(E_i) = DE_i = \frac{E_i}{\Delta E} = \frac{i\Delta E}{E} = i \tag{4.61}$$

For example, say the energy states are spaced 10 J apart. The density of states would be $1/10 = 0.1$ states per J. For an energy of say $E_2 = 20$ J, the number of states below this would be $20 \times 0.1 = 2$ states.

For some interval dE, we might ask how many states dn are within this interval? Note, it doesn't matter where dE is because in this case, the states are uniformly spaced. That is, D is a constant.

$$dn = DdE = \frac{1}{\Delta E} dE \tag{4.62}$$

Or, as we pass from $E = 0$ to $E = E_i$, the number of states below E_i increases at a fixed rate. $dn/dE = 1/\Delta E$ states per unit of energy.

For example, say we want to know how many states dn are within an interval of say $dE = 25$ J where $D = 0.1$ J^{-1}. $dn = 0.1 \times 25 = 2.5$.

In many cases, the density of states may not be a constant but a function of E. As shown in Figure 4.12(b), the rate of change of n with respect to E is now some function $D(E)$.

$$\frac{dn}{dE} = D(E) \tag{4.63}$$

Somewhat confusingly, we term the number of states within an interval dE as the degeneracy of the interval $g(E)$.

$$g(E) = D(E)dE \tag{4.64}$$

Previously, we said that the partition function is given by:

$$Z = \sum_{i=0}^{\infty} g(E_i)e^{-E_i/kT} \tag{4.65}$$

However, for a continuous distribution of energy levels, summing over all energy levels, the partition function, in the limit of small dE, becomes:

$$Z = \int_0^{\infty} D(E)e^{-E/kT}dE \tag{4.66}$$

Note, when we deal with a discrete sum, we identify each energy level as E_i, but when we deal with an integral, we just write the continuous variable E.

4.3.7 Energy in a Harmonic Oscillator System

Let us consider the special case of a continuum of energy levels and where the density of states $D(E) = D$ is a constant.

The fraction, or proportion, of molecules that have an energy in the range E to $E + dE$ will be given by the product of some probability density function $B(E)$ evaluated at E, times the width dE. The total area A under the curve of $B(E)$, as shown in Figure 4.13(a), is equal to 1. The area dA is equivalent to the probability of a molecule selected at random having an energy within the range $E + dE$.

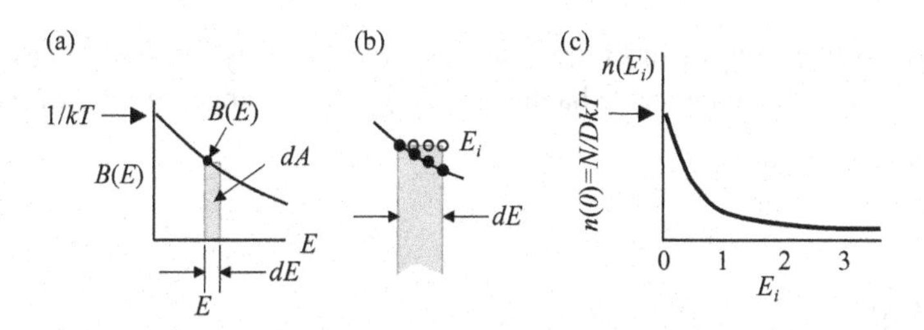

FIGURE 4.13 (a) Area under the curve from E to $E + dE$ gives the fraction of molecules with energy in the range E to $E + dE$. (b) Detail of the function $B(E)$ over that range. (c) Number of molecules with energy E_i as a function of E_i.

We ask: how many molecules are likely to occupy these states within this energy range dE? In the discrete case, the Boltzmann distribution shows the number of molecules with energy E_i:

$$n(E_i) = Ce^{-E_i/kT} \tag{4.67}$$

We can't use this directly because we don't have discrete energy levels. There will be no molecules with an energy precisely that of E_i. We must talk about an interval dE. We require the number of molecules $n_I(E_i)$ within that interval divided by the total number of molecules N. The subscript I denotes *interval*.

However, in this energy interval, there will be a number of available states $g(E)$ which can be calculated from the density of states.

$$g(E) = DdE \tag{4.68}$$

$n(E_i)$ is the number of molecules with an energy E_i. The number of states within the interval dE is $g(E)$. Thus, as shown in Figure 4.13(b), the number of molecules with an energy E_i within the *interval* is $n_I(Ei) = n(E_i) \times g(E)$.

That is, we assume that each *state* within the interval has an energy E_i, which is good enough if the interval is small.

And so:

$$n_I(E) = n(E)DdE = Ce^{-E/kT}DdE \tag{4.69}$$

We need to express this as a percentage of the total number of molecules N, not an absolute number $n_I(E)$.

$$\frac{n_I(E)}{N} = \frac{n(E)}{N}DdE = \frac{CD}{N}e^{-E/kT}dE \tag{4.70}$$

Since the sum over all energies must equal 1, we have:

$$1 = \frac{D}{N}\int_0^\infty n(E)dE = \frac{CD}{N}\int_0^\infty e^{-E/kT}dE \tag{4.71}$$

we find:

$$\frac{CD}{N} = \frac{1}{kT} \tag{4.72}$$

or:

$$C = \frac{N}{DkT} \tag{4.73}$$

Note, the units of D are J^{-1} and those of kT are J, so C is a number only as we would expect.

The probability of a selected molecule having an energy in the range E to $E + dE$ is thus given by:

$$\begin{aligned} dA &= \frac{n(E)}{N}dE \\ &= \frac{CD}{N}e^{-E/kT}dE \\ &= \frac{1}{kT}e^{-E/kT}dE \end{aligned} \tag{4.74}$$

At $E=0$, the value of $B(E)=1/kT$. This is shown in Figure 4.13(a).

It seems therefore that the function we are looking for, $B(E)$, is:

$$B(E)=\frac{1}{kT}e^{-E/kT} \tag{4.75}$$

This is the Boltzmann energy distribution.

$B(E)$ is a probability density function. The area dA under the curve gives the probability that a molecule selected at random will have an energy between E and $E + dE$:

$$dA = B(E)dE \tag{4.76}$$

Or:

$$P(E,E+dE)=\frac{1}{kT}e^{-E/kT}dE \tag{4.77}$$

Note that $B(E)$ has the units of J^{-1}. The constant $1/kT$ applies for this *special case of equally spaced energy levels* in the limit $dE \to 0$. This special case corresponds to the energy states for a classical simple harmonic oscillator, such as a pendulum, mass on a spring or a wave. It is a "classical" result – featuring a continuous distribution of energies with a constant density of states.

We can see that the quantity DkT is the value of the partition function Z. For the continuous case:

$$Z=\int_0^\infty De^{-E/kT}dE \tag{4.78}$$

And when D and $1/kT$ are both constant, we have:

$$\begin{aligned} Z &= D\int_0^\infty e^{-E/kT}dE \\ &= DkT \end{aligned} \tag{4.79}$$

The probability, expressed in terms of Z, becomes:

$$\begin{aligned} P(E,E+dE)dE &= \frac{De^{-E/kT}}{Z}dE \\ &= \frac{1}{kT}e^{-E/kT}dE \\ &= B(E)dE \end{aligned} \tag{4.80}$$

The number of molecules in the interval E_i is:

$$n_I(E)=Ce^{-E/kT}DdE \tag{4.81}$$

where:

$$C=\frac{N}{DkT} \tag{4.82}$$

And so:

$$n(0)=\frac{N}{DkT} \tag{4.83}$$

This relationship is shown in Figure 4.13(c).

4.3.8 Average Energy in a Harmonic Oscillator System

Boltzmann's energy distribution function *B*(*E*), times the interval *dE*, gives the probability of finding a molecule with an energy in the range *E* to *E* + *dE*. For molecules undergoing a harmonic oscillation, we have the distribution function:

$$B(E)=\frac{1}{kT}e^{-E/kT} \tag{4.84}$$

and the probability function:

$$P(E,E+dE)=B(E)dE=\frac{1}{kT}e^{-E/kT}dE \tag{4.85}$$

There must be some average energy E_{av} which could be assigned to each molecule, which when multiplied by the total number of molecules *N* gives the same total energy of the system.

To find this average energy for the discrete case, we could sum the energies at each state and divide by the number of molecules. The energy at each state is the number of molecules that exist at that state times the energy of the state. Thus,

$$\begin{aligned} E_{av} &= \frac{n(0)E_0+n(1)E_1+n(2)E_2+n(3)E_3+\ldots}{N} \\ &= \frac{1}{N}\sum_{i=0}^{\infty}(i\Delta E)n(0)e^{-i\Delta E/kT} \end{aligned} \tag{4.86}$$

For the continuum case, the probability of selecting a molecule with an energy between *E* and *dE* is:

$$P(E,E+dE)=B(E)dE=\frac{1}{kT}e^{-E/kT}dE \tag{4.87}$$

For each *dE*, if we multiply the value of *B*(*E*)*dE* times *N* we have the number of molecules with the energy of that state. We then multiply that number by the energy of that state, and we have the total energy of that state. We then sum up all these energies, over all the energy states, and divide by *N* to get the average energy per molecule.

Making use of a standard integral:

$$\int_0^{\infty} xe^{ax}dE=\frac{1}{a^2}(ax-1)e^{ax} \tag{4.88}$$

we obtain:

$$
\begin{aligned}
E_{av} &= \frac{1}{N}\int_0^{\infty} NEB(E)\,dE \\
&= \int_0^{\infty} EB(E)\,dE \\
&= \int_0^{\infty} E\frac{1}{kT}e^{-E/kT}\,dE \\
E_{av} &= kT
\end{aligned}
\tag{4.89}
$$

This is the average energy per molecule at the limit $dE \to 0$ (for the special case of equally spaced energy levels and one degree of freedom).

The partition function can be linked to various thermodynamic entities. For example, if we summed up all the total energies at each energy level (which would then be the total energy of the system), and then divided by the total number of molecules, we would have the average energy of each molecule.

This is the same as summing the product of the energies in the system and the probability of molecules having that energy, over all energies, divided by the total probability ($P_1 + P_2 + P_3 \ldots = 1$).

The average energy is therefore:

$$
\begin{aligned}
E_{av} &= \frac{E_0P_0 + E_1P_1 + E_2P_2 + E_3P_3\ldots}{P_0 + P_1 + P_2 + P_3\ldots} \\
&= E_0P_0 + E_1P_1 + E_2P_2 + E_3P_3\ldots \\
&= \sum_0^{\infty} P_iE_i
\end{aligned}
\tag{4.90}
$$

But, letting $\beta = 1/kT$ for convenience, we can write:

$$
P(E_i) = P_i = \frac{e^{-\beta E_i}}{\sum_{i=0}^{\infty} e^{-\beta Ei}} \tag{4.91}
$$

where the denominator is the partition function Z:

$$
Z = \sum_{i=0}^{\infty} e^{-\beta E_i} = e^{-\beta E_0} + e^{-\beta E_1}\ldots \tag{4.92}
$$

Thus:

$$
E_{av} = \frac{1}{Z}\sum_{i=0}^{\infty} E_ie^{-\beta E_i} \tag{4.93}
$$

But:

$$Z = \sum_{i=0}^{\infty} e^{-\beta E_i} = e^{-\beta E_0} + e^{-\beta E_1} \ldots$$

$$\frac{\partial Z}{\partial \beta} = -\sum_{i=0}^{\infty} E_i e^{-\beta E_i} \tag{4.94}$$

Thus:

$$E_{\text{av}} = -\frac{1}{Z}\frac{\partial Z}{\partial \beta} \tag{4.95}$$

But:

$$-\frac{1}{Z}\left(\frac{\partial Z}{\partial \beta}\right) = -\left(\frac{\partial \ln Z}{\partial Z}\frac{\partial Z}{\partial \beta}\right) \tag{4.96}$$

And so:

$$E_{\text{av}} = -\frac{\partial \ln Z}{\partial \beta} \tag{4.97}$$

By just knowing the value of the partition function, we can calculate the average energy of the molecules at temperature T. Indeed, with Z, we can calculate many thermodynamic properties of a system as long as the number of molecules N is fixed.

We can also determine the average energy for a harmonic oscillator using the partition function approach. For a continuum of energy states, and where the density of states is a constant $D(E) = D$, we have:

$$\begin{aligned} Z &= \int_0^{\infty} De^{-E/kT} dE \\ &= D\int_0^{\infty} e^{-E/kT} dE \\ &= DkT \end{aligned} \tag{4.98}$$

$D(E) = D$, a constant in this special case.

And so, in terms of the partition function:

$$
\begin{aligned}
E_{av} &= -\frac{\partial \ln Z}{\partial \beta} \\
Z &= DkT \\
\ln Z &= \ln\left(\frac{D}{\beta}\right) \\
&= \ln(D) - \ln(\beta) \\
\frac{\partial \ln Z}{\partial \beta} &= -\frac{1}{\beta} = E_{av} \\
E_{av} &= kT
\end{aligned}
\tag{4.99}
$$

For discrete energy states, and where the states are evenly spaced, $E_i = i\Delta E$, we have:

$$
E_{av} = \frac{\sum_{i=0}^{\infty}(i\Delta E)n(0)e^{-i\Delta E/kT}}{\sum_{i=0}^{\infty}n(0)e^{-i\Delta E/kT}} = \frac{\sum_{i=0}^{\infty}(i\Delta E)e^{-i\Delta E/kT}}{\sum_{i=0}^{\infty}e^{-i\Delta E/kT}} \tag{4.100}
$$

The numerator and denominator are convergent series, and so this leads to:

$$
E_{av} = \frac{\Delta E}{e^{\Delta E/kT} - 1} \tag{4.101}
$$

In the limit as $\Delta E \to 0$ we have:

$$
\begin{aligned}
E_{av} &= \lim_{\Delta E \to 0} \frac{\Delta E}{e^{\Delta E/kT} - 1} \\
&= kT
\end{aligned}
\tag{4.102}
$$

Quantum Theory 5

5.1 INTRODUCTION

Quantum theory had an unlikely start. By the middle of the 19th century, nearly everyone knew that light behaved like a wave, and by 1865, Maxwell had proved it. However, cracks began to appear. The radiation spectrum emitted by what is called a black body, a perfect radiator, was not as expected in the short wavelength regime. Then, a phenomenon which became known as the photoelectric effect could not be explained at all by the application of wave equations. It was in an effort to explain the black body radiation problem that gave the first clue to the notion that light was actually made up of particles – a belief held by Newton, but for different reasons, back in the early 1700s. Einstein put the idea on firm ground with the explanation of the photoelectric effect, and then the connection was made with the structure of the atom, followed by a probability interpretation of the nature of matter. To begin, we go back to Planck's problem with the radiation spectrum.

5.2 BLACK BODY RADIATION

A body whose temperature is above absolute zero emits radiant energy. The power P radiated can be calculated according to the Stefan–Boltzmann law:

$$P = e\sigma AT^4 \tag{5.1}$$

In this formula, A is the area over which emission of radiation occurs, T is the absolute temperature (K) and σ is the Stefan–Boltzmann constant where $\sigma = 5.67 \times 10^{-8}$ W m^{-2} K^{-4}. e is the emissivity of surface (varies between 0 and 1) and depends on:

- Nature of surface
- Temperature of surface
- Wavelength of the radiation being emitted or absorbed

Our interest here is the variation in intensity I (W m^{-2} in SI units) as a function of temperature and wavelength. To eliminate the "nature of the surface" effect, we will consider the radiation emission from what is called a black body.

A black body is a specially prepared radiative heat source that emits the maximum possible energy over different wavelengths without any self-absorption. Such a "body" can be constructed by heating a cavity with a small hole to allow the escape of radiation for measurement. At first, it might be thought that a black body would therefore emit a constant intensity over all wavelengths, but experiments show that this is not the case. What is observed is that the intensity from a black body varies as a function of wavelength, and the variation in intensity depends on the temperature of the body.

As the temperature of a black body increases, it is observed that:

- The emission spectrum does not depend on what the cavity is made of, but only the temperature inside.
- Total energy (area under curve) increases.
- Peak in emission spectrum shifts to shorter wavelengths.

These observations are shown in Figure 5.1(a).

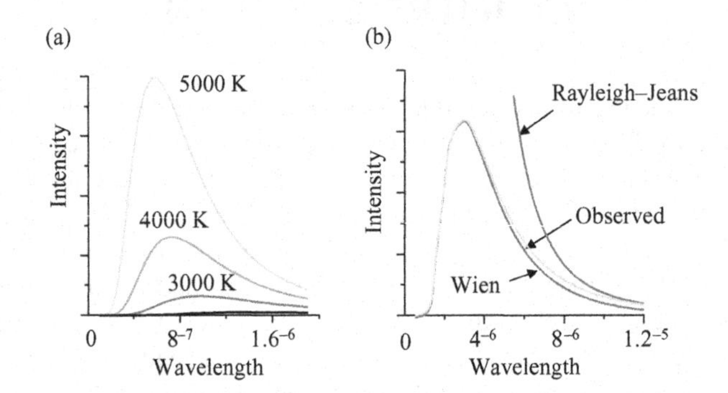

FIGURE 5.1 (a) Radiation emission spectrum showing variation of wavelength distribution with frequency. (b) Radiation distribution as calculated by Wien, and also Rayleigh and Jeans.

In 1893, Wilhem Wien argued that since the molecules within a heated body are vibrating in simple harmonic motion with thermal energy, then the resulting changes in velocities of these molecules would result in the acceleration of charges within the molecules, thus leading to the emission of radiation with a characteristic intensity spectrum.

Wien's radiation law agreed well with experimental results at high frequencies (short wavelength) but did not fit well at low frequencies (long wavelength). It was an empirical law. According to Wien, the energy density, or energy per unit volume is given by:

$$\frac{E}{V}=\frac{C_1 f^3}{e^{C_2 f/kT}} \tag{5.2}$$

where E is the energy, f is the frequency of the radiation, k is Boltzmann's constant and T is the absolute temperature. The comparison between the theory and the experiment is shown in Figure 5.1(b).

Rayleigh and Jeans applied the Boltzmann energy distribution for a harmonic oscillator to the radiation emitted and found that the predicted emission spectrum agreed well at low frequencies but had an ever-increasing intensity at higher frequencies – a feature which became known as the ultraviolet catastrophe. The Rayleigh–Jeans radiation law is:

$$\frac{E}{V}=\frac{8\pi}{c^3}f^2kT \tag{5.3}$$

It was a catastrophe because the formula was based upon the principle of equipartition of energy in vibrating systems and was thought to be absolutely correct, yet clearly was not observed. If radiation was emitted according to Rayleigh and Jeans, then the intensity would approach infinity at high frequencies with unfortunate consequences for any observer of black body radiation.

5.3 CAVITY RADIATION

Rayleigh and James Jeans approached the calculation of black body radiation by examining the standing waves formed by the radiation reflecting backwards and forwards inside a metal-walled, heated cavity. A small hole is provided to allow the wavelength and intensity of the radiation to be measured. Such a black body is shown in Figure 5.2(a).

We will assume that the cavity has no heat loss by conduction to the outside. The walls of the cavity are fed with a certain amount of energy, and the temperature of the walls increases. The energy input is then turned off. The temperature remains constant (no heat loss).

When the metallic walls of our cavity are heated, the weakly bound conduction electrons are thermally agitated and undergo random collisions with other electrons. These accelerations and decelerations produce radiant energy which is fed into the interior of the cavity. Much like the distribution of velocities in a gas, there is therefore a relationship between the temperature of the walls and the energy distribution of the electromagnetic radiation inside the cavity.

Looked at in one dimension, the radiated waves travel from one wall to the other, where they are reflected back on themselves. The incident and reflected waves combine to form standing waves (or modes, or wave patterns) with nodes at each wall.

Because the walls of the cavity are metal, the amplitude of the electric field at the walls is zero due to the redistribution of charge that occurs there. That is, for the oscillating electric field, there are nodes at the walls. Therefore, for a standing wave to be produced, the size L of the cavity must be equal to an integral number of half-wavelengths. This requirement is shown in Figure 5.2(b).

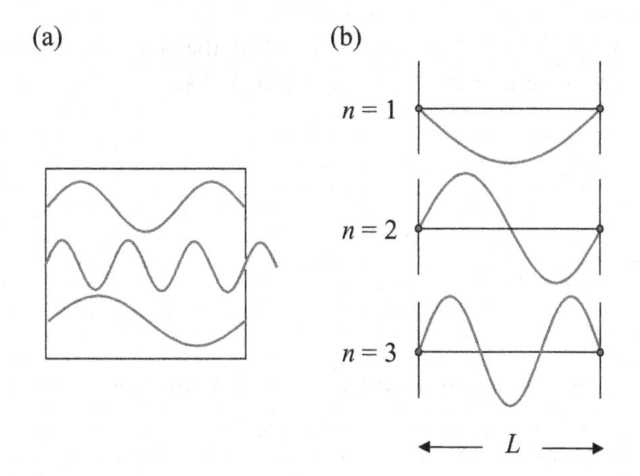

FIGURE 5.2 (a) Schematic of a black body comprising a cavity. (b) Standing waves within the cavity.

$$L = n\frac{\lambda}{2} \quad \text{where } n = 1,2,3,\ldots \tag{5.4}$$

In terms of frequency f, we have:

$$c = f\lambda$$

$$L = n\frac{c}{2f} \quad \text{where } n = 1,2,3,\ldots \tag{5.5}$$

$$f = n\frac{c}{2L}$$

Our ultimate aim is to determine the energy in the electromagnetic waves per unit volume of the cavity (the energy density) as a function of frequency. We do this by counting the number of standing waves $n(f)$ that exist in a frequency interval f to $f+df$. This number, times the average energy of these waves and then divided by the volume of the cavity, gives us the energy density per unit volume at that frequency.

5.4 FREQUENCY DENSITY OF STATES

5.4.1 Density of States – 1D

A travelling wave can be represented by a trigonometric cosine function:

$$u(x,t) = A\cos(\omega t - kx + \varphi) \tag{5.6}$$

where:

$$k = \frac{2\pi}{\lambda} \tag{5.7}$$

We can set $t=0$ when we are not interested in time-dependent effects, such as for a standing wave. k is called the wave number. The wave number describes the spatial characteristics of the wave (much like the frequency ω describes the time characteristics of a wave).

As shown in Figure 5.3 (a), let L be the length over which the boundary conditions are to apply. For a standing wave, L must be an integral number of half-wavelengths:

$$L = n\frac{\lambda}{2} \quad \text{where } n = 1,2,3\ldots \tag{5.8}$$

Or:

$$k = n\frac{\pi}{L} \tag{5.9}$$

In 1D k space, we can represent the allowed values of k for standing waves as shown in Figure 5.3 (b).

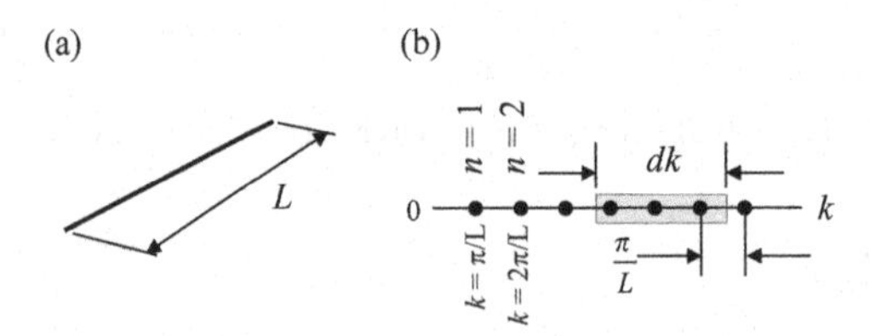

FIGURE 5.3 (a) 1D k space. (b) Allowed values of k.

Each point in k space represents a particular mode, or frequency, of vibration.

For an interval dk in k space, the number of modes dn within this interval is:

$$dn = \frac{L}{\pi}dk \tag{5.10}$$

but:

$$k = \frac{2\pi}{\lambda} = \frac{2\pi}{c} f \tag{5.11}$$

therefore:

$$dk = \frac{2\pi}{c} df \tag{5.12}$$

The number of modes (i.e. standing wave patterns) per unit of frequency is called the density of states $D(f) = dn/df$. Therefore, for this 1D case, with equally spaced states:

$$dn = \frac{L}{\pi} \frac{2\pi}{c} df \tag{5.13}$$

Thus:

$$D(f) = \frac{dn}{df} = \frac{2L}{c} \tag{5.14}$$

Note: for this 1D case, the density of states is a constant and independent of f.

As L gets larger, the spacings between the allowable values of k in k space decreases, and so for a particular interval dk, there are more allowable values of k possible.

The unit for the frequency density of states is $1/f$ (or sec).

In the example above, we focussed on an interval dk in k space. However, it is possible to look at the same situation and consider the number of states n less than or equal to a particular value of k.

5.4.2 Density of States – 1D Revisited

Consider again a 1D space of length L over which standing waves exist. The spacing between each state, or wave pattern, in k space is:

$$\frac{\pi}{L} \tag{5.15}$$

That is,

$$k = n\frac{\pi}{L} \tag{5.16}$$

Each standing wave has a frequency associated with it. We wish to determine how many states (i.e. modes, or standing wave patterns) dn exist with a frequency within an interval of $f + df$.

The number of states n less than or equal to a particular value of k is:

$$n = \frac{L}{\pi} k \tag{5.17}$$

So, we can express k as:

$$k = \frac{2\pi}{\lambda} = \frac{2\pi}{c} f \tag{5.18}$$

Since:

$$\lambda = \frac{f}{c} \tag{5.19}$$

Note, for electromagnetic waves, including polarisation would introduce a factor of 2 into the equations; this is not included here, but we will include it later.

The number of states with frequency less than or equal to f is thus:

$$\begin{aligned} n &= \frac{L}{\pi}\frac{2\pi}{c} f \\ &= \frac{2L}{c} f \end{aligned} \tag{5.20}$$

For a small interval df, there will be dn states present.

$$dn = \frac{2L}{c} df \tag{5.21}$$

The equations show that if we double f, then there are twice as many states with frequency less than f.

The rate of change of n with respect to f is called the density of states $D(f)$. $D(f)$ represents the number of available states per unit of frequency at a frequency f.

$$\frac{dn}{df} = D(F) = \frac{2L}{c} \tag{5.22}$$

In this case, $D(f)$ is independent of frequency and only depends on L.

The number of states per unit of frequency, the density of states, is a constant. This is plotted in Figure 5.4.

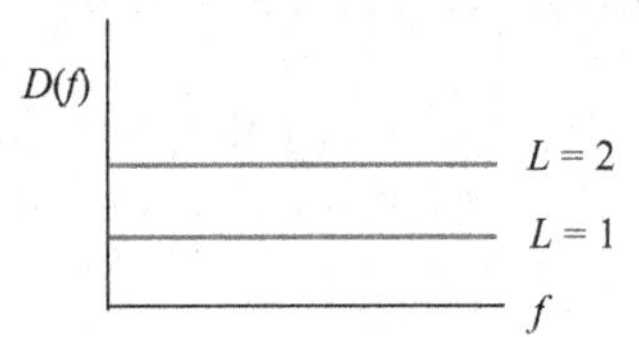

FIGURE 5.4 Density of states vs frequency for 1D space.

5.4.3 Density of States – 2D

Consider a 2D surface of length L and area A over which standing waves are present. This is shown in Figure 5.5(a). The spacing between each state in k space is shown in Figure 5.5(b).

The area per point (i.e. per state) in k space is thus:

$$\left(\frac{\pi}{L}\right)^2 \tag{5.23}$$

But, L^2 is the area A in real space, and so the number of states per unit area in k space is thus:

$$\frac{1}{(\pi/L)^2} = \left(\frac{L}{\pi}\right)^2 = \frac{A}{\pi^2} \tag{5.24}$$

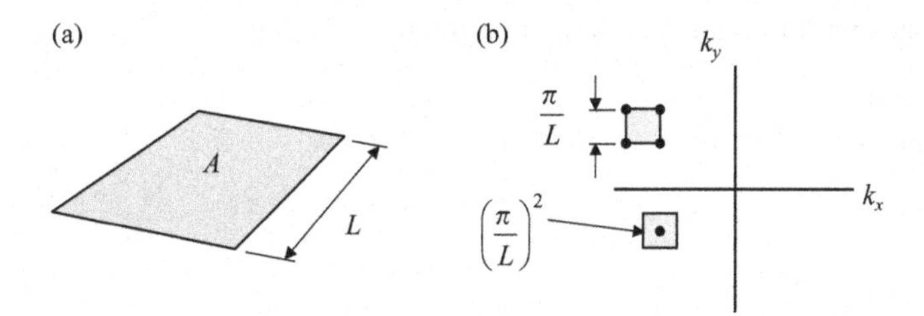

FIGURE 5.5 (a) 2D space. (b) Allowed values for k in 2D k space.

Now, each value of k represents a particular state, or mode of vibration, or frequency, so a circle of radius k in k space draws out a contour of a constant value of f. This is drawn in Figure 5.6(a). The area of the circle is πk^2.

Note, this is the area in k space, and is not A, the area of the surface in real space.

We can see that, as shown in Figure 5.6(b), when $k=2$, the circle encompasses more states compared to $k=1$. Thus, the number of states n within the circle is the number of states with a frequency less than or equal to a selected value of f, and is:

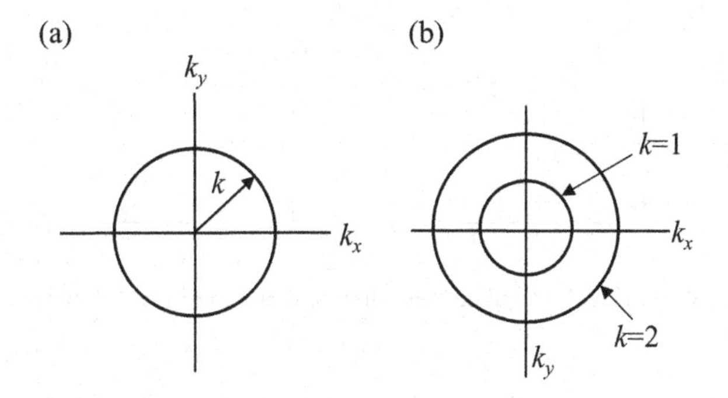

FIGURE 5.6 (a) Circle of radius $k=1$ in k space. (b) Circles in k space for $k=1$ and $k=2$.

$$
\begin{aligned}
n &= \pi k^2\left(\frac{A}{\pi^2}\right)\frac{1}{4} \\
&= \frac{k^2 A}{4\pi}
\end{aligned}
\tag{5.25}
$$

We divide by 4 here because we are only considering positive values of kx and ky.

But:

$$
k = \frac{2\pi}{\lambda} = \frac{2\pi}{c} f
\tag{5.26}
$$

since:

$$
\lambda = \frac{f}{c}
\tag{5.27}
$$

So, the number of states with frequency less than or equal to f is thus:

$$
\begin{aligned}
n &= \frac{A}{4\pi}\left(\frac{2\pi}{c}\right)^2 f^2 \\
&= \frac{\pi A}{c^2} f^2
\end{aligned}
\tag{5.28}
$$

This quadratic function is shown in Figure 5.7(a).

If we double f, then there are four times as many states with frequency less than f.

The density of states is thus:

$$
\frac{dn}{df} = \frac{\pi A}{c^2} f \tag{5.29}
$$

As can be seen in Figure 5.7(b), the number of states per unit of frequency, the density of states, increases at a constant rate with increasing frequency.

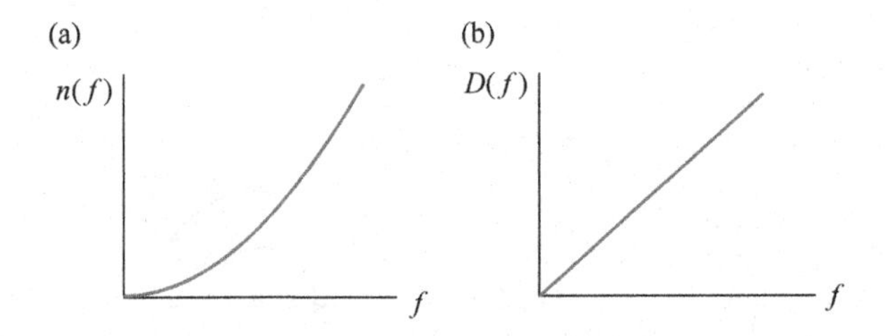

FIGURE 5.7 (a) Number of states with frequency less than or equal to f in two-dimensional space. (b) Density of states vs frequency.

The essential feature in this case is that the density of states for a 2D surface is a function of f as well as L.

Let's pick $k_x = 1$, and then look at k_y. As shown in Figure 5.8, the plate can vibrate in many different ways in the y direction for a single mode in the x direction. Hence, there are more ways to fit standing waves into the area as the frequency increases compared to the 1D case.

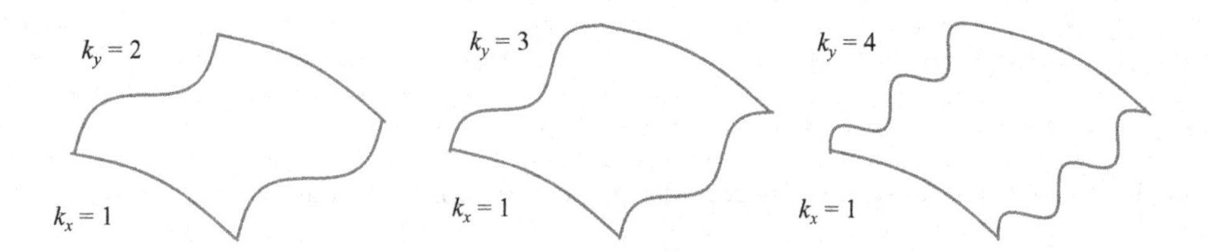

FIGURE 5.8 Different modes of vibration in two-dimensional space.

5.4.4 Density of States – 3D

Consider a 3D space of volume of length L and volume V as shown in Figure 5.9

Inside this volume, in k space, allowed states are uniformly distributed with a spacing π/L as illustrated in Figure 5.9(b).

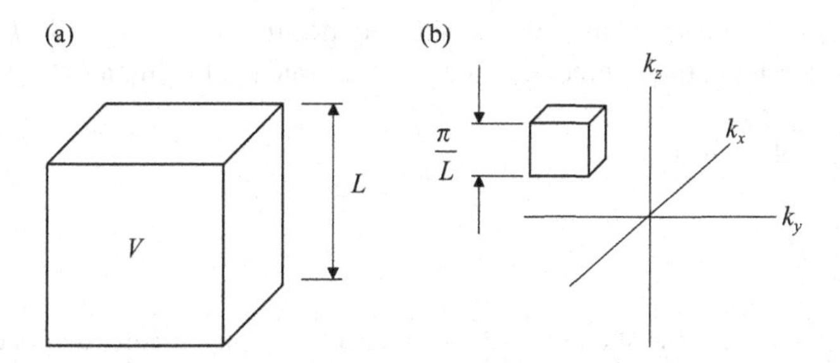

FIGURE 5.9 (a) 3D space (b) Density of states vs frequency for three-dimensional space.

The volume (in k space) per state is:

$$\left(\frac{\pi}{L}\right)^3 \tag{5.30}$$

But, $L^3 = V$ in real space, and so the states per unit volume in k space can be expressed:

$$\frac{V}{\pi^3} \tag{5.31}$$

where V is the volume in real space.

The surface of a sphere drawn out by a particular value of k represents all the states that have the same frequency. This is shown in Figure 5.10. Thus, the number of states n within the sphere is the number of states with a frequency less than or equal to a selected value of f.

The number of states within the volume is:

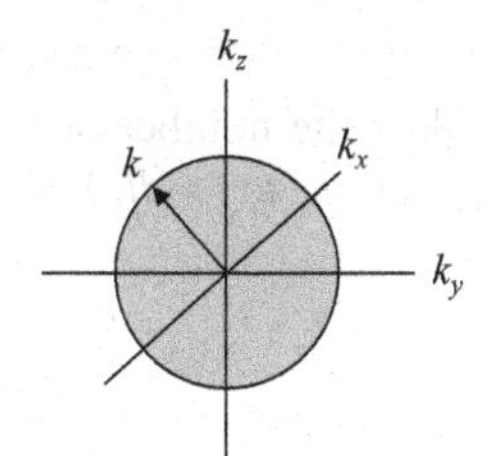

FIGURE 5.10 Surface of a sphere in three-dimensional k space representing those states with the same frequency.

$$n = \frac{4}{3}\pi k^3 \frac{V}{\pi^3}\frac{1}{8} \tag{5.32}$$

Since:

$$k = \frac{2\pi}{\lambda} = \frac{2\pi}{c} f \tag{5.33}$$

Then, the number of states with frequency less than or equal to f is thus:

$$\begin{aligned} n &= \frac{4\pi}{3}\frac{8\pi^3}{c^3}\frac{V}{\pi^3}\frac{1}{8} f^3 \\ &= \frac{4\pi}{3}\frac{V}{c^3} f^3 \end{aligned} \tag{5.34}$$

Note, we divide by 8 here because we are only considering positive values of k_x, k_y, k_z.

As shown in Figure 5.11(a), the number of states with frequency less than f increases dramatically as the frequency becomes larger.

And so, the density of states is:

$$\frac{dn}{df} = D(f) = \frac{4\pi V}{c^3} f^2 \tag{5.35}$$

As shown in Figure 5.11(b), the rate of change of $D(f)$ with f becomes greater as f becomes larger.

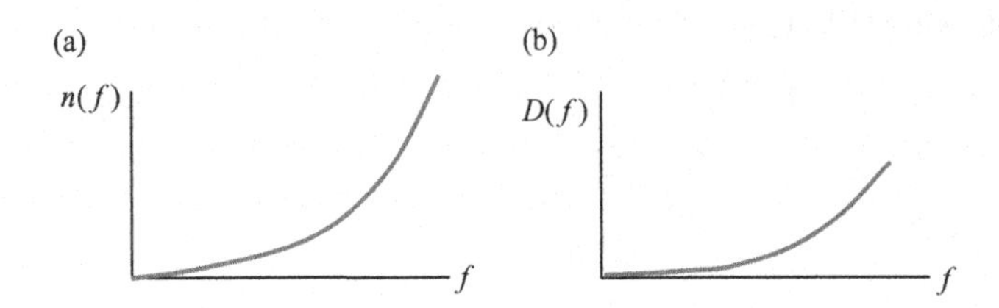

FIGURE 5.11 (a) Number of states vs frequency for three-dimensional space. (b) Density of states for three-dimensional space.

What this says is that (like the case of 2D, but unlike the case of 1D), in three dimensions, there are dramatically more ways of fitting a standing wave into the confines of the volume as the frequency gets higher.

5.5 RAYLEIGH–JEANS RADIATION LAW

The frequency density of states for cavity radiation allows us to calculate the number of standing wave patterns or modes within the confines of the cavity walls for a frequency range f to $f + df$. Thus, the number of modes is calculated from:

$$\begin{aligned} n(f)dn &= D(f)df \\ &= \frac{4\pi V}{c^3} f^2 df \end{aligned} \tag{5.36}$$

The significance of this is that for a certain interval df, we can encompass more standing wave patterns when the frequency is higher compared to low frequency (long wavelength) waves.

For electromagnetic waves, we have the possibility of two modes oriented at 90° (polarised). In a similar sense that we treated degeneracy in statistical mechanics, we can say that each mode is two-fold degenerate, and so:

$$D(f) = \frac{8\pi V}{c^3} f^2 \tag{5.37}$$

in the limit as $df \to 0$

The energy of a gas molecule is proportional to the square of the velocity. Individual molecules will have a range of velocities and hence a range of energies. In much the same way, for the electromagnetic waves in our cavity radiation, the waves can have a range of electric field amplitudes which carry the energy. But, in a *standing* wave, we have two waves: one travelling in one direction superimposed on one travelling in the other direction. The two waves together form a standing wave. The energy bound up in a standing wave is therefore twice that we would normally expect for a single travelling wave.

For freely moving gas molecules, the average energy per molecule is $1/2kT$ per degree of freedom. In the case of a body undergoing simple harmonic motion (such as if the body were subject to a restoring force) with one degree of freedom (e.g. movement in the positive and negative x direction), then because there is energy stored in the system as *both* kinetic and potential energy, the average total energy is twice that of the kinetic energy component, and so the average energy of the body is $E_{av} = kT$.

In the case of an electromagnetic field, we do not have "kinetic" and "potential" energies as such, but rather the energy associated with the E and B fields which are varying in a sinusoidal manner. We usually just express the total energy in terms of the magnitude of the E field since the total energy is equally shared between the E and B fields. That is, there is essentially just one degree of freedom here. For a *standing* wave we have twice the energy of a single travelling wave. Thus, the average energy over all those amplitudes, in the limit as $\Delta E \rightarrow 0$, at frequency f is $E_{av} = kT$.

Due to the principle of equipartition of energy, this same average energy should then be found in all the modes of vibration (or frequencies).

The energy E per unit volume (i.e. energy density) at frequency f is thus $D(f)/V$ times the average E_{av}:

$$\frac{E}{V} = \frac{D(f)}{V} = \frac{8\pi}{c^3} f^2 kT \tag{5.38}$$

Note that $D(f)$, the *frequency* density of states, depends on f and is not a constant for the three-dimensional volume V. The *energy* density of states $D(E)$is a constant for a harmonic oscillator system at some frequency f for the special case of equally spaced energy levels.

Rayleigh and Jeans calculated that the energy density (which is used to calculate the observable intensity) of the radiation in a cavity varies as a function of the square of the frequency. For a small frequency range df at frequency f, we have:

$$\frac{E}{V} df = \frac{8\pi}{c^3} f^2 kT df \tag{5.39}$$

The frequency density of states is greater at the higher frequencies (because there are more ways of filling the volume V with high frequency standing waves than with low frequency ones).

If we scan through the frequencies of black body radiation at intervals df, we find there is a relatively low number of standing waves $n(f)$ at low frequencies, but as we move up into the higher frequencies, we encounter more standing waves in the frequency interval. For a given frequency, the standing waves carry a distribution of energies (amplitudes). At a given frequency, these standing waves have the average energy kT, it's just that at higher frequencies, there are more of them, so the intensity of the emitted radiation is greater. The problem is that if we choose higher and higher frequencies, then the intensity should therefore be ever-increasing because $n(f)$ increases as f^2, and because of the equipartition of energy, those waves at higher frequencies should be carrying each an average energy of kT.

Comparison with observed experimental data shows that the Rayleigh–Jeans distribution equation gives the correct energy at low frequencies (long wavelength), but experimentally, the energy density starts to fall off as at higher frequencies rather than to continue to increase. This was shown earlier in Figure 5.2.

In the case of gases, the mass of the molecule puts an upper limit on the velocity of the molecule since each molecule has to carry its own $3/2kT$ of kinetic energy, and so the velocity adjusts to suit. Here, electromagnetic waves have no rest mass, their velocity is fixed at c, and so there is clearly something wrong with the application of these concepts when it comes to using them for electromagnetic waves at higher frequencies. Since the Rayleigh–Jeans formula was based on Maxwell–Boltzmann statistics and the equipartition of energy which could not be refuted, it was considered a catastrophe for what we now call "classical" physics.

5.6 THE BIRTH OF QUANTUM PHYSICS

5.6.1 Introduction

In 1900, Max Planck found that even though Wien's energy distribution was somewhat empirical, if he added a −1 to the denominator, he obtained a perfect fit to the experimental data. The conundrum was that this modified version of Wien's formula fitted the data, but the Rayleigh–Jeans law had physical significance and was based on Maxwell–Boltzmann statistics, which in turn had significant experimental support for the behaviour of gases.

In looking at the Wien radiation law:

$$\frac{E}{V}df = \frac{C_1 f^3}{e^{C_2 f/kT}}df \tag{5.40}$$

it can be seen that although the intensity rises sharply with f^3 in the numerator, it gets dragged down again by the increasing influence of the exponential term in the denominator. This then acts as a dampener for the high frequency components and avoids the ultraviolet catastrophe. The challenge therefore was to find some physical explanation to get from the Rayleigh–Jeans law to the modified version of Wien's equation.

Essentially, the problem was that the equipartition of energy was not being upheld at high frequencies. Rather, as the frequency increased, the value of $E_{av} = kT$ started to decrease instead of remaining constant. But, the principle of equipartition of energy is a cornerstone of classical physics and cannot be lightly dismissed.

Planck decided to study in more detail the origin of the law of equipartition of energy, and this takes us back to the Boltzmann distribution for molecules in a gas.

$$n(E_i) = n(0)e^{-E_i/kT} \tag{5.41}$$

In statistical mechanics, for gas molecules, Boltzmann divided up a continuous physical entity ($E = 1/2mv^2$) into discrete steps for the sake of counting the possibilities – a way of managing the infinite. Once he had a formulation for the distribution in a discrete sense, he let the step size approach zero to get the final distribution for a continuous quantity E. That is, the process of partitioning a continuous process up into countable steps allows us to use probability and statistics to calculate the most likely state of physical systems.

The partitioning of continuous energy distributions into steps was an ingenious tool in classical physics, and the integral of the result brings us back into the continuum regime. This was done well before the concept of quantum physics and was only done for the purpose of calculating the continuous quantities associated with the movement of gas molecules in a manageable way.

5.6.2 Boltzmann Statistics

In a gas, individual gas molecules have a range of velocities, and so there is a distribution of energies described by Boltzmann statistics. For cavity radiation, we have not gas molecules, but standing waves (i.e. harmonic oscillators). Standing waves continuously change energy (their amplitudes change) within the cavity.

Consider a particular mode, or standing wave pattern, or resonant frequency, of vibration. We assume that energy levels for that mode are equally spaced by an amount ΔE so that we have levels at 0, $\Delta 1E$, $2\Delta E$, $3\Delta E$, etc. That is, $E_i = i\Delta E$ where i is an integer. The number of standing waves of that frequency at an energy level $i\Delta E$ can be calculated in terms of the number of standing waves at the zero energy level, the Boltzmann distribution:

$$n(i\Delta E) = n(0)e^{-i\Delta E/kT} \tag{5.42}$$

Or:

$$n(E_i) = Ce^{-E_i/kT} \tag{5.43}$$

At an energy level $i\Delta E$, for these $n(i\Delta E)$ standing waves, the total energy at this energy level is the number of waves times their energy:

$$(i\Delta E)n(i\Delta E) = (i\Delta E)n(0)e^{-i\Delta E/kT} \tag{5.44}$$

If we summed up all the energies at each energy level (which would then be the total energy for that mode) and then divided by the number of standing waves, we would have the average energy for each standing wave for that mode of vibration.

$$E_{av} = \frac{\sum_{i=0}^{\infty}(i\Delta E)n(0)e^{-i\Delta E/kT}}{\sum_{i=0}^{\infty}n(0)e^{-i\Delta E/kT}} = \frac{\sum_{i=0}^{\infty}(i\Delta E)e^{-i\Delta E/kT}}{\sum_{i=0}^{\infty}e^{-i\Delta E/kT}} \tag{5.45}$$

The energy E per unit volume (i.e. energy density) at frequency f is $D(f)/V$ times the average E_{av}. This gives us the distribution of energies as the frequency is varied:

$$\frac{E}{V}df = \frac{8\pi}{c^3}f^2E_{av}df \tag{5.46}$$

How to find E_{av}? Is E_{av} a constant for all frequencies? It is a function of the frequency? Is it a function of the temperature T? For the case of our model gas, we found that $E_{av} = kT$ in the limit as $\Delta E \to 0$, and this was the way Rayleigh and Jeans approached the problem.

5.6.3 Rayleigh–Jeans Radiation Law

We can calculate the average energy if we know the partition function. For the discrete and the continuous cases of energy distributions, we have:

$$Z = \sum_{i=0}^{\infty} g(E_i) e^{-E_i/kT} = \int_0^{\infty} D(E) e^{-E/kT} dE \tag{5.47}$$

Rayleigh and Jeans then let $\Delta E \to 0$ since there was no reason to think that the energy levels (the amplitudes of the standing waves) could not take on any value and so vary continuously. So, the partition function, for the special case of a constant energy density of states D, becomes:

$$Z = \int_0^{\infty} De^{-E/kT} dE = DkT \tag{5.48}$$

And so:

$$\begin{aligned} E_{av} &= -\frac{\partial \ln Z}{\partial \beta} \\ Z &= DkT \\ \ln Z &= \ln\left(\frac{D}{\beta}\right) \\ &= \ln(D) - \ln(\beta) \\ \frac{\partial \ln Z}{\partial \beta} &= -\frac{1}{\beta} \\ E_{av} &= kT \end{aligned} \tag{5.49}$$

Thus, the energy density is:

$$\frac{E}{V} df = \frac{8\pi}{c^3} f^2 kT df \tag{5.50}$$

The significance of the Rayleigh–Jeans treatment is that since each mode of vibration (i.e. each frequency) carries an average of kT of energy, and the density of frequency states increases as f^2, then the equation predicts that the energy density should increase without limit as f becomes larger (or λ decreases). The underlying assumption is that the amplitudes can take on any value in a continuous manner.

5.6.4 Planck's Radiation Law

The average energy for each standing wave at a particular frequency is:

$$E_{av} = \frac{\sum_{i=0}^{\infty} (i\Delta E) e^{-i\Delta E/kT}}{\sum_{i=0}^{\infty} e^{-i\Delta E/kT}} \tag{5.51}$$

The sums in the numerator and denominator converge (by application of the binomial theorem):

$$\sum_{i=0}^{\infty} i\Delta E e^{-i\Delta E/kT} = \Delta E e^{-\Delta E/kT}\left(1+2\left(e^{-\Delta E/kT}\right)+3\left(e^{-\Delta E/kT}\right)^{\wedge 2}+\ldots\right)=\frac{\Delta E e^{-\Delta E/kT}}{\left(1-e^{-\Delta E/kT}\right)^{2}}$$

$$\sum_{i=0}^{\infty} e^{-i\Delta E/kT} = \left(1+\left(e^{-\Delta E/kT}\right)+\left(e^{-\Delta E/kT}\right)^{\wedge 2}+\ldots\right)=\frac{1}{1-e^{-\Delta E/kT}} \tag{5.52}$$

$$\therefore E_{av} = \frac{\sum_{i=0}^{\infty} i\Delta E e^{-i\Delta E/kT}}{\sum_{i=0}^{\infty} e^{-i\Delta E/kT}} = \frac{\dfrac{\Delta E e^{-\Delta E/kT}}{\left(1-e^{-\Delta E/kT}\right)^{2}}}{\dfrac{1}{1-e^{-\Delta E/kT}}} = \frac{\Delta E}{e^{\Delta E/kT}-1}$$

The energy E per unit volume in the frequency range f to $f + df$ is thus $D(f)/V$ times E_{av}. This gives us the distribution of energies as the frequency is varied:

$$\frac{E}{V}df = \frac{8\pi}{c^3}f^2\frac{\Delta E}{e^{\Delta E/kT}-1}df \tag{5.53}$$

Rayleigh and Jeans took the limit as $\Delta E \rightarrow 0$ and obtained:

$$E_{av} = \lim_{\Delta E \rightarrow 0}\frac{\Delta E}{e^{\Delta E/kT}-1} = kT \tag{5.54}$$

But this leads to the ultraviolet catastrophe. The only way to get an f^3 factor in the numerator for the energy density and a -1 in the denominator (as in Wien's formula) is to *not* take the limit and have the term ΔE be a linear function of f. So, and with a sense of desperation to save the phenomena, Planck set $\Delta E = hf$:

$$\frac{E}{V}df = \frac{8\pi}{c^3}f^2\frac{hf}{e^{hf/kT}-1}df \tag{5.55}$$

Where h is a new constant, Planck's constant, $h = 6.6256 \times 10^{-34}$ J s.

Planck's radiation law tells us that the energy levels (the amplitudes), for any one frequency, are evenly spaced and are discrete levels – but also, the spacing between these levels depends on the frequency. This is the key idea.

Thus was born the field of quantum physics.

5.6.5 Forms of the Radiation Laws

The various radiation law can be expressed in a variety of ways, as a function of either wavelength or frequency (Table 5.1).

The wavelength formulation arises from the frequency equations since:

$$f = \frac{c}{\lambda}$$
$$df = \frac{c}{\lambda^2}d\lambda \tag{5.56}$$

Energy density E/V has units J m^{-3} in the SI system. Often, especially when doing experimental work, it is convenient to measure intensity I in W m^{-2}. That is the rate of energy passing out from a unit area per

TABLE 5.1 Different expressions of the radiation law

LAW	*FREQUENCY*	*WAVELENGTH*
Planck	$\frac{E}{V}df = \frac{8\pi}{c^3}f^2\frac{hf}{e^{hf/kT}-1}df$	$\frac{E}{V}d\lambda = \frac{8\pi}{\lambda^5}\frac{hc}{e^{hc/\lambda kT}-1}d\lambda$
Wien	$\frac{E}{V}df = \frac{C_1 f^3}{e^{C_2 f/kT}}df$	$\frac{E}{V}d\lambda = \frac{C_1 c^4}{\lambda^5 e^{C_2 c/\lambda kT}}d\lambda$
Rayleigh–Jeans	$\frac{E}{V}df = \frac{8\pi}{c^3}f^2 kTdf$	$\frac{E}{V}d\lambda = \frac{8\pi}{\lambda^4}kTd\lambda$

unit time from the cavity. To convert from energy density to intensity we therefore multiply by c since this gives the energy per unit time.

However, the energy density is for the standing waves inside the cavity, being reflected backwards and forwards from the walls where the walls alternately absorb and emit radiation. The intensity actually observed is half this value (the emitted part), so we divide by a factor of 2.

The standing waves are distributed inside the cavity in three dimensions, and so for a particular surface over which the intensity is to be measured, radiation from various angles and directions will be incident on this surface. When this is taken into account, we need to divide the energy density by another factor of 2, and so, in terms of intensity, we obtain the Planck, Wien and Rayleigh-Jeans radiation laws as: Planck:

$$Id\lambda = \frac{2\pi}{\lambda^5}\frac{hc^2}{e^{hc/\lambda kT}-1}d\lambda \tag{5.57}$$

Wien:

$$Id\lambda = \frac{C_1 c^5}{4\lambda^5 e^{C_2 c/\lambda kT}}d\lambda \tag{5.58}$$

Rayleigh–Jeans:

$$d\lambda = \frac{2\pi c}{\lambda^4}kTd\lambda \tag{5.59}$$

We've expressed these quantities as energy density times the frequency interval, or wavelength interval. If we divide by $d\lambda$, we obtain intensity per unit wavelength, which is usually plotted against λ.

5.6.6 Stefan–Boltzmann Radiation Law

The Stefan–Boltzmann law gives the total radiated power as a function of temperature and area. For a black body, the emissivity $e=1$, and so we have:

$$\dot{Q} = \sigma AT^4 \tag{5.60}$$

The Planck radiation law gives the radiated power over a unit area for a wavelength interval $d\lambda$ or frequency interval dv. Thus, if we integrate the Planck law over all wavelengths, we arrive at the Stefan–Boltzmann law:

$$I d\nu = \frac{2\pi}{c^2} \nu^2 \frac{h\nu}{e^{h\nu/kT} - 1} d\nu$$

$$\dot{Q} = \int_0^\infty I d\nu = \int_0^\infty \frac{2\pi}{c^2} \nu^2 \frac{h\nu}{e^{h\nu/kT} - 1} d\nu$$

$$= \frac{2\pi h}{c^2} \int_0^\infty \frac{\nu^3}{e^{h\nu/kT} - 1} d\nu \tag{5.61}$$

$$= \frac{2\pi h}{c^2} \frac{k^3 T^3}{h^3} \int_0^\infty \frac{h^3}{k^3 T^3} \frac{\nu^3}{e^{h\nu/kT} - 1} d\nu$$

Letting:

$$x = \frac{h\nu}{kT}$$

$$\nu = \frac{kTx}{h} \tag{5.62}$$

we have:

$$d\nu = \frac{kT}{h} dx$$

$$\dot{Q} = \frac{2\pi h}{c^2} \frac{k^3 T^3}{h^3} \frac{kT}{h} \int_0^\infty \frac{x^3}{e^x - 1} dx \tag{5.63}$$

$$= \frac{2\pi}{c^2} \frac{k^4 T^4}{h^3} \int_0^\infty \frac{x^3}{e^x - 1} dx$$

$$\text{Now, } \int_0^\infty \frac{x^3}{e^x - 1} dx = \frac{\pi^4}{15} \tag{5.64}$$

And so, $\dot{Q} = \sigma T^4$

where the Stefan–Boltzmann constant is:

$$\sigma = \frac{2\pi k^4}{c^2 h^3} \frac{\pi^4}{15} \tag{5.65}$$

5.6.7 Wien Displacement Law

Wien found that when the temperature of a black body increased, the energy distribution maintained its general shape, but the maximum energy output shifted to higher frequencies.

When any object is above 0 K, it is emitting radiation according to Planck's radiation distribution law. That is, even an object at room temperature is emitting some radiation at very high frequencies, but at room temperature, something like an x-ray would rarely, if ever, occur. However, if we increase the temperature, we find that eventually there is significant radiation in the infrared portion of the electromagnetic spectrum, and then as the temperature is increased further, the body might start to glow "red hot", then white hot, and so on.

Not only might we see this radiation, but as the temperature rises, the intensity of the radiation also rises. Wien noticed that the wavelength of the maximum intensity varied as the inverse of the temperature. If the temperature doubled, the wavelength at maximum temperature halved.

To find an expression for the wavelength corresponding to maximum intensity, we differentiate Planck's radiation law with respect to λ and set this to 0:

$$0=\frac{d}{d\lambda}\frac{8\pi}{\lambda^5}\frac{hc}{e^{hc/\lambda kT}-1}$$

$$=8\pi hc\left[\frac{1}{\lambda^5}\left(\frac{-hc}{kT\lambda^2}\frac{e^{hc/\lambda kT}}{\left(e^{hc/\lambda kT}-1\right)^2}\right)+\frac{1}{e^{hc/\lambda kT}-1}\left(\frac{-5}{\lambda^6}\right)\right]$$

$$\frac{hc}{kT\lambda}\frac{e^{hc/\lambda kT}}{\left(e^{hc/\lambda kT}-1\right)^2}=\frac{5}{e^{hc/\lambda kT}-1} \tag{5.66}$$

$$\frac{hc}{5k}\frac{e^{hc/\lambda kT}}{\left(e^{hc/\lambda kT}-1\right)}\frac{1}{\lambda}=T$$

$$\frac{hc}{5k}\frac{1}{\left(1-e^{-hc/\lambda kT}\right)}\frac{1}{\lambda}=T$$

When this is solved numerically, we find that, with the wavelength in metres:

$$T=\frac{2.9\times10^{-3}}{\lambda}\text{ K} \tag{5.67}$$

which is one form of Wien's displacement law.

The Bohr Atom

6

6.1 INTRODUCTION

Although the phenomenon of electricity had been widely studied since the days of Benjamin Franklin in 1752, the quantitative nature of electric current was determined as a result of the study of electrolysis some 80 years later by Michael Faraday. Even then, the true nature of electricity remained unknown until the study of glow discharges in vacuum tubes. It was noticed that in a glass tube evacuated to a very low pressure, and with a high voltage applied across the plates positioned in the tube, the glass near the anode would glow brightly where it appeared a ray, coming from the cathode, struck it. In 1879, William Crookes showed these cathode rays to be deflected by electric and magnetic fields directed across the path, and it was John J. Thomson who, in 1897, deduced that the cathode rays were a stream of negatively charged particles which he called cathode corpuscles, but which came to be known as electrons.

Thomson was able to measure the ratio of the electric charge to the mass for these electrons. He found that this ratio was independent of the cathode material, and thus was something that was common to all atoms. Robert A. Millikan, in 1912, measured the charge on an electron from measurements of the behaviour of oil drops sprayed as a mist into the space between two horizontally charged plates. Millikan found the charge on an electron to be always an exact multiple of 1.6×10^{-19} C.

6.2 THE PHOTOELECTRIC EFFECT

In thermionic emission, (discovered by Thomas A. Edison in 1883), electrons can be ejected from a hot filament in a vacuum as a result of the kinetic energy imparted to them from the passage of the current in the filament. These electrons can form an electric current if an external field is applied.

In 1887, Hertz observed that an electric current could also be created if a metal was illuminated by light of a sufficiently high frequency. This is illustrated in Figure 6.1. Experiments showed that the current in this case could only be produced if the frequency of the light was above a critical value (the threshold frequency). If the frequency was below this value, no current was produced even if the light was of very high intensity. This effect was called the photoelectric effect and for many years remained unexplained.

It is found that even with no applied potential at the cathode, there is still a very small current. A small reverse voltage (the stopping potential) is needed to stop the most energetic of electrons from leaving the surface. The results of a typical experiment are shown in Figure 6.2(a).

The explanation of the photoelectric effect was given by Einstein in 1905. Einstein postulated that light consisted of energy quanta in accordance with Planck's equation. If the energy of the quanta of incoming light, which we now call photons, was greater than the work function W of the metal surface, then the excess energy would be imparted to the electron as kinetic energy. The maximum kinetic energy is given by:

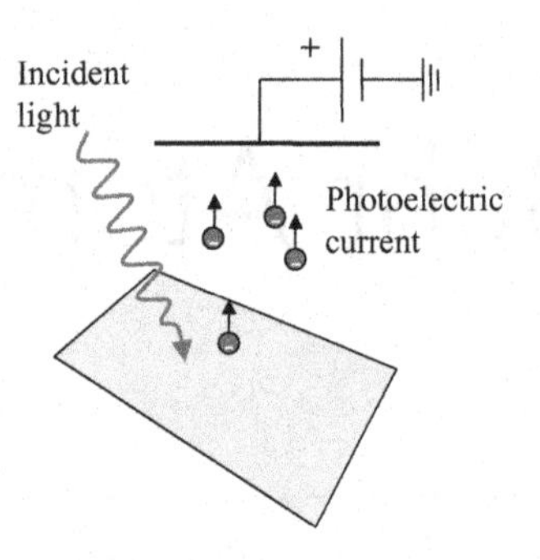

FIGURE 6.1 Schematic of the photoelectric effect.

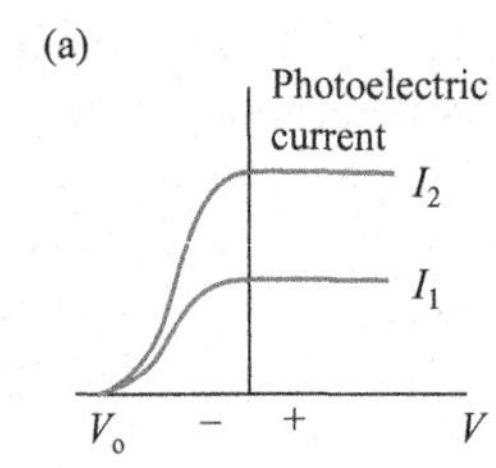

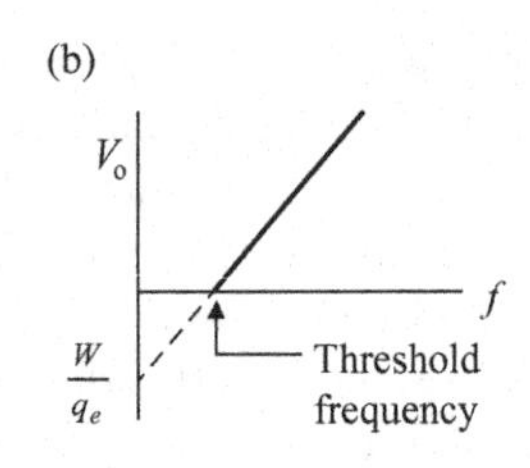

FIGURE 6.2 (a) Photoelectric current vs applied potential. (b) Stopping potential vs frequency of incoming radiation.

$$\begin{aligned}\frac{1}{2}mv^2 &= hf - W \\ &= q_e V_0\end{aligned} \tag{6.1}$$

where q_e is the charge on an electron.

The stopping potential is a measure of the maximum kinetic energy of the ejected electrons. As shown in Figure 6.2(b), if the stopping potential is plotted against frequency, the slope of the resulting linear function is Planck's constant, and the intercept is the work function.

The work function is a measure of the surface energy potential. That is, an electron needs a certain amount of energy to escape the surface. It is on the order of a few eV for most metals.

Note that Planck was still thinking of waves whose amplitude could only change in discrete steps according to the frequency of the wave. Einstein was thinking of particles of light, each particle carrying its own quanta of energy the value of which depended upon the frequency of the radiation.

6.3 LINE SPECTRA

Ever since the 18th century, it was known that the emission spectrum from a heated gas consisted of a series of lines instead of a continuous rainbow of colours. The position (or wavelength) of spectral lines was known to be unique to each type of element.

When white light is shone through a cool gas, it is found that dark lines appear in the resulting spectrum, the positions of which correspond exactly with the bright line spectra obtained when the gas is heated. A schematic of the observed dark lines is shown in Figure 6.3.

In 1885, Johann Balmer formulated an empirical equation that accounted for the position of the lines in the visible part of the hydrogen spectrum.

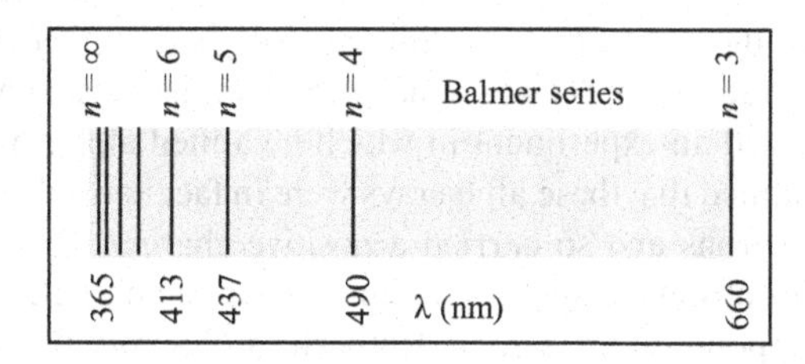

FIGURE 6.3 Radiation absorption spectrum for a gas.

$$\frac{1}{\lambda} = R\left(\frac{1}{2^2} - \frac{1}{n^2}\right) \tag{6.2}$$

where R is the Rydberg constant = 1.0973731×10^7 m^{-1}. The Balmer formula applies only to the hydrogen atom. Rydberg proposed a more general formula for the heavier elements.

In this formula, n is the line number and takes on the values 3, 4, 5,... The formula predicts an infinite number of spectral lines which become closer together with increasing value of n. At $n = \infty$, the wavelength is 364.6 nm, the limit of the Balmer series. Other series were discovered in the hydrogen spectrum by letting the first term in the brackets equal 1, 3, 4, etc.

Lyman series (ultraviolet) $n = 2, 3, 4\ldots$

$$\frac{1}{\lambda} = R\left(\frac{1}{1^2} - \frac{1}{n^2}\right) \tag{6.3}$$

Balmer series (visible) $n = 3, 4, 5\ldots$

$$\frac{1}{\lambda} = R\left(\frac{1}{2^2} - \frac{1}{n^2}\right) \tag{6.4}$$

Paschen series (infrared) $n = 4, 5, 6\ldots$

$$\frac{1}{\lambda} = R\left(\frac{1}{3^2} - \frac{1}{n^2}\right) \tag{6.5}$$

Brackett series (infrared) $n = 5, 6, 7\ldots$

$$\frac{1}{\lambda} = R\left(\frac{1}{4^2} - \frac{1}{n^2}\right) \tag{6.6}$$

The existence of spectral lines could not be explained by classical physics. Balmer's equation demonstrated an order, and an integral order at that, to the position of lines within the frequency spectrum. Balmer did not propose any physical explanation for his formula, but simply noted that it described the phenomena almost exactly. The importance of the equation is that it provides a test for what was to become a model for atomic structure.

6.4 THE BOHR ATOM

By 1910, the best model for the structure of the atom was one put forward by Thomson, the discoverer of the electron. In Thomson's model, negatively charged electrons were thought to be embedded within a continuous sphere of material of positive charge, the plum-pudding model. From Millikan's oil drop experiment, it was known that the mass of an electron was very small in comparison to the mass of the whole atom, and so most of the atomic mass was concentrated in the positively charged material.

Ernest Rutherford performed an experiment in which he aimed alpha "rays" at a thin sheet of gold foil. It had by then been determined that these alpha rays were in fact ionised helium atoms (that is, helium atoms which had lost two electrons and so carried a positive charge). These particles were emitted at high velocity from radioactive substances such as radium. When the particles strike a fluorescent screen, a flash of light appears at that point on the screen. Rutherford observed that most of the alpha particles went straight through the foil with very little deviation. This was as expected, but the reasoning is subtle. As we've seen previously, the electric field surrounding an isolated charge can be represented by lines of force which extend radially outwards from the charge. The strength of the field decreases as $1/r^2$ from the charge. Near the charge, the field lines are close together (strong electric field) compared to the lines further out where they are more widely spaced.

Now, with a Thomson atom, the positive charge is distributed throughout the size of the atom, which in the case of gold, is fairly large. An incoming positively charged alpha particle is very small in comparison. The alpha particle is not greatly affected by the electrons since the electrons are extremely small in comparison to the alpha particles. The alpha particles will just push them out of the way. It is the relatively heavy positively charged sphere that is important. Although the positively charged portion of the atom may exert a Coulomb force on the alpha particle, the field around the atom is relatively diffuse (due to its size), and so any deviation of the path of the alpha particle averages out to be very small by the time the particle leaves the foil before going off to strike the screen.

To his surprise, Rutherford found that a very small number of particles were scattered over a very wide angle – some by even greater than 90°, coming back off the front surface of the foil. Such an alpha particle would have to be acted upon by an enormous force to cause this, and this would only occur as a result of a near-head-on collision with something very small and positively charged in the foil. That is, the positively charged part of the atom had to be extremely small so as to produce an electric field intense enough to cause an appreciable deviation of the path of the alpha particle.

In 1911, Rutherford therefore proposed that the electrons orbited at some considerable distance from a central, heavy and positively charged nucleus. That is, most of an atom was empty space.

Initially, it was thought that the electrostatic attraction between the nucleus and an outer electron was balanced by the centrifugal force arising from the orbital motion. However, if this were the case, then the electron (being accelerated inwards towards the centre of rotation) would continuously radiate all its energy as electromagnetic waves and very quickly fall into the nucleus.

In 1913, Niels Bohr postulated two important additions to Rutherford's theory of atomic structure. The first postulate was that electrons can orbit the nucleus in a circular path in what are called stationary states in which no emission of radiation occurs. Further, noting that Planck's constant had units of angular momentum, Bohr stipulated that in these stationary states, the orbital angular momentum is constrained to have values:

$$L = m_e vr = \frac{nh}{2\pi} \tag{6.7}$$

Note, the 2π appears because we customarily express the energy in terms of f (i.e. $E=hf$) rather than ω. In many cases, it is convenient to express Planck's constant as $\hbar = h/2\pi$ in which case Bohr's postulate becomes $L = n\hbar$.

The second postulate was that electrons can make transitions from one state to another accompanied by the emission or absorption of a single photon of energy $E=hf$ thus leading to absorption and emission spectra.

This was very much a mechanical model of the atom, but the model proved to be amazingly productive.

For a single electron orbiting a positively charged nucleus, as shown in Figure 6.4 for the hydrogen atom, we see that the centrifugal force is balanced by Coulomb attraction:

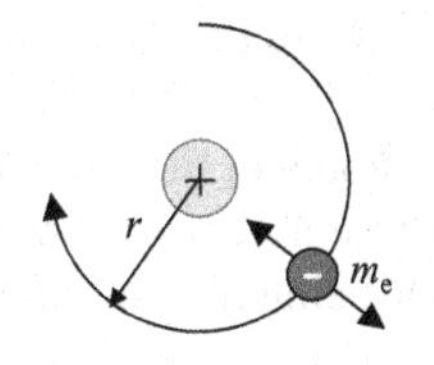

FIGURE 6.4 Bohr's model for the hydrogen atom.

$$\frac{1}{4\pi\varepsilon_0}\frac{q_e^{\,2}}{r^2}=\frac{m_e v^2}{r} \tag{6.8}$$

with the additional condition that the orbital angular momentum can have values such that:

$$m_e vr=\frac{nh}{2\pi} \tag{6.9}$$

where n is an integer, 1, 2, 3…

The kinetic energy of rotation of the electron is

$$\begin{aligned}KE&=\frac{1}{2}m_e v^2\\&=\frac{1}{2}m_e\frac{r}{me}\frac{1}{4\pi\varepsilon_0}\frac{q_e^{\,2}}{r^2}\\&=\frac{1}{8\pi\varepsilon_0}\frac{q_e^{\,2}}{r}\end{aligned} \tag{6.10}$$

The potential energy W at a distance r from the nucleus is:

$$W=-\frac{1}{4\pi\varepsilon_0}\frac{q_e^{\,2}}{r} \tag{6.11}$$

The total energy of an electron at a given energy level is therefore:

$$\begin{aligned}E&=\frac{1}{8\pi\varepsilon_0}\frac{q_e^{\,2}}{r}-\frac{1}{4\pi\varepsilon_0}\frac{q_e^{\,2}}{r}\\&=\frac{1}{8\pi\varepsilon_0}\frac{q_e^{\,2}}{r}\end{aligned} \tag{6.12}$$

But r can only have values according to:

$$r=\frac{nh}{2\pi}\frac{1}{m_e v} \tag{6.13}$$

And so, the allowed energy levels are:

$$E_n=-\frac{m_e Z^2 q_e^{\,4}}{8\varepsilon_0^{\,2}h^2n^2} \tag{6.14}$$

where we have included Z, the number of electrons in a multi-electron atom. For $Z=1$, for the hydrogen atom, the energy of the ground state, the innermost level, is 13.6 eV. The energy levels for each state n rise as Z^2, thus, the energy level of the innermost shell for multi-electron atoms can be several thousand eV.

The stationary states or energy levels allowed by the Bohr model of the atom are observed to consist of sub-levels (evidenced by a fine splitting of spectral lines). These groups of sub-levels are conveniently called electron shells, and are numbered K, L, M, N, etc., with K being the innermost shell corresponding to $n=1$. The number n is called the principal quantum number and describes how energy is quantised.

Energies are customarily measured in electron-volts for convenience, because of the magnitude of the numbers. 1 eV $= 1.602 \times 10^{-19}$ J.

The energy required to move an electron from an electron shell to infinity is called the ionisation energy. It is convenient to assign the energy at infinity as being 0 since as an electron moves closer to the nucleus (which is positively charged) its potential to do work is less; thus, the energy levels for each shell shown are negative. For hydrogen, the ionisation energy is −13.6 eV. Thus, for the hydrogen atom, the energies for the higher energy levels are given by:

$$E = -\frac{13.6}{n^2} \tag{6.15}$$

At each value of n (i.e. at each energy level) the angular momentum L can take on several distinct values. Later, we will see that the number of values of angular momentum is described by a second quantum number l. The allowed values of l are 0, 1…(n − 1). Each value of l is indicated by a letter, *s, p, d, f, g, h* for $l=0$, 1, 2, 3, 4, 5 respectively where:

$$L = \frac{h}{2\pi}\sqrt{l(l+1)} \tag{6.16}$$

We might notice that, according to the above, for $l=0$, we have $L=0$ indicating that the electron does not have any angular momentum which is contrary to the Bohr postulate ($L=nh/2\pi$). The Bohr atom is a model – an orbiting electron. The wave mechanics treatment which we will come to shortly makes the correct prediction.

A third quantum number m describes the allowable changes in the angle of the angular momentum vector in the presence of a magnetic field. It takes the values −1 to 0 to +1. It was later shown that a fourth quantum number was needed to describe the spin of an electron where the spin can be either −1/2 or +1/2. According to the Pauli exclusion principle, no electron in any one atom can have the same combination of quantum numbers. This provides the basis for the filling of energy levels. For example, the 3*d* energy level can hold up to ten electrons. With $n=3$ we have $l=0$, 1, 2, or *s*, *p* or *d*, giving values for m as −2, −1, 0, 1, 2. That is, the 3*d* energy level, with five values of m times two for spin, can accommodate ten electrons. When all the electrons in an atom are in the lowest possible energy levels, the atom is said to be in its ground state.

6.5 THE RYDBERG CONSTANT

Can the Bohr model explain the emission spectra of the hydrogen atom? Let's consider the transition of an electron from one energy level to another – it doesn't matter which, but we can choose $n=3$ to $n=2$ for something to work with. Figure 6.5 shows a transition from a higher energy state to a lower energy state with the emission of a photon.

The resulting emission of a photon is in the visible region (it is one of the Balmer series of lines) with $\lambda=656$ nm.

According to the Bohr model, the energy level at the $n=2$ is:

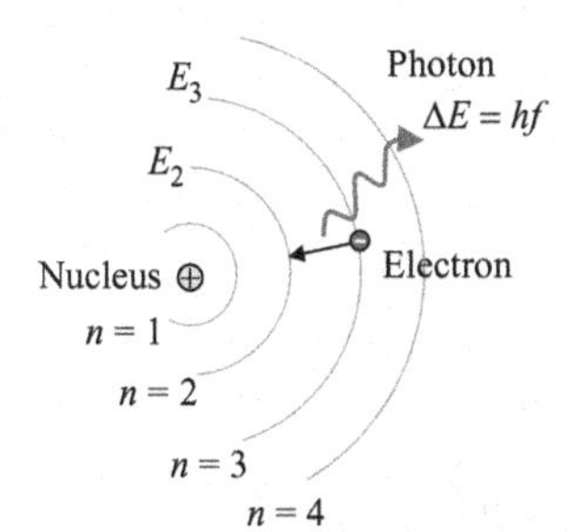

FIGURE 6.5 Transition from a higher energy state to a lower energy state with the emission of a photon.

$$E_2 = -\frac{m_e q_e^{\,4}}{8\varepsilon_0^{\,2} h^2 2^2} \tag{6.17}$$

The energy level at $n=3$ is:

$$E_3 = -\frac{m_e q_e^{\,4}}{8\varepsilon_0^{\,2} h^2 3^2} \tag{6.18}$$

The energy difference between $n=3$ and $n=2$.

$$\begin{aligned} \Delta E = E_3 - E_2 &= -\frac{m_e q_e^{\,4}}{8\varepsilon_0^{\,2} h^2 3^2} + \frac{m_e q_e^{\,4}}{8\varepsilon_0^{\,2} h^2 2^2} \\ &= \frac{m_e q_e^{\,4}}{8\varepsilon_0^{\,2} h^2}\left(\frac{1}{2^2} - \frac{1}{3^2}\right) \end{aligned} \tag{6.19}$$

But:

$$\begin{aligned} \Delta E &= hf \\ c &= f\lambda \\ f &= \frac{c}{\lambda} \\ \Delta E &= \frac{hc}{\lambda} \end{aligned} \tag{6.20}$$

So:

$$\begin{aligned} \Delta E &= \frac{hc}{\lambda} = \frac{m_e q_e^{\,4}}{8\varepsilon_0^{\,2} h^2}\left(\frac{1}{2^2} - \frac{1}{3^2}\right) \\ \frac{1}{\lambda} &= \frac{m_e q_e^{\,4}}{8\varepsilon_0^{\,2} h^3 c}\left(\frac{1}{2^2} - \frac{1}{3^2}\right) \end{aligned} \tag{6.21}$$

And so, with: $q_e = -1.6\times10^{-19}$ C; $m_e = 9.1\times10^{-31}$ kg; $\varepsilon_0 = 8.85\times10^{-12}$ F m^{-1}; $h = 6.63\times10^{-34}$ J s and $c = 3\times10^{8}$ m s^{-1}.

Comparing with Balmer's formula:

$$\frac{1}{\lambda} = R\left(\frac{1}{2^2} - \frac{1}{3^2}\right) \tag{6.22}$$

the Rydberg constant is predicted to be:

$$R = \frac{9.1\times10^{-31}\left(1.6\times10^{-19}\right)^4}{8\left(8.85\times10^{-12}\right)^2\left(6.63\times10^{-34}\right)^3\left(3\times10^8\right)} \tag{6.23}$$
$$= 1.09\times10^7 \text{ m}^{-1}$$

which compares very favourably with the empirical value found by Rydberg.

The resulting wavelength from $n=2$ to $n=3$ is thus:

$$\frac{1}{\lambda} = 1.09\times10^7\left(\frac{1}{2^2}-\frac{1}{3^2}\right) \tag{6.24}$$
$$\lambda = 660.5 \text{ nm}$$

which is consistent with the experimentally observed value. Alternately, we could take the difference in measured energy levels of the hydrogen atom for $n=2$ and $n=3$ to get $-1.51-(-3.39)=1.88$ eV and use $\Delta E=hf$ to obtain the same result.

6.6 MATTER WAVES

The Bohr model of the atom applies only to a single electron orbiting a nucleus and ignores interactions between electrons and other neighbouring atoms. Further, the theory does not offer any explanation as to why the angular momentum is to be quantised, nor the relative intensities of the spectral lines. Such explanations and treatments can only be explained in terms of wave mechanics.

In 1924, Louis de Broglie postulated that matter exhibited a dual nature (just as did electromagnetic radiation) and proposed that the wavelength of a particular body of mass m with velocity v is found from:

$$\lambda = \frac{h}{mv} \tag{6.25}$$

where mv is the momentum p of the mass.

This then explained the orbits of the Bohr atom. If an electron were a wave, and the wavelength of the electron was an integer multiple of the circumference of the orbit (that is, if $2\pi r$ is the circumference of the orbit, then that orbit contains n wavelengths, i.e. $\lambda n=2\pi r$), then if we were able to observe that wave, we would see the crests and troughs of the wave fixed in place around the circumference of the orbit as shown in Figure 6.6(a). If the wavelength were slightly off, then the crests and troughs would slowly revolve around the axis of the orbit as shown in Figure 6.6(b). The situation is similar to that of a standing wave but this time, there are no reflections such as in our earlier example of a stretched string between two supports. Here, we have the crests and the troughs of the matter wave remaining stationary. That is, the electron is in a stationary state. We might call these waves circular standing waves.

Or, put another way, the allowable wavelengths associated with a particular orbit for a stationary state are:

$$\lambda = \frac{2\pi r}{n} \tag{6.26}$$

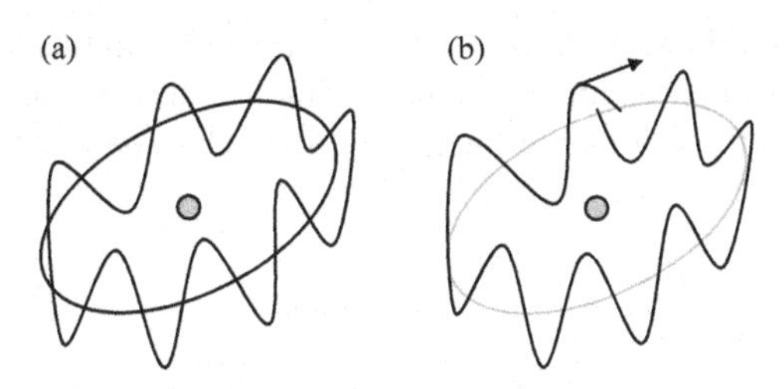

FIGURE 6.6 (a) Modes of vibration in a circular path for a standing wave. (b) Modes of vibration where the wavelength is not an integer multiple of the circumference.

with $n = 1, 2, 3\ldots$ Thus, in terms of angular momentum:

$$\frac{h}{mv} = \frac{2\pi r}{n}$$
$$mvr = \frac{nh}{2\pi} \tag{6.27}$$

These waves are called matter waves. The matter waves for electrons are the standing waves that correspond to particular electron orbits. Because h is a very small number, the wavelength of large bodies is very small. For small bodies, e.g. electrons, the wavelength is comparable to atomic dimensions.

We might ask ourselves: what is the velocity of these matter waves? For standing waves of a stretched string, the velocity of the wave, which passes backwards and forwards along the string, is given by:

$$v = \sqrt{\frac{T}{\rho}} \tag{6.28}$$

Where T is the tension in the string and ρ is the mass per unit length. For electrons in an atom, say the hydrogen atom, the velocity of the matter wave can be found from:

$$v = \frac{nh}{2\pi mr} \tag{6.29}$$

Returning for a moment to the balance of the Coulomb and centrifugal forces for the Bohr hydrogen atom, but this time solving for r, we have:

$$\frac{1}{4\pi\varepsilon_0}\frac{q_e^2}{r^2} = \frac{m_e v^2}{r} \tag{6.30}$$
$$m_e vr = \frac{nh}{2\pi}$$
$$v = \frac{nh}{2\pi m_e r}$$
$$\frac{1}{4\pi\varepsilon_0}\frac{q_e^2}{r^2} = \frac{m_e}{r}\frac{n^2h^2}{4\pi^2 m_e^2 r^2}$$
$$r = \frac{\varepsilon_0 n^2 h^2}{\pi m_e q_e^2}$$

For $n = 1$, and inserting known values for h, ε_0, m_e and q_e, we obtain a radius for the electron in the ground state for a hydrogen atom of about 0.05 nm. Note that the radius increases as the square of n.

Substituting back, we find the velocity of the ground-state electron in its orbit for hydrogen is:

$$v = \frac{nh}{2\pi m_e r}$$

$$v = \frac{nh}{2\pi m_e} \frac{\pi m_e q_e^2}{\varepsilon_0 n^2 h^2} \tag{6.31}$$

$$= \frac{q_e^2}{2\varepsilon_0 nh}$$

which is of the order of 2×10^6 ms^{-1}, much less than the speed of light. Note that the velocity increases as r decreases. For multi-electron atoms, the radii of the innermost electron orbits are much smaller than that of the hydrogen atom, and the velocity becomes comparable to the speed of light –for which relativistic considerations have to be included.

Although we have computed the velocity of the orbiting Bohr electron, we might ask ourselves, is this the same as the velocity of the de Broglie matter wave? That is, the velocity of the standing wave associated with the electron. In the case of the stretched string, we have waves travelling backwards and forwards being reflected of each fixed end of the string. What so for the matter wave for an electron? The only requirement is that the wavelength of the electron is a multiple of the circumference of the orbit. That is, $\lambda n = 2\pi r$. For $n = 1$, this implies that the wavelength is the same as the circumference. The wave wraps itself around the circumference of the orbit. For $n = 2$, the wavelength is half of the circumference (which now has a radius four times larger than the case for $n = 1$) – as if there are two complete wavelengths fitting around the circumference. Unlike the case of a string, the matter waves do not reflect off anything, but clearly have some velocity. It is the allowed orbitals (the allowed values of angular momentum) that "fix" the ends of the wave much like the end supports in the case of a stretched string. There is a connection between the velocity of the matter wave and the velocity of the orbiting Bohr electron, but the connection is a subtle one and not essential for our understanding – since we will leave the Bohr atom shortly and focus on the waves rather than the electron.

The wave–particle duality of matter means that, inherently, an electron is a particle, but its motion can be quantified using the mathematical equations appropriate to waves. The wave nature of matter is often interpreted as being one of probabilities. The square of the amplitude of a matter wave represents the probability of finding the associated particle at a particular position.

Since momentum and kinetic energy are related by:

$$E = \frac{1}{2}mv^2$$
$$p = mv = \frac{2E}{v} \tag{6.32}$$

and:

$$\lambda = \frac{h}{mv} \tag{6.33}$$

Then we can express the de Broglie relationship in terms of energy and momentum:

$$E = \frac{1}{2}mv^2$$

$$v = \sqrt{\frac{2E}{m}} \tag{6.34}$$

$$p = mv$$

$$= \sqrt{2Em}$$

Since matter is described in terms of a probability, there becomes an inherent limitation in what we can know about the motion and position of a particle such as an electron. The Heisenberg uncertainty principle quantifies these uncertainties. For momentum and position, the requirement is:

$$\Delta p \Delta x \geq \frac{h}{2\pi} \tag{6.35}$$

where Δp and Δx are the uncertainties associated with these quantities. The more we reduce the uncertainty in one, the more the uncertainty in the other increases.

6.7 THE PHOTON

We might ask ourselves at this point, exactly what is a photon?

In 1900, Planck envisaged that electromagnetic radiation in a cavity consisted of standing waves whose amplitude could vary only in discrete steps, and that the step size depended upon the frequency of the wave. That is, it was the presence of the cavity and the circumstances of the standing waves within it that gave rise to the discreteness and so accounted for the emission spectrum.

In 1905, Albert Einstein proposed that this discreteness was a feature of light itself, and not a product of the circumstances of its emission. The discrete nature of light was said to be a result of light being composed of particles, photons, each with its own energy which depended upon that particle's frequency. Einstein then showed that a photon has zero rest mass, but it does have a relativistic mass and thus momentum.

$$E = hf = mc^2$$

$$\frac{h}{\lambda} = mc = p \tag{6.36}$$

According to Bohr, in 1913, when an electron makes a transition from one energy state to another, a photon is emitted or absorbed, the photon having an energy that depends on the change of energy levels.

In 1924, de Broglie proposed that all matter exhibits a wave-like character just as waves exhibit a particle-like behaviour.

Returning to our concept of cavity radiation, if we scan through the frequencies of black body radiation at intervals df, we find there is a relatively low number of standing waves at low frequencies, but as we move up into the higher frequencies, we encounter more standing waves in the same frequency interval. For a given frequency interval, the standing waves carry a distribution of energies (amplitudes). These standing waves have an average energy kT, it's just that at higher frequencies, there are more of them, so the intensity of the emitted radiation is greater at those higher frequencies – up to a point. If the frequency interval we are looking at is too high, then the step up in amplitude for a wave to exist at that frequency is so great that not very many can be excited to that level, and so the intensity falls off, thus avoiding the ultraviolet catastrophe. That is, if $hf > kT$, then fewer and fewer waves can exist at those higher energy (amplitude) levels. If the frequency is very high, such that $hf >> kT$, then hardly any waves will be found at that energy level despite the density of states being very large. Those states are largely unfilled.

At one particular frequency, there will be a number of electromagnetic waves of a range of amplitudes (energies). Those waves can increase or decrease their energies (amplitudes) in steps according to $E=hf$. That is, each energy state (amplitude) can be $0hf$, $1hf$, $2hf$, $3hf$... but any *change* in energy state can only happen one hf at a time.

Imagine now there are no electromagnetic waves, but there are, in their place, particles which we now call photons. Each photon can carry a certain amount of energy, but some photons carry more energy than others. All photons move at the speed of light, but since they can have different energies, they have different relativistic masses and different momenta.

Higher energy photons are produced by higher frequency oscillations of an electron. Such an electron may be oscillating as a result of thermal agitation or the oscillations between two states of an atom (beat frequency) in a transition from one energy state to another.

A photon has an energy $E=hf$. It cannot have energy $2hf$, only hf. Another photon may carry an energy $2hf$, but that photon would have been produced at a different frequency f' and have its own energy hf'. In terms of electromagnetic radiation, the number of photons per unit area per unit time is related to the intensity of the radiation. High intensity, low wavelength electromagnetic waves are associated with a large number of low energy photons.

In 1925, Bose and Einstein re-derived Planck's radiation law for cavity radiation by examining the statistics of photon energies, a "photon gas", within the cavity.

We must now therefore think of light as consisting of photons. Later we shall see that this can be reconciled with the apparent wave nature of light. We shall see that the photon is the quanta of the electromagnetic field.

The New Quantum Theory

7

7.1 INTRODUCTION

At this point in our discussion, we recognise the utility of the Bohr model of the atom. The notable feature is the mechanistic description of electrons orbiting the nucleus with the added conditions of quantisation of the energy levels and stationary states as originally proposed by Planck and given further interpretation by Einstein. These ideas are known now as the old quantum theory. While describing the phenomena, they don't really provide much in the way of explanation. Just why are there stationary states? Why does the energy not just radiate away when the electron is orbiting in an allowed orbit? Why is the angular momentum quantised?

The orbital, or circular, motion of the electron together with the stationary states foretell a connection to wave phenomena and standing waves. Further, we have seen that circular motion can be expressed in terms of complex numbers, and so it will not be surprising if we find that a mathematical description of the energy levels in an atom involve wave equations and complex numbers. This leads us into the realm of quantum mechanics.

7.2 THE SCHRÖDINGER EQUATION

The total energy of a particle (e.g. an electron in an atom, a free electron, an electron in an electric field, a conduction electron in a solid) is the sum of the potential and kinetic energies. Expressed in terms of momentum p, and mass m, this is stated:

$$E = \frac{1}{2}mv^2 + V \tag{7.1}$$

Since $p = mv$, then:

$$\begin{aligned} E &= \frac{1}{2}pv + V \\ &= \frac{p^2}{2m} + V \end{aligned} \tag{7.2}$$

The factor $h/2\pi$ occurs often in the equations to follow, and so we let:

$$\hbar = \frac{h}{2\pi} \tag{7.3}$$

and also:

$$\omega = 2\pi f$$
$$k = \frac{2\pi}{\lambda} \tag{7.4}$$
$$E = hf = \hbar\omega$$

The potential energy, V, may depend upon the position and time, and so in general, $V = V(x,t)$.

Returning for a moment to the complex representation of a wave:

$$y(x,t) = Ae^{i(kx-\omega t)} \tag{7.5}$$

Let us take the derivative of y with respect to t:

$$\begin{aligned}\frac{\partial y}{\partial t} &= -Ae^{i(kx-\omega t)}i\omega \\ &= -i\omega y\end{aligned} \tag{7.6}$$

Now, letting:

$$\begin{aligned}E &= \hbar\omega \\ \omega &= \frac{E}{\hbar}\end{aligned} \tag{7.7}$$

Thus:

$$\begin{aligned}\frac{\partial y}{\partial t} &= -i\frac{E}{\hbar}y \\ i\hbar\frac{\partial y}{\partial t} &= Ey\end{aligned} \tag{7.8}$$

We can thus define the energy operator, $\hat{E}$:

$$\hat{E} = i\hbar\frac{\partial}{\partial t} \tag{7.9}$$

Now, taking the derivative of y with respect to x:

$$\begin{aligned}\frac{\partial y}{\partial x} &= Ae^{i(kx-\omega t)}ik \\ &= iky\end{aligned} \tag{7.10}$$

From de Broglie, we have:

$$p = \hbar k \tag{7.11}$$

and so:

$$\frac{\partial y}{\partial x} = i\frac{p}{\hbar}y$$
$$\frac{\hbar}{i}\frac{\partial y}{\partial x} = py \tag{7.12}$$

The momentum operator is thus expressed:

$$\hat{p} = \frac{\hbar}{i}\frac{\partial}{\partial x}$$
$$= -i\hbar\frac{\partial}{\partial x} \tag{7.13}$$

That is, the differential operator $\hat{p}$ is equivalent to momentum, and the differential operator $\hat{E}$ is equivalent to energy. Note, this is quite a different thing to the "velocity operator" in Section 1.7. There, the operator $\hat{D}$ acted upon the distance s, and the result was the velocity. Note that the operator $\hat{D}$ has units of t^{-1}, and after being applied to the distance s, the result is velocity v. The operator $\hat{D}$ is an algebraic operator.

$$\hat{D}s = v \tag{7.14}$$

In the case of the momentum and energy operators above, the physical quantity p or E *becomes the operator.* The units of the operators $\hat{p}$ and $\hat{E}$ are momentum and energy, respectively. They are, until they act upon a function, "unsatisfied" operators. They will be used to operate on the wave function. These operators are called quantum mechanical operators to distinguish them from algebraic operators.

The energy equation thus becomes a differential operator equation.*

$$\frac{p^2}{2m} + V = E$$
$$\frac{1}{2m}\left(-i\hbar\frac{\partial}{\partial x}\right)^2 + V(x,t) = i\hbar\frac{\partial}{\partial t} \tag{7.15}$$
$$-\frac{\hbar^2}{2m}\frac{\partial^2}{\partial x^2} + V(x,t) = i\hbar\frac{\partial}{\partial t}$$

In this scheme, physical quantities which can vary in space and time, such as energy and momentum, are represented by operators.

In 1926, Erwin Schrödinger was very impressed by the concept of de Broglie's matter waves and so sought to explain the structure of the Bohr atom in terms of wave mechanics. But what was the nature of the wave?

We have seen previously that a travelling wave can be described by a wave function $y(x,t)$. But say we have a wave function $\Psi(x,t)$ that we do not yet know the form of. Whatever it is, we now let the energy differential operator equation operate on the function Ψ.

$$-\frac{\hbar^2}{2m}\frac{\partial^2\Psi(x,t)}{\partial x^2} + V(x,t)\Psi(x,t) = i\hbar\frac{\partial\Psi(x,t)}{\partial t} \tag{7.16}$$

* Note the subtlety in proceeding to the last line of the equation. By the laws of operators: $DD = D^2$ That is, an operator squared is equivalent to the second derivative. We are not saying that the second derivative is the square of the first derivative. We are saying that the second derivative *operator* is the square of the first derivative *operator*. When the operator comes to actually operate on the function, we take the second derivative.

This is an energy equation. It is a differential equation. It is a complex differential equation. It is called the Schrödinger equation in one dimension. The solution to the Schrödinger equation is the wave function $\Psi(x,t)$.

We might ask, just what is the physical significance of the wave function Ψ? In classical physics, a wave function describes the displacement of a particle, or the amplitude of an electric field, or some other phenomenon, at a given coordinate x at some time t. For example, the magnitude E of the electric field in an electromagnetic wave at some location x at time t can be expressed:

$$E(x,t) = E_o \sin(kx - \omega t) \tag{7.17}$$

The energy density is the energy contained within the electric field per unit volume (J m^{-3} in SI units). The intensity of the field is a measurement of the power (i.e. the rate of energy transfer) transmitted per unit area by the field. The average (or rms) power of an electromagnetic wave is:

$$I_{av} = \varepsilon_0 c\left[\frac{1}{2}E_o^{\ 2}\right] \tag{7.18}$$

The important feature here is that the energy carried by a wave is proportional to the square of the amplitude of the wave. In the case of electromagnetic waves, what we are really measuring as energy is the density of photons (i.e. number per unit volume) – since each photon carries its own quanta of energy. The wave function gives us some physical or experimental observable.

By analogy to the case of photons, the wave function for an electron has a connection with the energy carried by it since Schrödinger's equation is an energy equation. Max Born postulated that the square of the amplitude of the wave function is a measure of the probability density of the position of an electron. Since Ψ is complex, to obtain a physical value for this probability, we use the product:

$$P(x,t) = |\Psi|^2 = \Psi^*\Psi \tag{7.19}$$

where Ψ^* is the complex conjugate of Ψ. $|\Psi|^2$ is interpreted as a probability density function. For example, the probability that an electron is located within a small increment Δx around x at time t is:

$$P(x,t)\Delta x \tag{7.20}$$

When a small particle, such as an electron (or a proton), travels from place to place, it does so as a particle. The electron is not smeared out into some kind of wave – it retains its particle-like nature. The amplitude of the wave function provides information about the probability of finding an electron at some particular place. The wave function, in this case, does not provide a physical observable. Instead, the quantum mechanical wave function provides probability information about physical observables.

The Schrödinger wave equation is a complex equation. The wave functions that are solutions to the Schrödinger equation will, in general, be complex functions. However, as we shall see, it is the amplitude of the complex quantities that are of interest.

7.3 SOLUTIONS TO THE SCHRÖDINGER EQUATION

7.3.1 Separation of Variables

The solution to the Schrödinger wave equation is the wave function Ψ. For many cases of interest, the potential function V is a function of x only, that is, the potential is static (independent of time). This allows the wave equation to be separated into time-independent and time-dependent equations that can be readily solved independently.

We start with the full Schrödinger equation:

$$-\frac{\hbar^2}{2m}\frac{\partial^2\Psi(x,t)}{\partial x^2}+V(x,t)\Psi(x,t)=i\hbar\frac{\partial\Psi(x,t)}{\partial t} \tag{7.21}$$

Let the solution $\Psi(x,t)$ be the product of two equations, one a function of x, the other a function of t.

$$\Psi=\psi(x)\varphi(t) \tag{7.22}$$

Thus:

$$-\frac{\hbar^2}{2m}\frac{\partial^2\psi\varphi}{\partial x^2}+V\psi\varphi=i\hbar\frac{\partial\psi\varphi}{\partial t}$$

$$-\frac{\hbar^2}{2m}\frac{\partial^2\psi\varphi}{\partial x^2}\frac{1}{\psi\varphi}+V=i\hbar\frac{\partial\psi\varphi}{\partial t}\frac{1}{\psi\varphi} \tag{7.23}$$

$$\frac{1}{\psi}\left[-\frac{\hbar^2}{2m}\frac{\partial^2\psi}{\partial x^2}+V\psi\right]=i\hbar\frac{\partial\varphi}{\partial t}\frac{1}{\varphi}$$

Each of the two sides of this equation can be thought to be equal to some constant G which will be later shown to be the energy E. Setting the left-hand side equal to G and multiplying by ψ, we obtain:

$$-\frac{\hbar^2}{2m}\frac{\partial^2\psi}{\partial x^2}+V\psi=G\psi \tag{7.24}$$

This is called the time-independent Schrödinger equation.

The right-hand side is called the time-dependent Schrödinger equation, and this is also set equal to G:

$$i\hbar\frac{\partial\varphi}{\partial t}\frac{1}{\varphi}=G \tag{7.25}$$

G is a constant that just connects the two equations. It is termed the separation constant because it allows the variables x and t to be separated into two equations. Its significance will be examined shortly.

The resulting solutions of time-independent and time-dependent equations are functions, one a function of x, the other a function of t. When these two functions are multiplied together, we obtain the wave function. The wave function $\Psi(x,t)$ is the solution to the original Schrödinger differential wave equation. That is:

$$\Psi(x,t)=\psi(x)\varphi(t) \tag{7.26}$$

$\psi(x)$ is the solution to the time-independent equation. $\phi(t)$ is the solution to the time-dependent equation.

When solving a linear differential equation with constant coefficients, we make use of an auxiliary equation. The solution to the original differential equation involves finding the roots of the auxiliary equation which is a polynomial in m. For example, say we wish to solve:

$$2\frac{d^2y}{dx^2}+\frac{dy}{dx}-y=0. \tag{7.27}$$

This equation can be written in operator form:

$$\left(2D^2+D-1\right)y=0 \tag{7.28}$$

Where D is the differential operator d/dx. The solution has the form $y = e^{mx}$. The auxiliary equation is written:

$$2m^2 + m - 1 = 0$$

$$(2m-1)(m+1) = 0 \tag{7.29}$$

$$m = +\frac{1}{2}, -1$$

Thus, two solutions for the differential equation are:

$$y_1 = e^{\frac{1}{2}x}$$

$$y_2 = e^{-x} \tag{7.30}$$

In general, the solution to the differential equation is written as a linear combination of these:

$$y(x) = C_1 e^{\frac{1}{2}x} + C_2 e^{-x} \tag{7.31}$$

In general, there may be many solutions to the time-independent equation each differing by a multiplicative constant. The collection of functions $\psi(x)$ that are solutions are called eigenfunctions, eigenvectors or characteristic functions. The eigenfunctions for the time-independent equation $\psi(x)$ determine the space dependence of the wave function Ψ. The quantum state associated with an eigenfunction is called an eigenstate. When a physical quantity, such as energy or momentum, is extracted from the eigenfunction, these are called eigenvalues.

7.3.2 Solution to the Time-Dependent Schrödinger Equation

The solution to the time-dependent Schrödinger equation involves the use of an auxiliary equation. We proceed as follows:

$$i\hbar \frac{\partial \varphi}{\partial t} \frac{1}{\varphi} = G$$

$$\frac{\partial \varphi}{\partial t} = -G\varphi \frac{i}{\hbar} \tag{7.32}$$

The auxiliary equation in m is:

$$\mathrm{m} = -G\frac{i}{\hbar} \tag{7.33}$$

Therefore, the solution has the form:

$$\varphi = e^{-\frac{Gi}{\hbar}t}$$

$$= \cos\frac{Gt}{\hbar} - i\sin\frac{Gt}{\hbar} \tag{7.34}$$

Using Euler's formula. $\phi(t)$ has a frequency:

$$\omega = \frac{G}{\hbar} \tag{7.35}$$

But:

$$\omega = \frac{E}{\hbar} \tag{7.36}$$

And so:

$$G = E$$
$$\varphi(t) = e^{-i\frac{Et}{\hbar}} \tag{7.37}$$

That is, G is equivalent to the energy E.

7.3.3 The Wave Function

Note that the de Broglie relationship is:

$$\lambda = \frac{h}{mv} \tag{7.38}$$

Thus, where we would normally write kx in the wave equation, k becomes, in terms of momentum:

$$k = \frac{2\pi}{\lambda} = \frac{2\pi mv}{h} = \frac{p}{\hbar} \tag{7.39}$$

And so, in terms of momentum and energy $E = \hbar\omega$, we write:

$$\Psi(x,t) = Ae^{i\left(\frac{px}{\hbar}x - \frac{E}{\hbar}t\right)} \tag{7.40}$$

The factor A is a normalisation constant which we will discuss in the next section.

Separating the time and space components:

$$\Psi(x,t) = \left[Ae^{i\left(\frac{px}{\hbar}x\right)}\right]e^{-i\left(\frac{E}{\hbar}t\right)} \tag{7.41}$$

The time-dependent part of the wave function represents the phase of the matter wave, and the time-independent part, the square-bracketed term in the above, represents the amplitude of the wave. The amplitude term is thus the time-independent equation:

$$\psi(x) = Ae^{i\left(\frac{px}{\hbar}x\right)} \tag{7.42}$$

The interpretation of the wave function $\Psi(x,t)$ for an electron is one of probabilities. The amplitude of the wave function (squared) provides information about the probable location x of an electron in space at some particular time t. Note, we are not saying that the value of the wave function (the matter wave) is a measure of the probability, but the *amplitude* of the wave function is a measure of probability. The actual value of the wave function oscillates between $-A$ and $+A$ with a frequency $\omega = E/\hbar$ whereas the amplitude A may remain constant in time but may be a function of position x.

The probability amplitude is given by the square of the amplitude of the matter wave, which is equal to the magnitude of the time-independent equation ψ:

$$\begin{aligned} P(x) &= |\Psi|^2 = \Psi^*\Psi \\ &= \psi(x)e^{-i\frac{Et}{\hbar}}\psi(x)e^{+i\frac{Et}{\hbar}} \\ &= |\psi|^2 \end{aligned} \quad (7.43)$$

But, ψ is a complex number in spatial terms, and so we must find the magnitude by multiplying by the complex conjugate ψ^*:

$$\begin{aligned} |\psi|^2 &= \psi^*\psi = Ae^{i\left(\frac{px}{\hbar}x\right)}Ae^{-i\left(\frac{px}{\hbar}x\right)} \\ &= A^2 \end{aligned} \quad (7.44)$$

That is, to determine the probability, we find the value of A. In this case, A is a constant, and therefore the probability of finding an electron which has a matter wave Ψ is a constant for all space. In this case, the electron could be anywhere, a result which does not appear to be very useful. We will see however that when matter waves combine, A is not a constant but varies in space. A is called the normalisation factor.

7.3.4 Normalisation

In order for the positional probability of an electron to have physical meaning, the electron must be somewhere within the range between $-\infty$ and $+\infty$. That is:

$$\int_{-\infty}^{\infty} \Psi^*\Psi dx = 1 \quad (7.45)$$

The amplitude of the wave function is found from the solution to the time-independent equation. We shall see that the general solution $\psi(x)$ to this equation contains a constant A. For a particular solution, the value of A depends upon the boundary conditions of the problem. The most general situation is that the electron must be somewhere. That is, the total probability of finding the electron between $-\infty$ and $+\infty$ is 1. When the value of A has been found from the boundary conditions, the wave function is said to be normalised.

The value of the wave function, $\Psi(x,t)$ is not a physically observable quantity. It provides probability information about physically observable quantities. For example, we might ask what then is the expected, or average, location of a particle, such as an electron? Since the electron must be somewhere between $-\infty$ and $+\infty$, the expected value $\langle x \rangle$ is the weighted sum of the individual probabilities over all values of x – much like what we did with the partition function for the energies of molecules in a gas:

$$\langle x \rangle = \frac{\int_{-\infty}^{\infty} xP(x,t)dx}{\int_{-\infty}^{\infty} P(x,t)dx} \quad (7.46)$$

where the numerator is the weighted sum, and the denominator is equal to 1.

$$\langle x \rangle = \int_{-\infty}^{\infty} xP(x,t)\,dx$$
$$= \int_{-\infty}^{\infty} \Psi^* x \Psi\, dx \tag{7.47}$$

$\langle x \rangle$ is the expectation value of the electron's position. This is not necessarily the most likely value of x. The most likely value of x for a given measurement is given by the maximum (or maxima) in the probability density function. The expectation value is the *average* value of x that would be obtained if x were measured many times. As shown in Figure 7.1, when the probability density is symmetric about $x=0$, the expectation value for $x=0$ even though the maximum value of the probability density function might be elsewhere.

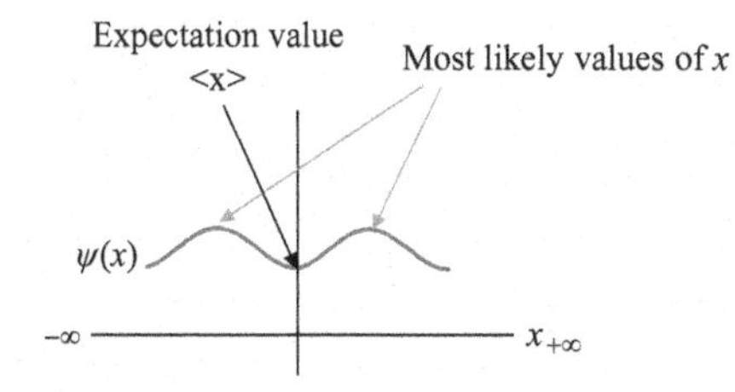

FIGURE 7.1 Comparison between the expected value and most likely value of position x.

Expectation values for energy and momentum, which are functions of x, may thus be calculated from:

$$\langle E \rangle = \int_{-\infty}^{\infty} \Psi^* \hat{E} \Psi\, dx$$
$$\langle p \rangle = \int_{-\infty}^{\infty} \Psi^* \hat{p} \Psi\, dx \tag{7.48}$$

The terms inside the integral sign are customarily written in the order shown here to reflect a style of notation introduced by Dirac.

7.3.5 Solutions to the Time-Independent Schrödinger Equation

7.3.5.1 The Zero-Potential

Solutions to the time-independent part of the Schrödinger equation attract most interest because of the information they provide about the location of a particle such as an electron. One example might be the application of quantum mechanics to the motion of conduction electrons in a solid, where the dimensions of the solid are far greater than the dimensions of an electron. Where there is no electrostatic force acting, the electrons are free to move. Under these conditions, an electron may be either at rest, or moving with a constant velocity. There is no potential energy associated with the electron, $V(x)=0$. We have an electron of mass m which is moving with a constant velocity v and has some energy E. The Schrödinger equation is thus written:

$$-\frac{\hbar^2}{2m}\frac{\partial^2\psi}{\partial x^2} = E\psi$$

$$\frac{\partial^2\psi}{\partial x^2} = \frac{2Em}{-\hbar^2}\psi \tag{7.49}$$

$$\frac{\partial^2\psi}{\partial x^2} + \frac{2Em}{-\hbar^2}\psi = 0$$

Using the technique of separation of variables, the auxiliary equation is:

$$m^2 + \frac{2Em}{\hbar^2} = 0 \tag{7.50}$$

The roots of the auxiliary equation are:

$$m = \pm i\frac{\sqrt{2Em}}{\hbar} \tag{7.51}$$

Since the solution must have the form e^{mx}, then we can let:

$$k = \frac{\sqrt{2Em}}{\hbar} \tag{7.52}$$

and so, we obtain:

$$\psi(x) = C_1 e^{+ikx} + C_2 e^{-ikx}. \tag{7.53}$$

Converting to trigonometric form using Euler's formula:

$$\begin{aligned}\psi(x) &= C_1 \cos kx + C_1 i \sin kx + C_2 \cos(-kx) + C_2 i \sin(-kx) \\ &= (C_1 + C_2)\cos kx + (C_1 - C_2) i \sin kx\end{aligned} \tag{7.54}$$

The eigenfunctions become:

$$\psi(x) = A\cos kx + Bi\sin kx \tag{7.55}$$

where:

$$\begin{aligned}A &= C_1 + C_2 \\ B &= C_1 - C_2\end{aligned} \tag{7.56}$$

Now, since:

$$\frac{E}{\hbar} = \omega \tag{7.57}$$

Then the wave function is:

$$\begin{aligned}\Psi(x,t) &= \psi(x)\varphi(t) \\ &= \left(A\cos kx + Bi\sin kx\right)e^{-i\omega t}\end{aligned} \tag{7.58}$$

Or in exponential terms:

$$\begin{aligned}\Psi(x,t) &= \left(C_1e^{ikx} + C_2e^{-ikx}\right)e^{-i\omega t} \\ &= C_1e^{i(kx-\omega t)} + C_2e^{-i(kx+\omega t)}\end{aligned} \tag{7.59}$$

This is a general solution that describes the superposition of a wave travelling to the right ($+kx$) and one travelling to the left ($-kx$) with amplitudes C_1 and C_2, respectively.

A particular solution for the case of a probability wave travelling in the $+x$ direction can be found by setting $C_2=0$ and so $C_1=A=B$ and hence:

$$\begin{aligned}\Psi(x,t) &= \left(A\cos kx + Ai\sin kx\right)e^{-i\omega t} \\ &= \left(Ae^{ikx}\right)e^{-i\omega t}\end{aligned} \tag{7.60}$$

or:

$$\Psi(x,t) = Ae^{i(kx-\omega t)} = A\cos\left(kx-\omega t\right) + Ai\sin\left(kx-\omega t\right) \tag{7.61}$$

The amplitude squared of the wave function is:

$$\begin{aligned}\psi(x) &= \left(A\cos kx + Bi\sin kx\right) \\ \psi^*\psi &= \left(A\cos kx + Bi\sin kx\right)\left(A\cos kx - Bi\sin kx\right) \\ &= A^2\cos^2 kx - A\cos kxBi\sin kx + Bi\sin kxA\cos kx + B^2\sin^2 kx \\ &= A^2\cos^2 kx + B^2\sin^2 kx \\ &= A^2\end{aligned} \tag{7.62}$$

Or:

$$\begin{aligned}|\Psi|^2 &= \Psi^*\Psi = Ae^{-i(\omega t-kx)}Ae^{i(\omega t-kx)} \\ &= A^2\end{aligned} \tag{7.63}$$

For our free electron, we therefore take one solution (say $C_2=0$) as the case where the electron moves in the $+x$ direction, *or* $C_1=0$ being the case of the electron moving in the $-x$ direction.

Just because at some position x, and a particular time t, $\Psi(x,t)=0$, this does not mean the electron will not be found at that position. It is the *amplitude* (squared) $|\psi|^2$ of the wave function, not the *value* of the wave function $\Psi(x,t)$ that determines the likely position of the electron at some time t. Because $\Psi(x,t)$ is complex, to find this amplitude we must multiply the function by its complex conjugate. As we've seen, for this case, the amplitude is a constant – the value of which we will work out in a moment. The significance is that the probability of finding the electron is the same no matter what the value of x. The electron could be anywhere.

Normalisation of the wave function provides the value of A and is found from:

$$
\begin{aligned}
1 &= \int_{-\infty}^{\infty} \Psi^* \Psi dx \\
&= \int_{-\infty}^{\infty} A e^{-i(kx-\omega t)} A e^{i(kx-\omega t)} dx \\
&= A^2 \int_{-\infty}^{\infty} dx
\end{aligned}
\tag{7.64}
$$

$$A = 0$$

The difficulty here is that the limits of integration are far larger than those which would ordinarily apply in a real physical situation. If the electron is bound by a large, but finite boundary, then (as in the case of a square well potential as discussed in the next section) a non-zero value of A may be calculated while retaining an approximation to the ideal case of infinite range. We will discuss this point further in Section 7.4.

We might well ask what is the average value for the position x of the electron? The average value is called the expectation value.

For a given value of energy E, the amplitude of the resulting wave function is a constant, A, independent of x (and t). That is, the amplitude squared of the wave function is the same no matter where in x or when in t we look. Therefore, for a free electron, the electron can be equally likely to be anywhere in x with a velocity v (and momentum p). The uncertainty in x is infinite. The energy E is not quantised for a completely free electron.

More formally, the average value of x is given by the weighted sum of the individual values:

$$
\begin{aligned}
\langle x \rangle &= \int_{-\infty}^{\infty} \Psi^* x \Psi dx \\
&= \int_{-\infty}^{\infty} A e^{-i(kt-\omega x)} x A e^{i(kt-\omega x)} dx \\
&= A^2 \int_{-\infty}^{\infty} x dx \\
&= (0)(\infty)
\end{aligned}
\tag{7.65}
$$

That is, the average value of the position of the particle is undefined; it could be anywhere along the x axis. This is shown in Figure 7.2.

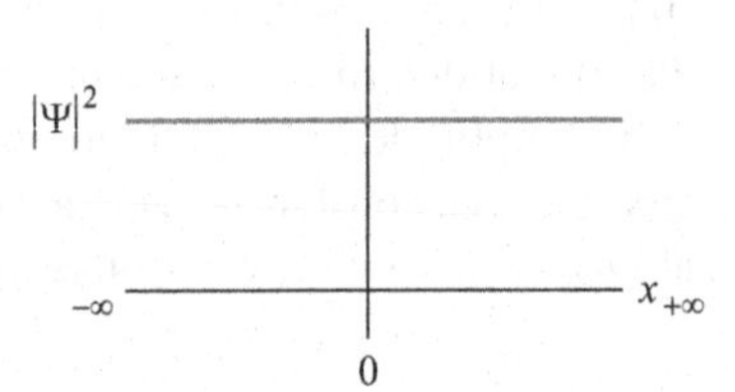

FIGURE 7.2 Average value of position of the particle in the zero-potential.

Remembering from above that:

$$1 = \int_{-\infty}^{\infty} \Psi^* \Psi dx \tag{7.66}$$

The expected, or average, value of momentum is found from:

$$\begin{aligned}
\langle p \rangle &= \int_{-\infty}^{\infty} \Psi^* \hat{p} \Psi dx \\
&= \int_{-\infty}^{\infty} \Psi^* (-i\hbar) \frac{d\Psi}{dx} dx \\
&= \int_{-\infty}^{\infty} \Psi^* (-i\hbar)(ik) A e^{i(kx - \omega t)} dx \\
&= \int_{-\infty}^{\infty} -i\hbar (ik) \Psi^* \Psi dx \\
&= \hbar k \int_{-\infty}^{\infty} \Psi^* \Psi dx \\
&= \hbar k \\
\langle p \rangle &= \sqrt{2Em}
\end{aligned} \tag{7.67}$$

This is the de Broglie relation expressed in terms of energy. The positive value of momentum here is consistent with the sign convention we adopted earlier, that is $(kx - \omega t)$ for an electron and wave moving in the positive x direction.

The expected, or average, value of energy E for a free electron is found from:

$$\begin{aligned}
\langle E \rangle &= \int_{-\infty}^{\infty} \Psi^* \hat{E} \Psi dx \\
&= \int_{-\infty}^{\infty} \Psi^* i\hbar \frac{d\Psi}{dt} dx \\
&= \int_{-\infty}^{\infty} A e^{-i(kx - \omega t)} i\hbar (-i\omega) A e^{i(kx - \omega t)} dx \\
&= \int_{-\infty}^{\infty} -i^2 \hbar \omega \Psi^* \Psi dx \\
&= \hbar \omega \\
\langle E \rangle &= hf
\end{aligned} \tag{7.68}$$

which is the Planck relation.

The de Broglie relation allows us to calculate the wavelength of the electron:

$$\begin{aligned}\lambda &= \frac{h}{p} \\ &= \frac{h}{\sqrt{2Em}}\end{aligned} \tag{7.69}$$

Since energy is not quantised in the case of a free electron, the de Broglie wavelength can take on any value in this case.

This is an example of Heisenberg's uncertainty principle, since in this case, the momentum (and hence the velocity) can be precisely calculated, but the position is completely undetermined. It might seem a trivial result for so much computation but has significance when compared to the case where the electron is bound by some potential.

7.3.5.2 The Infinite Square Well Potential

Consider the case of an infinite square well potential. This is also known as the particle-in-a-box problem. It is depicted in Figure 7.3.

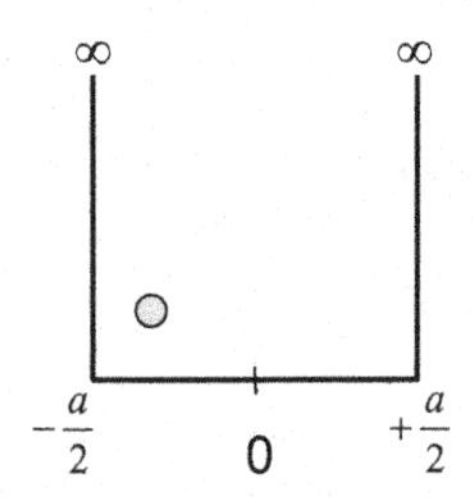

FIGURE 7.3 Schematic of the infinite square well potential.

The boundary conditions are:

$$V(x) = \infty \text{ where } x \leq -\frac{a}{2}; x \geq \frac{a}{2} \tag{7.70}$$

and:

$$V(x) = 0 \text{ where } -\frac{a}{2} < x < \frac{a}{2} \tag{7.71}$$

The particle, say an electron, is confined to the region inside the well. In this region, $V(x)=0$, and so:

$$-\frac{\hbar^2}{2m}\frac{\delta^2\psi}{\delta x^2} = E\psi \tag{7.72}$$

which is the same as the zero-potential case. Inside the well, the electron can move freely. For the moment, we will say that it travels with constant velocity until it hits one side and then rebounds, travelling with constant velocity in the other direction. Note, this is *not* simple harmonic motion between the walls. It is, in the first approximation, constant velocity between the walls.

Note that as with the zero-potential:

$$k = \frac{\sqrt{2Em}}{\hbar} \tag{7.73}$$

The eigenfunctions are:

$$\psi(x) = C_1 e^{+ikx} + C_2 e^{-ikx}. \tag{7.74}$$

This is a general solution to the wave equation for the case of $V(x) = 0$ which is the superposition of two travelling waves in opposite directions. In the present case, the electron might be free to travel within the walls of the container, but when it gets to one wall, it bounces back in the other direction. So, here, we do not select one or the other direction as we did in the zero-potential case. Instead, both directions must be considered together, and further, the travelling waves must have the same amplitude.

In trigonometric form, we obtain (as in the zero-potential case):

$$\begin{aligned}\psi(x) &= C_1 \cos kx + C_1 i \sin kx + C_2 \cos(-kx) + C_2 i \sin(-kx) \\ &= (C_1 + C_2)\cos kx + (C_1 - C_2) i \sin kx\end{aligned} \tag{7.75}$$

However, unlike the case of zero-potential, the travelling waves, in opposite directions, have the same amplitude. Thus, $C_1 = C_2$ and so:

$$\psi(x) = A\cos kx \quad \text{where } A = 2C_1 \tag{7.76}$$

Or, alternately, $C_2 = -C_1$ and so:

$$\psi(x) = Bi \sin kx \quad \text{where } B = 2C_1 \tag{7.77}$$

The wave function $\Psi(x,t)$ is written:

$$\Psi(x,t) = [A\cos kx]\, e^{-i\omega t} \text{ or } \Psi(x,t) = [Bi \sin kx]\, e^{-i\omega t} \tag{7.78}$$

Now, compare this with the zero-potential, free electron case:

$$\begin{aligned}\Psi(x,t) &= Ae^{i(kx-\omega t)} \\ &= Ae^{ikx}e^{-i\omega t} \\ &= \left(A\cos(kx) + Ai\sin(kx)\right)e^{-i\omega t}\end{aligned} \tag{7.79}$$

where we can choose either the real or imaginary parts. We might say that at this point, there is no difference at all between the wave functions for the case of the free electron and inside the potential well. However, the difference will become apparent when we come to examine the value of A.

The boundary conditions for the infinite square well are the restriction that the electron has zero probability of being at the walls of the well. At the walls, the electron will never have enough energy to surmount the barrier $V(x) = \infty$. That is, the square of the amplitude of the wave function will be zero at the walls.

Separating the space and time components, we have an amplitude function $\psi(x)$ which varies periodically in x with an amplitude A but, unlike the case of a free electron, the value of $\psi(x)$ is fixed at zero at the walls. The wave $\psi(x)$ is therefore a standing wave. This situation only occurs when there are constraints on the range of the electron. Standing waves are formed from the superposition of two waves which are travelling in opposite directions. The value of the combined wave function is still oscillating, but the amplitude of that oscillation depends on x. The amplitude of the oscillation at the nodes is 0, and the position of these nodes does not change with time. These nodes occur whenever $kx - \omega t = \pi/2, 3\pi/2$, etc.

The most general form of solution for a standing wave is the superposition of the two solutions so that:

$$\psi(x) = A\cos kx + Bi\sin kx \text{ where } A = 2C_1,\ B = 2C_1 \tag{7.80}$$

At the wall where $x=a/2$, the wave function $\psi(x)=0$, and so:

$$\begin{aligned}\psi(x) &= A\cos\frac{ka}{2} + Bi\sin\frac{ka}{2} \\ &= 0\end{aligned} \tag{7.81}$$

Similarly, where $x=-a/2$:

$$\begin{aligned}\psi(x) &= A\cos\frac{-ka}{2} + Bi\sin\frac{-ka}{2} \\ &= A\cos\frac{ka}{2} - Bi\sin\frac{ka}{2} \\ &= 0\end{aligned} \tag{7.82}$$

Thus:

$$A = 0;\ Bi\sin\frac{ka}{2} = 0 \tag{7.83}$$

Or:

$$B = 0;\ A\cos\frac{ka}{2} = 0 \tag{7.84}$$

$A=B=0$ is a trivial solution – that is, the particle is not inside the well. Non-trivial eigenfunctions are found by letting, say, $A=0$ (or $B=0$) and letting k take on values such that:

$$A = 0;\ Bi\sin\frac{ka}{2} = 0 \tag{7.85}$$

therefore:

$$\begin{aligned}\frac{ka}{2} &= n\pi \\ k &= \frac{2n\pi}{a} \quad \text{where } n = 1,2,3,4\ldots\end{aligned} \tag{7.86}$$

or:

$$k = \frac{n\pi}{a} \quad \text{where } n = 2,4,6\ldots \tag{7.87}$$

Now, let:

$$B = 0;\ A\cos\frac{ka}{2} = 0 \tag{7.88}$$

we have:

$$\frac{ka}{2} = n\frac{\pi}{2}$$
$$k = \frac{n\pi}{a} \text{ where } n = 1,3,5,7\ldots \tag{7.89}$$

Thus:

$$\psi_n(x) = A_n \cos k_n x \text{ where } n = 1,3,5,7\ldots \tag{7.90}$$

or:

$$\psi_n(x) = B_n i \sin k_n x \text{ where } n = 2,4,6,8\ldots \text{ and } k_n = \frac{n\pi}{a} \tag{7.91}$$

To determine the values of the constants A and B, the eigenfunctions are normalised. For the odd n case:

$$1 = \int_{-a/2}^{a/2} \psi^* \psi dx$$
$$= A^2 \int_{-a/2}^{a/2} \cos^2 \frac{n\pi}{a} x dx$$
$$= A^2 \int_{-a/2}^{a/2} \frac{1}{2} + \frac{1}{2}\cos\frac{2n\pi}{a} x dx \tag{7.92}$$
$$= A^2 \left[\frac{1}{2}x - \frac{a}{4n\pi}\sin\frac{2n\pi}{a}x\right]_{-a/2}^{a/2}$$
$$= A^2\left[\frac{a}{2}\right]$$
$$A^2 = \frac{2}{a}$$

Similarly, for the even case:

$$B^2 = \frac{2}{a} \tag{7.93}$$

The normalised eigenfunctions are thus:

$$\psi_n(x) = \sqrt{\frac{2}{a}} \cos k_n x \text{ where } n = 1,3,5,7\ldots \text{ and} \tag{7.94}$$

$$\psi_n(x) = \sqrt{\frac{2}{a}} \sin k_n x \text{ where } n = 2,4,6,8\ldots \tag{7.95}$$

A plot of the first three eigenfunctions is shown in Figure 7.4(a).

The full wave function is written:

$$\Psi_n(x,t) = \left[\sqrt{\frac{2}{a}}\cos k_n x\right] e^{-i\omega t} \text{ where } n = 1,3,5,7\ldots \tag{7.96}$$

or:

$$\Psi_n(x,t) = \left[\sqrt{\frac{2}{a}}\sin k_n x\right] e^{-i\omega t} \text{ where } n = 2,4,6,8\ldots \tag{7.97}$$

Figure 7.4(b) shows a plot of the probability distributions, the positional probability of the location of the electron for each of the allowable energy levels. In all cases, the probability is of course zero at the walls where the energy barrier is infinitely high.

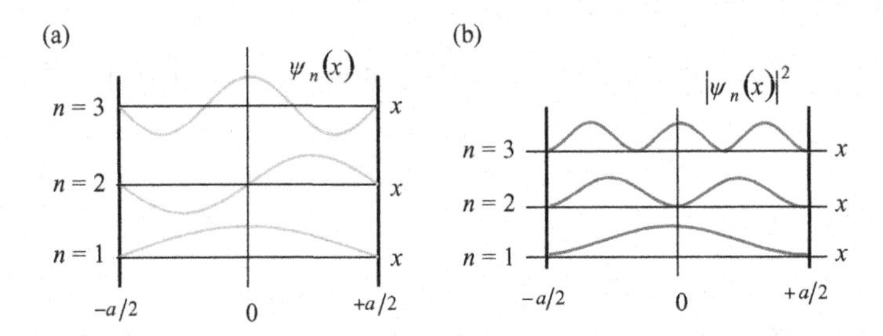

FIGURE 7.4 (a) First three eigenfunctions for the infinite square well potential. (b) First three probability distributions.

Note that because of symmetry around $x=0$, the expectation value, or average value, of x is 0.

The energies associated with the particle are given by the allowed values of k. These values of k, and hence E, are the eigenvalues:

$$\begin{aligned} k &= \frac{n\pi}{a} = \frac{\sqrt{2Em}}{\hbar} \\ \frac{n^2\pi^2}{a^2} &= \frac{2Em}{\hbar^2} \quad \text{where } n = 1,2,3,4\ldots \\ E &= \frac{n^2\pi^2\hbar^2}{2a^2m} \end{aligned} \tag{7.98}$$

A plot of the energy levels is shown in Figure 7.5.

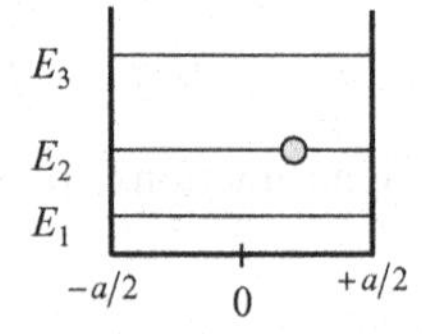

FIGURE 7.5 First three energy levels, eigenvalues, for the infinite square well potential.

The energy $n=1$ corresponds to the zero-point energy:

$$E_1 = \frac{\pi^2\hbar^2}{2a^2m} \tag{7.99}$$

Energy is quantised. Unlike the case of a free electron (such as in the zero-potential case), energy is quantised in this case as a consequence of it being bound (between the walls of the well).

We might ask ourselves just what is this energy? It is the energy of the electron that it possesses by virtue of its velocity in between the walls of the potential. But note carefully that the probability distributions are periodic in space. The electron therefore spends more time where the probability is greater compared to where it is smaller. The velocity of the electron is therefore not constant. For example, for $n=1$, the electron spends more time towards the middle of the well compared to near the walls indicating it is moving with a lower velocity in the middle of the well and speeding up as it approaches the walls. For $n=3$, the electron appears to increase and decrease its velocity in a periodic manner depending on where it is between the walls. Any increase in energy must occur by a discrete step up to the next mode of spatial vibration.

Each mode of spatial variation carries with it an oscillation in time according to the time-dependent part of the wave function but this is not a consideration here. Here, our attention is on the spatial variation of the probability density.

When the particle is in the well, $\Delta x \approx a$. The uncertainty in the momentum is thus, for E_1:

$$\begin{aligned}\Delta p &= \frac{\hbar}{2\Delta x} = \frac{\hbar}{2a} \\ p_1 &= \sqrt{2mE_1} = \frac{\pi\hbar}{a} \\ \Delta p &= 2p_1 = \frac{2\pi\hbar}{a} \\ \Delta x \Delta p &= \frac{2\pi\hbar}{a}a = 2\pi\hbar\end{aligned} \tag{7.100}$$

which is consistent with Heisenberg's uncertainty principle.

Standing waves can be found in the solutions for other potentials as well, such as the finite square well potential, the step potential, the Coulomb potential, the harmonic oscillator potential and more. In all of these cases, the standing waves arise from the superposition of waves where the values of their wave functions are constrained – the electron is bound.

7.4 SIGNIFICANCE OF THE BOUNDARIES

7.4.1 Free Electron

In complex exponential form, the electron wave function $\Psi(x,t)$ is written in much the same way as any other wave:

$$\Psi(x,t) = Ae^{i(kx-\omega t)} \tag{7.101}$$

Using the de Broglie relationship, we write the wave function in terms of momentum and energy:

$$\begin{aligned}\Psi(x,t) &= Ae^{i\left(\frac{p_x}{\hbar}x-\frac{E}{\hbar}t\right)} \\ &= \left[Ae^{i\frac{p_x}{\hbar}x}\right]e^{-i\frac{E}{\hbar}t}\end{aligned} \tag{7.102}$$

For a free electron, we found that the factor A was a constant (even though it was calculated to be zero for a completely free electron in infinite space). While the amplitude A of the wave function might remain constant, the value of the wave function of some point is dependent on the time t and x. That is, whenever $kx-\omega t=\pi/2, 3\pi/2$, etc., the function $\Psi=0$. If we fix a time t (e.g. by taking a snapshot of $\Psi(x,t)$) then these zero positions (we should not call them nodes to avoid confusion with the nodes of a standing wave) will appear at periodic intervals of x. As t increases, the positions x for the zero positions must also increase, and so they "travel" along in the x direction – that is, the whole probability wave travels, and we might imagine that the electron is carried along with it. This is shown in Figure 7.6.

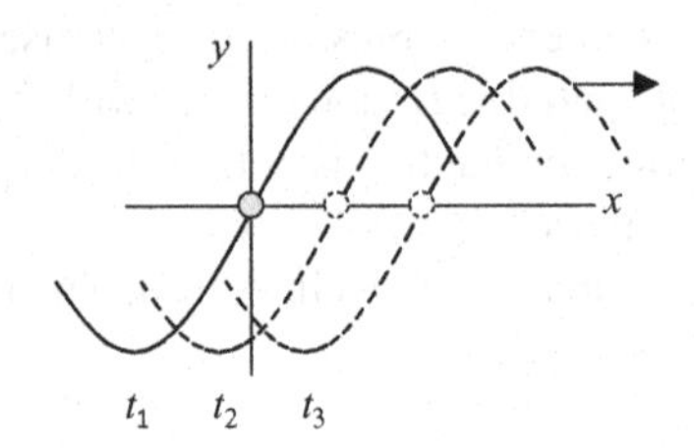

FIGURE 7.6 Travelling wave for the zero-potential case.

But we might say, if the electron has a de Broglie wavelength given by:

$$\lambda = \frac{h}{mv} \tag{7.103}$$

then the amplitude of the wave function cannot really be zero, even if the electron is completely free. The wave function itself must have some value at some place to indicate the presence of the electron even if we don't know where that location might be.

Imposing some kind of constraint on the range of motion (e.g. say the diameter of the solar system) lifts the value of A a little at the expense of introducing a slight ripple in the amplitude of the wave function. But this only happens if the value of the wave function was fixed at those constraints as in the infinite square well. For a free electron, this need not be the case. Much like the circular wave pattern shown in Figure 6.6(b), if the radius of this orbit were to be made very large, then the electron is virtually free, but contained within the circumference $2\pi R$. This leads to a converging normalisation condition with a normalisation factor given by:

$$\begin{aligned} 1 &= A^2 \int_0^{2\pi R} e^{-i(kx-\omega t)} e^{i(kx-\omega t)} dx \\ A &= \frac{1}{\sqrt{2\pi R}} \end{aligned} \tag{7.104}$$

Note that even with a free electron, there is a space-like periodic character to the amplitude function $\psi(x)$ even if the amplitude of that periodic variation in $\psi(x)$ is very small. This point has particular significance. It means that the amplitude function $\psi(x)$ of the wave function $\Psi(x,t)$ is a complex number, even for a "free" electron.

7.4.2 Bound Electron

The presence of the walls in the infinite square well potential leads to restrictions on the value of $\psi(x)$, which are manifested as standing waves forming between the boundaries.

The form of the equations is very similar to that of the zero-potential, where the amplitude could also be seen to have a periodic character, but there, the amplitude of the variation in amplitude was zero (i.e. $A=0$ in the limiting case of an infinite range) or made very small by constraining the motion in some way. The moment we impose a fixed boundary on the value of the wave function at the extreme range of x, even if that range is large, we obtain quantum discreteness, a ripple in $\psi(x)$ in spatial terms.

Note that the amplitude of $\psi(x)$ is inversely proportional to the square root of a, the distance between the walls. As a decreases, the amplitude of the space-like variation (i.e. the amplitude of the standing waves) increases.*

Let's imagine for a moment that we thought of an electron as a localised ripple in a special type of field which pervades all of space. We can imagine the value of the field, where the electron is located, is oscillating with a magnitude from $-A$ to $+A$ and a frequency ω. The magnitude of this oscillation, the amplitude (squared), tells us where the electron is most likely to be. Note carefully, we are not saying the value of the field says where the electron might be, we are saying that the amplitude of the oscillation of the field tells us where the electron could be. If at another place, the amplitude of oscillation is zero, then there is no electron there.

Now, for the case of a moving electron with no constraints, if we look at a certain location x_1, as shown in Figure 7.7(a), we have the same probability of finding the electron there compared to if we looked at another place x_2. The probability of finding the electron is the same everywhere. This does not mean the electron *is* everywhere, but it means we have the same chance of finding it anywhere. The moving electron is described by a travelling wave in the electron field of constant amplitude, rather more like a wave packet, with a de Broglie wavelength and a velocity $\lambda = h/p$. It's just that it could be anywhere.

For the bound electron, the probability of finding an electron depends on where in x we look. For a standing wave, if the frequency of the oscillation is, say, ω, the strength of the oscillation, that is, the amplitude, depends on the cosine of kx, which in turn depends upon a, the spacing of the outer bounds since $k=2\pi/\lambda$ and $\lambda=a/2$ is the spacing between the bounds for the first mode of vibration. If the frequency were to increase to a new mode of vibration,† as the result of the electron taking on more energy, then the shape of the standing wave would change. Figure 7.7(b) shows the shape for the second mode of vibration. The amplitude (squared) tells us where the electron appears to spend most time. That is, if we look at some time t, we are more likely to find the electron at x_1 compared to the position x_2, as shown in Figure 7.7(b). In this case, the de Broglie wavelength of the electron is the spacing a.‡

* Imagine we hold our hands out far apart and the electron is in the space between. In the limit of an infinite spacing of our hands, the "amplitude" $\sqrt{2/a}$ of the amplitude function $\psi(x)$ drops to zero, and the matter wave disappears. There are no standing waves because in the limit of infinite spacing, there are no reflections. As we bring our hands closer together, the amplitude term increases, nodes begin to appear. As we bring our hands very close together, the amplitude term becomes very large, and so does the size of the ripples in the wave function. Being a quantum system, the change in amplitudes occurs in a series of steps, the step size being dependent on the frequency of the standing waves. We might ask, well why are there a series of difference standing waves – that is, what gives rise to the harmonics above the fundamental frequency? The answer is that for an electron, the amplitude depends upon the bounds of the potential. In an atom, this is usually fixed by the Coulomb potential. In this system, the amplitude is a probability. The probability amplitude does not store energy like a stretched string or a mass on a spring. Any more energy coming in must result in the electron moving from one mode of vibration to another where that energy can be accommodated as a change of position in the Coulomb potential. Further, it is a feature of quantum physics that except for electrons with opposite spins, one cannot have more than one electron occupying the same energy state. This is the Pauli exclusion principle. Once that state (amplitude) is filled, any more electrons in the vicinity have to go into energy states with a different mode of spatial vibration.

† Remembering that the amplitude at any one frequency is determined by the spatial limits of the potential and so cannot change.

‡ There is an interesting physical interpretation to be considered in Figure 7.8(b). For the nodes, where the amplitude of the oscillation is 0, there is no chance of finding the electron there. We might ask, well what is happening at the nodes? How does the electron get from point x_1 to x_2 if it cannot pass through the zero point at the node? The electron is not a little ball which has to go from x_1 to x_2 via some path. It is the matter *wave* that is travelling between the two walls. If an electron appears at x_2, it is because it disappeared from position x_1. For example, say you had some marbles in a box. There is a layer of identical marbles covering the floor of the box, every space taken up by a marble. The layer of marbles is covered by a special sheet that make them invisible if they are under the sheet. You take one more marble and put it on top of the sheet which is on top of the layer of marbles. You can see this marble; it is above the sheet. You then push down on that marble, and what happens? It goes into the layer below the sheet, and another marble somewhere else pops up. You see one marble disappear and another one appear in another location. The marble you pushed down on did not travel under the sheet and pop up in the other location. The space above the sheet is visible, we do not see what is going on underneath, but to us, it looks like the marble we pressed down into the sheet popped up somewhere else. If the sheet has different thicknesses, then the likelihood of the electron popping up at some place where the sheet is thinnest is greater compared to somewhere else where the thickness of the sheet is greater. This sheet could have thin patches all over it, arranged in a regular pattern, so it might look to us like a marble will pop up in a regular pattern if we keep pushing the one that pops up down again.

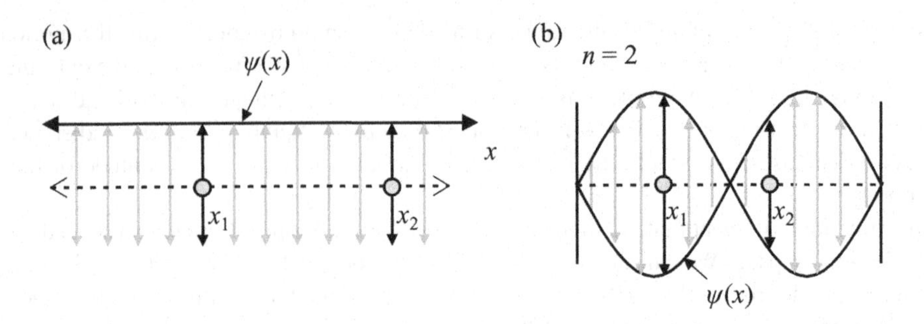

FIGURE 7.7 (a) Probability distribution along x for the zero-potential. (b) Probability distribution for $n=2$ for the infinite square well potential.

7.5 WAVE FUNCTIONS AND PHOTONS

Quantum mechanics describes how an electron behaves in an atom – its position. The actual wave function Ψ does not describe the shape of the probability distribution. The wave function Ψ is a matter wave whose amplitude (squared) $|\Psi|^2$ is a measure of the probability of an electron's position in an atom. The wave function Ψ itself cannot be observed. In some circumstances, such as the bound state potentials, the shape of the probability distribution function can appear wave-like in that the amplitude (squared) of the matter wave changes with position in a regular manner with distance. In other circumstances, e.g. the free electron, the shape of the probability distribution function approaches a constant value independent of position.

When an electron makes a transition, say from a higher energy state to a lower energy state, the shape of the probability distribution changes from one shape to another, and the "centre of charge" may be at a different radius from the positively charged nucleus as a result – the electron has moved within the Coulomb potential. The amplitude of the wave function does not store energy; it only stores probability information. For the energy to be accounted for, the electron has to go to another probability distribution where the centre of charge has moved in the Coulomb potential.

In three dimensions, the shape of the probability distribution depends on which energy level is being considered. For example, "s" energy levels have a probability distribution of a spherical shape. "p" energy levels have a probability distribution of a dumbbell shape. These shapes also determine the allowable values of the orbital angular momentum for each state.

Stationary quantum states occur when the potential V is a function of position only. When an electron is in a stationary state, the amplitude of the probability function is simply:

$$|\Psi|^2 = \Psi^*\Psi = \psi^* e^{+i\frac{E_n}{\hbar}t} \psi e^{-i\frac{E_n}{\hbar}t} = \psi^*\psi \tag{7.105}$$

That is, $|\Psi|^2$ is independent of time. The shape of the amplitude of the wave function is in the nature of a standing wave. That is, the centre of charge associated with the electron is static, and so, unlike the case of the Bohr atom, there is no emission of radiation, and the electron does not spiral into the nucleus.

The time taken for an electron to make a transition from one state to another is of the order of 10^{-9} seconds. During this time, the wave functions of the two states temporarily combine, forming a mixed-state wave function. During the transition, the amplitude of the first state wave function gradually decreases, and the amplitude of the second increases.

The wave function for the two states m and n might be written as:

$$\Psi_m = \psi_m(x)e^{-i\frac{E_m}{\hbar}t}$$
$$\Psi_n = \psi_n(x)e^{-i\frac{E_n}{\hbar}t} \tag{7.106}$$

During a transition, the combined wave function is:

$$\Psi_m + \Psi_n = \psi_m(x)e^{-i\frac{E_m}{\hbar}t} + \psi_n(x)e^{-i\frac{E_n}{\hbar}t} \tag{7.107}$$

The probability density function for the combined state is therefore:

$$\Psi^*\Psi = \left(\psi_m^* e^{+i\frac{E_m}{\hbar}t} + \psi_n^* e^{+i\frac{E_n}{\hbar}t}\right)\left(\psi_m e^{-i\frac{E_m}{\hbar}t} + \psi_n e^{-i\frac{E_n}{\hbar}t}\right)$$

$$\begin{aligned} &= \psi_m^*\psi_m e^{+i\frac{E_m}{\hbar}t}e^{-i\frac{E_m}{\hbar}t} + \psi_m^*\psi_n e^{+i\frac{E_m}{\hbar}t}e^{-i\frac{E_n}{\hbar}t} + \psi_n^*\psi_m e^{+i\frac{E_n}{\hbar}t}e^{-i\frac{E_m}{\hbar}t} + \psi_n^*\psi_n e^{+i\frac{E_n}{\hbar}t}e^{-i\frac{E_n}{\hbar}t} \\ &= \psi_m^*\psi_m + \psi_m^*\psi_n e^{+i\frac{E_m}{\hbar}t}e^{-i\frac{E_n}{\hbar}t} + \psi_n^*\psi_m e^{+i\frac{E_n}{\hbar}t}e^{-i\frac{E_m}{\hbar}t} + \psi_n^*\psi_n \\ &= \psi_m^*\psi_m + \psi_n^*\psi_n + \psi_m^*\psi_n e^{+i\frac{E_m}{\hbar}t}e^{-i\frac{E_n}{\hbar}t} + \psi_n^*\psi_m e^{+i\frac{E_n}{\hbar}t}e^{-i\frac{E_m}{\hbar}t} \end{aligned} \tag{7.108}$$

The first two terms are functions of position only, and do not concern us for the moment. The last two terms have a time dependence. These two terms show that the probability density function is, during the transition, a function of time and more importantly, is periodic in time. That is, the amplitude of the combined wave function, when squared, remains a function of time (whereas in all our examples before, the amplitude of a single wave function had periodicity in space, not time). This is in the manner of a beat frequency.

$$\psi_m^*\psi_n e^{+i\frac{E_m}{\hbar}t}e^{-i\frac{E_n}{\hbar}t} + \psi_n^*\psi_m e^{+i\frac{E_n}{\hbar}t}e^{-i\frac{E_m}{\hbar}t} = Ae^{+i\left(\frac{E_m-E_n}{\hbar}\right)t} + Be^{+i\left(\frac{E_n-E_m}{\hbar}\right)t} \tag{7.109}$$

So, letting:

$$\omega_1 = \frac{E_m - E_n}{\hbar}$$
$$\omega_2 = \frac{E_n - E_m}{\hbar} \tag{7.110}$$

we obtain a combined wave of the form:

$$\begin{aligned} y &= Ae^{i\omega_1 t} + Be^{i\omega_2 t} \\ &= A\cos\omega_1 t + iA\sin\omega_1 t + B\cos\omega_2 t + Bi\sin\omega_2 t \end{aligned} \tag{7.111}$$

Taking the imaginary part of the trigonometric forms (to align with Section 2.10) and since $A = B$:

$$\begin{aligned} y &= A\sin(\omega_1 t) + A\sin(\omega_2 t) \\ &= 2A\cos\left[\frac{\omega_1 - \omega_2}{2}t\right]\sin\left[\frac{\omega_1 + \omega_2}{2}t\right] \end{aligned} \tag{7.112}$$

since:

$$\sin X + \sin Y = 2\cos\left(\frac{X - Y}{2}\right)\sin\left(\frac{X + Y}{2}\right). \tag{7.113}$$

The sine term is the frequency of the resultant and is the average of the two component frequencies. This does not concern us at present. It is the amplitude of the resultant, the cosine term, that is important. This amplitude term oscillates with a frequency of:

$$\begin{aligned} \frac{\omega_1 - \omega_2}{2} &= \frac{1}{2}\left(\frac{E_m - E_n}{\hbar} - \frac{E_n - E_m}{\hbar}\right) \\ &= \frac{1}{2}\frac{E_m - E_n - E_n + E_m}{\hbar} \\ &= \frac{E_m - E_n}{2\hbar} \\ f &= \frac{E_m - E_n}{h}. \end{aligned} \tag{7.114}$$

That is:

$$E_m - E_n = hf \tag{7.115}$$

Where E_n and E_m are the energies of the two states. This value of f is precisely that of the frequency predicted by Bohr and the observed emission spectra of the hydrogen atom. During the transition time, a photon energy hf is emitted, carrying away the excess energy $\Delta E = hf = E_n - E_m$ where f is the beat frequency.

The combined mixed-state wave function has a beat frequency due to the difference in the de Broglie wavelengths of the states, that is, it is the *beat* frequency that is the photon frequency, not the frequency of the matter waves.

So, what makes an electron make a transition from one state to another? As with most physical systems, there is a natural tendency for a system to assume the lowest energy state, i.e. the ground state. In an atom, say the hydrogen atom, the single electron is usually always in the ground state. If the electron makes a transition to a higher energy state as a result of absorbing an incoming photon, then it will very quickly return to its ground state, if not in one step, then in a series of steps. There are certain restrictions, or selection rules, which dictate which transitions are possible on the way back to the ground state.

The intensity of the spectral lines that we might see in an experiment arises from the rate of transitions between the excited state and a lower energy state. That is, the intensity is related to the number of transitions per second.

In this chapter, we have had a lot to say about the wave functions associated with a particle (usually taken to be an electron) bound by various forms of a potential. However, although the new quantum theory was born out of the necessity of accounting for the particulate nature of light – i.e. photons – we have not as yet had much to say about the wave function for a photon, the quanta of the electromagnetic field. We will return to this most interesting point in Part 2.

7.6 SPIN

7.6.1 Spin Angular Momentum

When two magnetic fields interact, they tend to align themselves. This often results in a torque being applied – for example, when a compass needle aligns itself with the earth's magnetic field.

To take a somewhat mechanistic view, we can say that all the magnetic properties of materials are due to the motion of electric charges within atoms. Such movement is usually due to the effect of what is interpreted as orbiting and spinning electrons around the nucleus.

Consider a single electron in orbit around a nucleus with an angular velocity ω (ignore the spinning motion of the electron just now).

The angular momentum L of the orbiting electron is given by:

$$L = m_e r^2 \omega \tag{7.116}$$

The area A of the path traced out by the electron is:

$$A = \pi r^2 \tag{7.117}$$

The electric current associated with the orbital motion of the electron is:

$$I = q\frac{\omega}{2\pi} \tag{7.118}$$

The orbital magnetic moment μ_M is given by the product of I and A, hence:

$$\begin{aligned} \mu_M &= q\frac{\omega}{2\pi}\pi r^2 \\ &= \left(\frac{-q_e}{2m_e}\right)L \end{aligned} \tag{7.119}$$

The physical significance of the negative sign means that μ in this case is in the opposite direction to L because for an electron, $q=-q_e=-1.6\times 10^{-19}$ C.

The above expression applies to the orbital motion of the electron around the nucleus. This is shown in Figure 7.8(a).

Now, let's consider the internal property of electron spin. As with most phenomena on a sub-atomic scale, comparison with larger scale phenomena is not often possible. Electrons do not "spin" in the ordinary sense of the word, but the property we call spin manifests itself as if the electron were spinning in the way in which electrons align with an externally applied magnetic field.

The angular momentum arising from the so-called spinning motion of the electron, the spin magnetic moment, is expressed:

$$\mu_M = \left(\frac{-q_e}{m_e}\right)S \tag{7.120}$$

where S is the spin angular momentum. These quantities are illustrated in Figure 7.8(b).

The magnetic moment is similar in a sense to the dipole moment formed by an electric dipole placed in an electric field. In magnetism, there are no magnetic charges to form dipoles (which is why when a bar

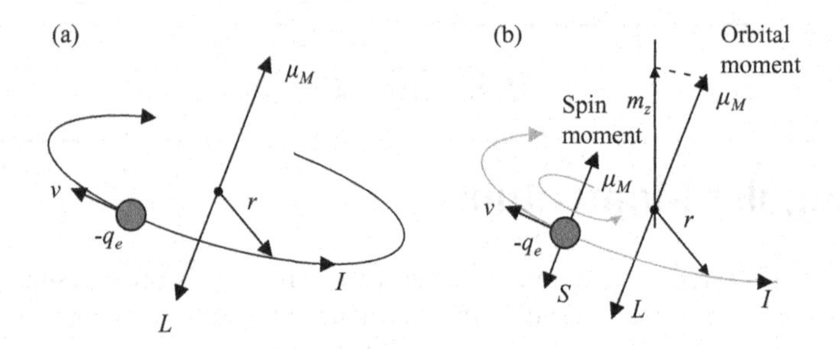

FIGURE 7.8 (a) Orbital magnetic moment. (b) Orbital magnetic moment and spin magnetic moment.

magnet is cut, each piece has its own N and S poles). The term magnetic *dipole* moment is therefore not applicable, and so we say magnetic moment instead.

The total magnetic moment for an atom has contributions from both orbital and spin motion of several electrons. In general, we have:

$$\boldsymbol{\mu}_{\mathbf{M}} = g\left(\frac{-q_e}{2m_e}\right)\mathbf{J} \tag{7.121}$$

g is the Lande factor (equal to 1 for pure orbital motion and 2 for pure spin), and in this formula, J is the total angular momentum from the combination of spin and orbital motions for all the electrons. It is a vector.

7.6.2 Quantum Numbers

As we have seen before, in an atom, at each value of the principal quantum number n corresponds to the allowable energy levels E for the surrounding electrons where $n = 1, 2, 3, \ldots$

The orbital angular momentum L can take on several distinct values. Each of the values is described by a second quantum number l. The allowed values of l are 0, 1, … $(n-1)$.

The allowable values of the orbital angular momentum are:

$$L = \sqrt{l(l+1)}\frac{h}{2\pi} = \sqrt{l(l+1)}\hbar \tag{7.122}$$

$$l = 0, 1, 2, \ldots n-1$$

A third quantum number m describes the allowable changes in the angle of the orbital angular momentum vector in the presence of a magnetic field. It takes the integer values –l to 0 to +l. What this means is that the z component of L is constrained to have values in increments such that:

$$L_z = m\frac{h}{2\pi} = m\hbar \tag{7.123}$$

$$m = -l, -l+1, \ldots l-1, l$$

Note that L_z cannot ever equal L. The largest component of L_z is $l\hbar$.

For a given l, there are $2l+1$ possible states or values of L_z. For $l=2$, we have:

$$L = \sqrt{2(2+1)}\hbar = \sqrt{6}\hbar \tag{7.124}$$

But the allowed values of L_z are $L_z = -2\hbar, -1\hbar, 0, +1\hbar, +2\hbar$.

These values are illustrated in Figure 7.9(a).

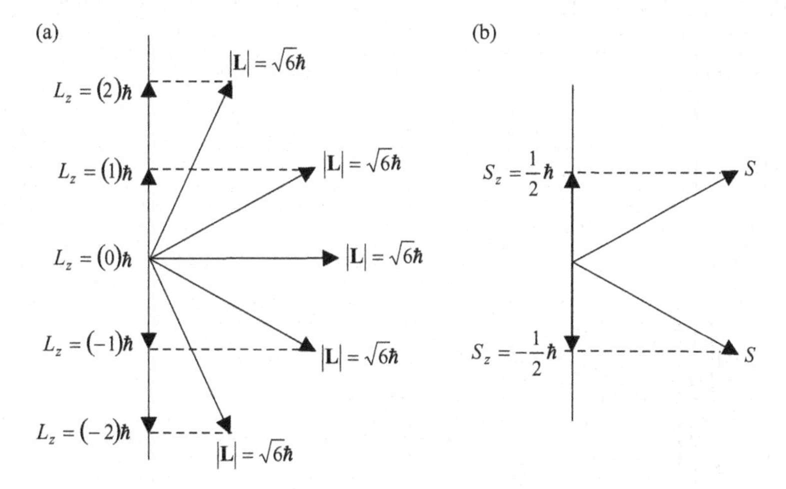

FIGURE 7.9 (a) Allowed values of orbital angular momentum L for an electron. (b) Two states of intrinsic spin for an electron.

It was postulated by Pauli that no two electrons could exist in the same quantum state – but experiments showed that there were often two electrons occupying the same energy state. Pauli proposed that another state, not included in Schrödinger's equation, was responsible. The properties of this state suggested it was in the nature of an intrinsic angular momentum, a spin.

The fourth quantum number s describes this intrinsic spin of an electron where $s = +½$ (up) or $-½$ (down) and indicates whether the z component of the spin angular momentum S_z is aligned with (up) or opposite (down) to any external magnetic field (usually oriented along the z axis). S_z can thus take on two values:

$$S_z = \pm s\hbar = \pm \frac{h}{4\pi} = \pm \frac{\hbar}{2} \tag{7.125}$$

These are illustrated in Figure 7.9(b).

The electron is therefore called a spin ½ particle. Spin ½ particles are grouped into a class of particles called fermions.

Each possible spin state is an eigenstate, but not one predicted or included in the eigenstates of the Schrödinger equation. It is an additional quantum state that can be associated with a particle, such as an electron. Other sub-atomic particles can also have a spin property.

We can represent spin in operator notation $\hat{S}$.

We shall see later that the spin operator is best expressed in matrix form. For example, in the z direction:

$$\hat{S}_z = \frac{\hbar}{2}\begin{bmatrix} 1 & 0 \\ 0 & -1 \end{bmatrix} \tag{7.126}$$

The factor $\hbar/2$ is often not written for convenience, and the operators in x, y and z directions become:

$$\hat{\sigma}_x = \begin{bmatrix} 0 & 1 \\ 1 & 0 \end{bmatrix}; \; \hat{\sigma}_y = \begin{bmatrix} 0 & -i \\ i & 0 \end{bmatrix}; \; \hat{\sigma}_z = \begin{bmatrix} 1 & 0 \\ 0 & -1 \end{bmatrix} \tag{7.127}$$

These are known as the Pauli matrices.

The spin-dependent part of the wave function can be incorporated into the quantum mechanics of Schrödinger's equation by expressing the wave function as a matrix of two components:

$$\Psi\uparrow=\begin{bmatrix}\Psi(x,y,z,t)\\0\end{bmatrix}$$
$$\Psi\downarrow=\begin{bmatrix}0\\\Psi(x,y,z,t)\end{bmatrix} \tag{7.128}$$

Or:

$$\Psi\uparrow=\Psi(x,y,z,t)\begin{bmatrix}1\\0\end{bmatrix}$$
$$\Psi\downarrow=\Psi(x,y,z,t)\begin{bmatrix}0\\1\end{bmatrix} \tag{7.129}$$

These two matrices are sometimes written in a short-hand notation as the "ket" vector:

$$\Psi\uparrow=\Psi(x,y,z,t)\begin{bmatrix}1\\0\end{bmatrix}=\Psi|\uparrow\rangle \tag{7.130}$$

$$\Psi\downarrow=\Psi(x,y,z,t)\begin{bmatrix}0\\1\end{bmatrix}=\Psi|\downarrow\rangle \tag{7.131}$$

A general expression for the two-component wave function is:

$$\Psi=\begin{bmatrix}\Psi_1(x,y,z,t)\\\Psi_2(x,y,z,t)\end{bmatrix} \tag{7.132}$$

This form of the wave function is called a spinor.

The amplitude (squared) of Ψ_1 gives the probability that the particle exists at some point in time t at location (x,y,z) in the spin-up state, while the amplitude (squared) of Ψ_2 gives the probability that the particle exists at some point in time t at location (x,y,z) in the spin-down state. The total probability over all time and space, and some spin (either up or down) is unity.

$$\int\left(|\Psi_1|^2+|\Psi_2|^2\right)dV=1 \tag{7.133}$$

7.7 SIGNIFICANCE OF THE SCHRÖDINGER EQUATION

Schrödinger originally imagined that the charge on the electron was smeared out within the atom and that the square of the amplitude of the matter wave determined the density of the charge from place to place. He did not care much for Born's probability interpretation of his wave function and devised the famous

Schrödinger's Cat scenario to illustrate the apparent absurdity of the notion. Einstein did not believe that nature would be probabilistic either and is quoted as saying that "God does not play dice".

At about the same time, Heisenberg also felt somewhat dissatisfied with the Bohr model, and believed there was no value in trying to interpret the behaviour of electrons if they could not be seen. Instead, he developed a mathematical framework centred around the notion of a simple harmonic oscillator. He envisaged that on a relatively large scale (for example, an electron orbiting at a very large orbit), conventional classical equations of motion could be applied. Working down to smaller length scales, he discovered that it became necessary to accept that the product of two numbers, say pq, did not equal the same product in reverse, qp. The difference between the results was Planck's constant h divided by 2π. This brought the required quantum discreteness into the phenomena. Heisenberg's analysis was put into a more formal form by Born and Pascal Jordan, who developed a matrix approach to the analysis, and then Born applied it to the hydrogen spectrum where it was found that both energies and intensities could be calculated. This *matrix mechanics* approach relied on no physical model such as electrons in orbit; it was purely mathematical.

In 1927, this led Heisenberg to a formal statement which was now called the uncertainty principle, in which it is said that both the position and momentum of a particle cannot be known at the same time. For example, if we measure the momentum of a particle precisely, we can never know where it is. The product of the uncertainty in position times the uncertainty in momentum is always greater than $h/2\pi$.

Schrödinger's wave *mechanics* could do everything that Bohr could do and as well, account for the intensity of spectral lines – a major achievement. The planetary model of Bohr was now replaced with a wave model which showed how photon emission was the result of the beat frequency between two states when an electron made a transition from one state to the other. We now are comfortable with the notion that the amplitude (squared) of the wave function was interpreted as a probability of an electron (regarded as a whole particle, and not spread out in some way) being somewhere travelling with some velocity according to the limits imposed by the uncertainty principle.

Successful as the Schrödinger equation is, there are significant limitations when it comes to dealing with the energies of multi-electron atoms. As well, the equation could not account for velocities of particles which might be a significant fraction of the speed of light, and as well could not account for what was thought to be an intrinsic, or internal, property of a particle called spin – by which time had been demonstrated experimentally. When later it was realised that particles could be created or destroyed during high energy collisions, the Schrödinger equation was just not applicable.

In order to go forward, we need to first know about Einstein's theory of relativity.

Relativity

8

8.1 INTRODUCTION

Although we tend to now take it for granted now that electromagnetic waves travel through empty space (from say the sun to the Earth), in the 1870s, when Maxwell's equations were developed to describe the nature of electric and magnetic fields, no one knew what medium supported these waves. That is, what was the medium whose particles underwent periodic motion so that a disturbance, the changing E and B fields, could travel?

Consider the required properties of this medium:

- Experiments showed that electromagnetic waves were found to be transverse waves, not compression or longitudinal waves.
- Thus, they could only be transmitted by a solid since a liquid cannot support any shear (transverse) stresses.
- The velocity of electromagnetic waves was found to be very great, which in turn requires the medium carrying the waves to be very dense.

Thus, the medium for electromagnetic waves was required to be almost an infinitely dense solid that filled all space, through which ordinary bodies could travel without any impediment (such as the Earth moving in orbit through space). Although the medium itself could not be identified, it was given the name: luminiferous aether.

Newton's laws of motion are derived from the point of view of a frame of reference that is either completely at rest or is moving with a constant velocity. Traditionally, the best frame of reference available to us on Earth is the "fixed" stars. Although the stars are moving, they are so far away that their movement, from an ordinary perspective, is undetectable.

The question then arises as to what exactly is the real fixed frame of reference that we should use for our Newtonian mechanics. It was thought that the aether would be a good candidate. This made sense because light from the sun and stars all moved through the aether. It would not make much sense if the aether itself was moving in some places and not others. Plus, the aether was "solid" and so if one part of it moved, all the other parts would also have to move. Considerable effort was put into identifying the aether, or even proving its existence.

8.2 SPECIAL RELATIVITY

8.2.1 The Michelson–Morley Experiment

In 1883, Albert Michelson and Edward Morley undertook to measure the Earth's velocity relative to the aether (the aether drift) by using a very sensitive interferometer to detect the change in time it took light rays to travel a distance thought to be perpendicular and parallel to the Earth's orbital motion around the sun.

The apparatus is shown in Figure 8.1.

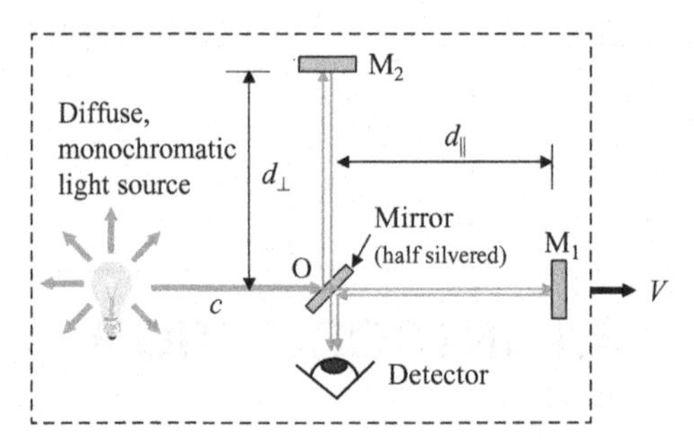

FIGURE 8.1 The Michelson–Morley experiment.

Because of the diffuse nature of the light source, many rays of light fall upon the half-silvered mirror and make the journey to the detector. The diagram shows the path taken by one of these light beams. Due to the differences in path lengths traversed by the different light beams, and because the light is monochromatic, there will be interference fringes visible at the detector. By changing the path length traversed by one of the light beams, say, by, moving one of the mirrors, the fringes appear to shift, and so a precise measurement of the distance the path length changes can be obtained by counting the number of fringes that move past a graticule in the field of view.* This is the working principle of an interferometer.

To simplify the use of symbols, we will assume in our discussion that the distance between the half-silvered mirror and the other mirrors is exactly the same. That is, $d_{\parallel} = d_{\perp}$. This does not lead to any loss of the principles involved. It is the *change* in the distances travelled by the light beams in the perpendicular and parallel directions that is measured, not their absolute value.

We begin with the apparatus on the earth which is itself moving to the right with velocity V with respect to the aether which is assumed to be fixed, or at absolute rest.

Beginning with the point O on the half-silvered mirror, a ray of light is launched into the aether going to the right with velocity c towards the mirror M_1. From the point of view of the fixed aether, during the time the light takes to reach the mirror, the mirror has moved a little bit to the right due to the velocity V of the earth. Compared to the case where the earth was not moving, the time taken for this outward trip is therefore a little greater than it would be if the apparatus were at rest. Since the actual distance d from O to M_1 has not changed during this time, and the time taken has increased, the velocity of the light, timed between two events over a distance d, measured by someone travelling with the apparatus, should be less than c. That is, it should be $c - V$. That is, the time taken to travel the distance d from O to M_1 from the point of view of someone travelling with the apparatus is therefore:

$$t_1 = \frac{d}{(c - V)} \tag{8.1}$$

Alternately, we can say that the path length travelled by the light in the aether for the outward-bound trip has increased because the mirror has moved to the right during the trip, and so the light, travelling at velocity c through the aether, has to travel further through the aether to catch up with it.

Either way, we can see that the velocity of light measured by someone at rest in the aether, c, is different to that measured by someone travelling with the apparatus $(c - V)$.

* The movement of the fringes is an illusion, much like an animated cartoon. What is happening is that as one of the mirrors moves, areas of bright bands in the field become dimmer while areas of dark bands become brighter as the regions of destructive and constructive interference change. The effect is as if the bands move sideways in the field of view.

During the time for the return trip from M_1 to O, the point O has moved to the right a little. This time, the light reaches the mirror a little before it would have compared to the case where the apparatus was at rest since the point O has moved forward to meet the returning light beam. From the point of view of someone travelling with the apparatus, the velocity of light would be somewhat greater than c. The time taken for the return trip is thus:

$$t_2 = \frac{d}{(c+V)} \tag{8.2}$$

Alternately, we can say that the path length of the light in the aether has decreased because the point O has moved forward to the right a little during the return trip and the light, travelling at speed c, does not have to travel as far through the aether to arrive there.

The total time for the round trip from O to M_1 and back to O is thus:

$$\begin{aligned} t_{\|} &= \frac{d}{c-V} + \frac{d}{c+V} \\ &= \frac{d\left((c+V)+(c-V)\right)}{(c+V)(c-V)} \\ &= \frac{2cd}{c^2-V^2} \\ &= \frac{2d}{c\left(1-V^2/c^2\right)} \end{aligned} \tag{8.3}$$

We can see that the total round-trip time reduces to $2d/c$ if the velocity, with respect to the fixed aether, approaches zero.

Now we consider the light beam which travels to the mirror M_2.

The path for this light beam is shown in Figure 8.2.

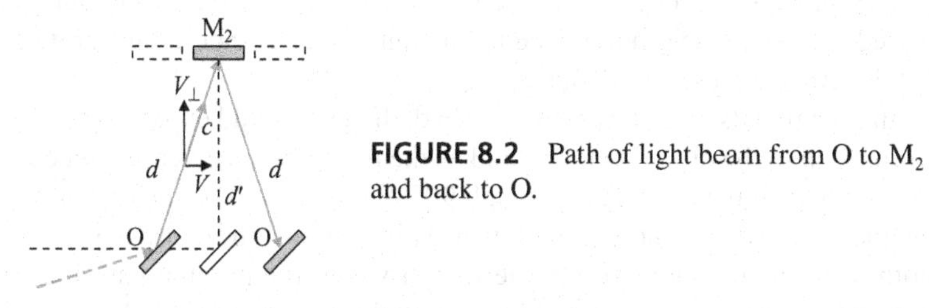

FIGURE 8.2 Path of light beam from O to M_2 and back to O.

From the point of view of the fixed aether, during the time the light is travelling from O to the top mirror M_2, the apparatus has moved to the right, and so to reflect off M_2 and reach the mirror at O again, the ray has to travel a greater total distance $2d$ compared to $2d'$ if the apparatus were not moving. Note that in this case, the ray starts from a slightly different angle, which is why a diffuse light source is required for this apparatus.

The light ray is moving at an angle with respect to the vertical direction. If the velocity of light within the fixed aether at that angle is c, then there is a component $V_\perp$ in the vertical direction and a component V in the horizontal direction where V is the velocity of the apparatus in the aether. From Pythagoras' theorem, we have:

$$V_\perp = \sqrt{c^2 - V^2} \tag{8.4}$$

Looking at the vertical components of V and d, the round-trip time becomes:

$$\begin{aligned} t_{\perp} &= \frac{2d'}{V_{\perp}} \\ &= \frac{2d'}{\sqrt{c^2 - V^2}} \\ &= \frac{2d'}{c\sqrt{1 - V^2/c^2}} \end{aligned} \tag{8.5}$$

That is, the time taken for the perpendicular trip has increased because the velocity has reduced to $V_{\perp}$ over the perpendicular distance $2d'$. We can see that this time reduces to $2d'/c$ when V goes to zero.

Alternately, we can say that the light, travelling at speed c through the aether, has to traverse a longer path length, $2d$, for the round trip, and so the time for the return trip has increased.

From the point of view of the aether, the light travels at speed c, but from the point of view from someone travelling with the apparatus, the speed is $V_{\perp}$.

Let's examine the ratio of the time taken for the light rays to travel the two paths where we remember that in our example here, the horizontal path d is equal to the perpendicular path d':

$$\begin{aligned} \frac{t_{\parallel}}{t_{\perp}} &= \frac{2d}{c\left(1 - V^2/c^2\right)} \frac{c\sqrt{1 - V^2/c^2}}{2d} \\ &= \frac{1}{\sqrt{1 - V^2/c^2}} \end{aligned} \tag{8.6}$$

When $V > 0$, this ratio is always > 1, which means that the time for the light to travel the parallel path should be always longer compared to the time taken to travel the perpendicular path. At velocities V approaching c, the difference in times becomes appreciable. Alternately we can say that from the point of view of the aether, the path length travelled by the light within the aether in the perpendicular direction is always longer compared to the path length travelled by the light in the aether in the parallel direction. The difference in the path lengths increases as V increases.

If we could stop the earth in its orbit, and begin at $V=0$, then we would expect to see the fringes move at the detector as V ramps up to the orbital velocity and as the path lengths diverge. Since we can't do this, what we can do is to slowly turn the apparatus through 90°. What used to be the distance for the perpendicular trip now becomes the distance for a parallel trip. This is shown in Figure 8.3.

When $V>0$, both path lengths increase but the rate at which the parallel path increases is greater compared to that of the perpendicular path, and so by turning the apparatus through 90°, we should see the fringes in the interference pattern shift as the paths involved swap from being parallel to perpendicular.

Michelson and Morley's interferometer had an effective path length of 11 m and was capable of resolving a difference in path length of about 3×10^{-9} m, which in the case of the orbital velocity of the Earth and the then known velocity of light, translated into a fringe shift of nearly half a fringe – easily detectable by the eye.

The result of the Michelson–Morley experiment was negative. The fringes did not move. There appeared to be no aether drift at all. It was as if the aether, if it existed, got carried along with the motion of the Earth so that $V=0$.

The consequence of this was that the velocity of light always measured at c no matter whether one was moving or not. This remarkable behaviour could not be explained by conventional classical physics.

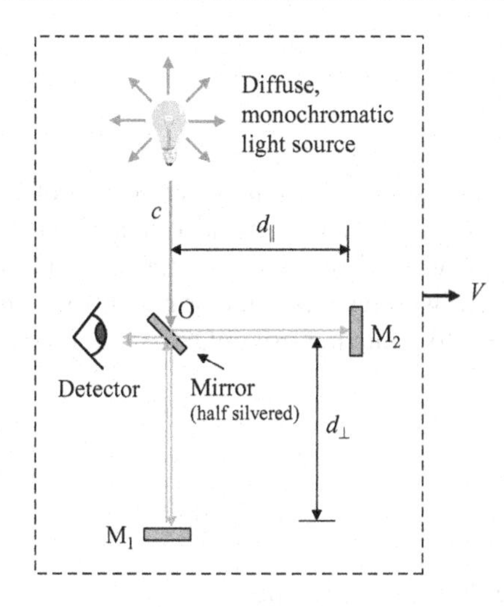

FIGURE 8.3 Path from O to M_2 and back to O with apparatus rotated.

8.2.2 The Principle of Relativity

Galileo asked us to imagine the following experiment:* "Take aboard a ship (which is initially tied up at port on the shore) some butterflies and a fishbowl (with fish). Go below deck and shut yourself up, along with the butterflies and fish, in a cabin and close the shutters on the windows. With the ship at port, notice how the butterflies fly around the cabin, and how the fish swim around in the bowl freely in all directions. Now, set the ship in motion. The motion is to be uniform velocity. Observe your butterflies and fish again. Do the butterflies all clump towards the stern? Do the fish get left behind and gather towards one end of the bowl? No. The butterflies fly around as before and the fish swim to and fro without any preference for being towards the front or the back of the bowl".

According to Galileo, there is no way of telling whether the ship is in motion from looking at anything within the cabin. The only way to know if the ship is moving is to look outside at the shore. That is, by itself, without any fixed frame of reference, uniform straight-line motion is indistinguishable from being at rest.

This is Galileo's *principle* of relativity. In its time, it explained why the motion of the Earth around the sun could not be detected by people on the Earth, since they were moving also along with the Earth.

In Galileo's example, the shore is a fixed, or stationary, reference frame, while the moving ship is a moving reference frame.

We specify the coordinates of an object in the cabin in terms of the local coordinates x', y', z'. These will be different to the global x, y, z coordinates of the object as measured from the point of view of the shore – because as time progresses, the local coordinates may remain the same while the global coordinates will be continuously changing due to the velocity of the ship.

To determine the absolute global position of any object on board the moving ship, we need to express the local coordinates (x', y', z') of a point within the cabin in terms of the global coordinates (x, y, z) on the shore and the velocity and time that the measurements of these positions are taken.

* Note, there were no trains at the time of Galileo.

8.2.3 Frames of Reference

8.2.3.1 Distance

Imagine now we wish to measure the width of Galileo's fishbowl from the point of view of being on the ship and on the shore. The ship is moving with velocity V. We make this measurement at some time t. During the time t, the ship has moved a distance Vt from where it started from, the origin O. At this time t, the observer on the shore takes measurements x_2 and x_1 with respect to the origin O(0, 0). The observer on the ship takes measurements x'_2 and x'_1 with respect to a moving origin O′(0, 0). The coordinate x is related to x' by:

$$\begin{aligned} x_1 &= x_1' + Vt \\ x_2 &= x_2' + Vt \end{aligned} \tag{8.7}$$

The width of the fishbowl is thus, as measured on the ship, Figure 8.4(a):

$$\Delta x = x_2' - x_1' \tag{8.8}$$

and as measured from the shore, Figure 8.4(b):

$$\begin{aligned} \Delta x &= x_2 - x_1 \\ &= \left(x_2' + Vt\right) - \left(x_1' + Vt\right) \\ &= x_2' - x_1' \end{aligned} \tag{8.9}$$

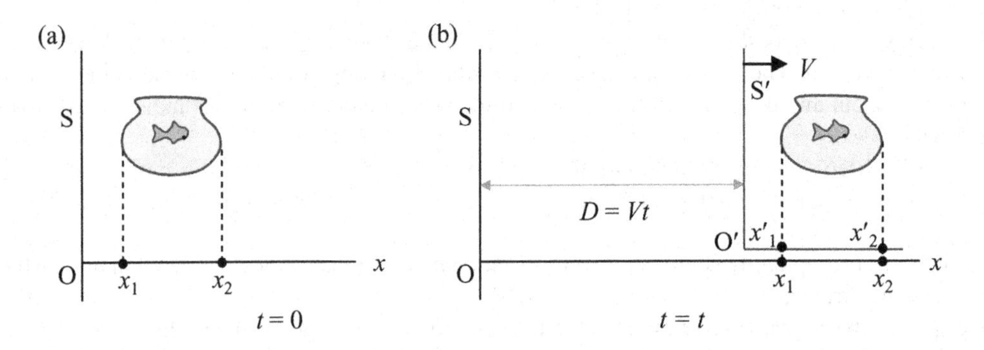

FIGURE 8.4 (a) Length of a body in stationary reference frame. (b) Non-relativistic length of body as it appears in both stationary and moving reference frames.

That is, both observers will get the same measurement of distance Δx. Although the actual time t doesn't need to be specified, it is important that the readings x_1 and x_2 are taken at the *same* time. If there were a delay between recording x_1 and x_2, then there would be an error in the distance Δx because the position of the edge of the moving fishbowl for the second measurement would have changed during the delay.

The two key points here are that the measurements x_1 and x_2 have to be simultaneous, and if that is so, then the distance between two points measured in one frame of reference is the same as that measured in another frame of reference. That is, for Galileo, the distance between two points in space is an invariant quantity.

8.2.3.2 Velocity and Acceleration

Let us imagine we now have a point $P(x, y, z)$ moving with velocity v with respect to a reference frame S. Let us now imagine there is a frame of reference S′ which is travelling with velocity V relative to S and V different to v. The coordinates of the point, in the S′ reference frame at some time t, might be $P(x', y', z')$. The values of x and x' at this time t are connected by:

$$x = x' + Vt \tag{8.10}$$

We might ask ourselves what will be the velocity of the point P with respect to the S′ reference frame? This velocity, v', will be somewhat less than v (if both V and v are in the same direction) since this reference frame is already moving with velocity V:

$$\begin{aligned} v &= \frac{dx}{dt} = \frac{d}{dt}(x' + Vt) \\ v &= \frac{dx'}{dt} + V \\ v' &= v - V \end{aligned} \tag{8.11}$$

- v is the velocity of a point P in the S reference frame.
- $v' = dx'/dt$ is the velocity of a point P as measured within the S′ reference frame.
- V is the velocity of the S′ reference frame.

So, according to Galileo, the velocity of the point P depends upon which reference frame we are using to measure it.

If the point P is accelerating with respect to the S reference frame, then we have:

$$\begin{aligned} a &= \frac{dv}{dt} = \frac{d}{dt}\left(\frac{dx'}{dt} + V\right) \\ &= \frac{dv'}{dt} + 0 \\ &= a' \end{aligned} \tag{8.12}$$

That is, we will get the same value of acceleration if we measure a in either the S reference frame or measure a' in the S′ reference frame, whereas previously we saw that the velocity measurement does depend on which reference frame is used.

Quantities which do not change when measured in different reference frames are called invariant. We have seen that the distance between two points is an invariant quantity; so is mass, and so a force ($=ma$) is also invariant. Velocity is not an invariant quantity, it depends on the frame of reference being used to measure it.

Galileo's principle of relativity appeared to be incompatible with the results of the Michelson–Morley experiment since Michelson and Morley showed that the velocity of light did *not* depend on which reference frame was used to make the measurement. It was Albert Einstein who resolved the paradox.

8.2.4 Postulates of Special Relativity

Consider a lamp inside a train carriage* and the carriage itself is moving with the speed of light c. Once the light from the lamp is launched, then it would travel with velocity c with respect to the fixed aether. That is, the light waves are on their own, travelling through the aether, with speed c.

* By the time of Einstein, trains were the marvel of the age, not ships as in the days of Galileo.

So, an observer at the other end of the carriage would not see the lamp because the waves cannot travel fast enough through the aether to catch up with him.

But, if the carriage slowed down a little, the observer would see the light since the waves could then reach him. So, the observer could tell if he was moving at velocity c or not without looking outside – which is in violation of Galileo's principle of relativity.

To satisfy the principle of relativity, the light would have to move through the fixed aether at $2c$ (if the carriage were moving at c already) so that the observer could not tell if he were moving or not and for the observer to measure a local velocity of light at c.

This is contrary to experimental evidence of Michelson and Morley where it is observed that the velocity of light is always c no matter whether the source or the observer is moving. That is, the person inside the carriage should measure the speed of light, relative to them, as c, and a person on the ground should measure the same value c relative to them.

A person in the carriage will observe the light to travel the distance d from one end of the carriage to the other in a time t so that $d/t=c$. A person on the ground, observing the light travelling from the mirror to the person on the train, will also measure the speed of this light, relative to the ground, as c. That is, this person will observe the light to travel the distance d from one end of the carriage to the other in a time t so that $d/t=c$.

Einstein's paper "On the Electrodynamics of Moving Bodies" published in 1905 explained how the principle of relativity could be applied to light while at the same time requiring that the velocity of light be a constant independent of the motion of any observer. He began with what was observed as being the starting point and worked backwards to determine how these observations could be maintained. His two postulates were:

- There is no such thing as a fixed frame of reference.
- Light travels in a vacuum with a fixed velocity which is independent of the velocity of the observer and the velocity of the source.

In the words of Sherlock Holmes, "when you have eliminated all which is impossible, then whatever remains, however improbable, must be the truth". It was Einstein who discovered that the awful truth was that our concepts of both distance and time required radical adjustment to accommodate the facts. To see how this is so, we return to the Michelson–Morley experiment.

8.2.5 Time Dilation

In the Michelson–Morley experiment, from the point of view of an observer in the S′ reference frame, the time taken for the ray to traverse the total distance $2d'$ is $t'=2d'/c$ as shown in Figure 8.5(a).

From the point of view of an observer in the S reference frame, as shown in Figure 8.5(b), the light has to travel a greater distance $2d$. But the velocity of light has to be the same for each observer, and so

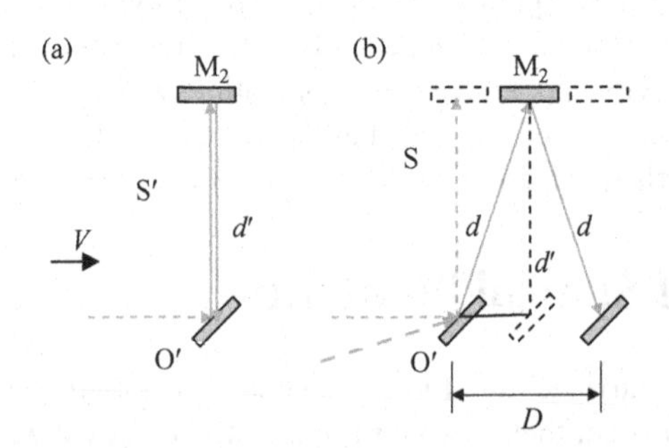

FIGURE 8.5 Comparison of light path for round trip between O and M_2 for both (a) moving and (b) stationary reference frames in the Michelson–Morley experiment.

the time measured by this observer is $t=2d/c$. That is, t must be different from (greater than) t' because $2d$ is greater than $2d'$.

As well, during the time t it takes for the light to make the journey from the point of view of the observer in the S reference frame, the apparatus itself has moved a distance $D=Vt$. In summary, we have:

$$
\begin{aligned}
t' &= \frac{2d'}{c} \\
t &= \frac{2d}{c} \\
D &= Vt
\end{aligned}
\tag{8.13}
$$

By Pythagoras, we have $d^2 = d'^2 + (D/2)^2$ and so:

$$t = \frac{t'}{\sqrt{1 - V^2/c^2}} \tag{8.14}$$

This expression is the same as that we obtained earlier for the differences in time that the light rays travel in the perpendicular and parallel paths when we were discussing the motion of the earth through the aether:

$$\frac{t_{\parallel}}{t_{\perp}} = \frac{1}{\sqrt{1 - V^2/c^2}} \tag{8.15}$$

where we saw that the time for the parallel path to the motion of the earth through the aether is always greater than the time taken for the perpendicular path. The principles involved are very similar, but this time we are taking the velocity c as being the constant factor for both reference frames. The equation shows that if the velocity of light is to be considered a constant, independent of the status of the observer, then the time between two events, that is the departure of the light ray from O′ and its return to the same position O′, is different depending on whether the measurement is made from the S reference frame or the S′ reference frame.

Let's say the Michelson–Morley interferometer was large enough so that the light travelling the perpendicular path, as measured in the moving reference frame, takes $t'=1$ second to make the trip from the source to the detector, and that the device itself had a velocity V of half the velocity of light. For an observer in the S reference frame, the time t taken would be:

$$t = \frac{1}{\sqrt{1 - (0.5c)^2/c^2}} = 1.155 \text{ s} \tag{8.16}$$

or:

$$\frac{t}{t'} = 1.155 \text{ s} \tag{8.17}$$

That is:

- If there was a clock on the apparatus, we, if we are travelling along with it, see the second hand move at one second intervals. $t'=1$ second.
- An observer outside in the S reference frame, with an identical clock, would also see that the light takes one second to make the round trip. $t=1$ second.

What this means is that by the time one second has passed in the S′ reference frame, more than one second (i.e. 1.155 s) has passed out there in the S reference frame relative to the time intervals in the S′ reference frame. Or, when one second has passed in the S reference frame, a shorter time (1/1.155 = 0.865 s) has passed in the S′ reference frame relative to the S reference frame. From the point of view of the person in the S reference frame, the clock in the S′ reference frame is running slow.

If the velocity of light has to be the same for both observers, but the time, previously thought to be an invariant quantity that ticks meticulously in the "background" at a steady and unchanging rate, has stretched or compressed (depending on your point of view), then the distances travelled must also be brought into question.

Before we consider the distances involved, there is an important feature here that deserves mentioning. The time for the round trip is in fact the time between two events. The first event is the leaving of the light ray from the mirror at point O′. The second event is the reception of the light wave back at point O′ after it bounces off the top mirror M_2. From the point of view of the moving reference frame, these two events happen at the same place, O′.

The observer in the moving reference frame S′ uses a clock which is at rest with respect to O′. The time interval t' between two events, *where the events occur at the same place*, measured by a clock at this place, is given the name: proper time. There is nothing improper about other time measurements in other reference frames, it is just a name to indicate the special condition that the events being timed occur at the same place. That is:

$$t' = t\sqrt{1 - V^2/c^2} \tag{8.18}$$

is the time interval between two events in the S′ reference frame where those events occur at the same place in that S′ reference frame. The observer in the S reference frame measures the time interval t between these same events but at two different places in the S reference frame since during the interval, point O′ has moved a distance D.

Note that:

$$\sqrt{1 - V^2/c^2} < 1 \text{ for } V < c \tag{8.19}$$

8.2.6 Length Contraction

From the point of view of the S reference frame, the vertical distance d' from O′ to the mirror M_2 in the Michelson–Morley experiment has not changed as the apparatus has moved horizontally a distance $D = Vt$. But we should not be so hasty making any conclusion about the horizontal component of the distance.

Let us now examine the light waves travelling parallel to the moving reference frame as shown in Figure 8.6(a).

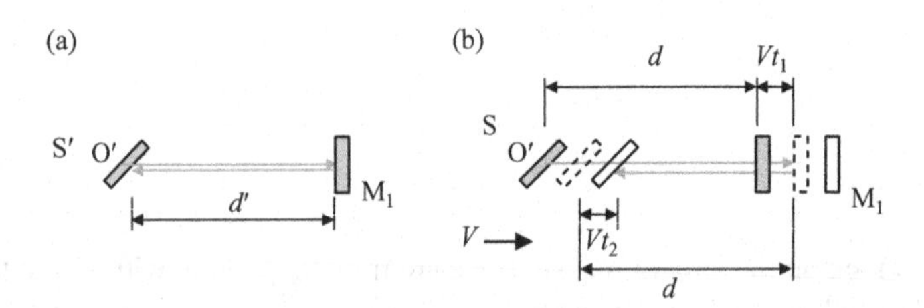

FIGURE 8.6 Comparison of light path for round trip from O to M_1 for (a) moving and (b) stationary reference frames.

From the point of view of the moving reference frame S′, the distance travelled is $2d'$. The time taken for the light waves to travel from O′ to the mirror M_1 and back to O′ is t':

$$t' = \frac{2d'}{c} \tag{8.20}$$

t' is a proper time interval because the two events, the departure and arrival of the light waves at O′, are at the same place and the clock timing this interval is at rest with respect to O′.

As shown in Figure 8.6(b), in the S reference frame, the point O′ is moving with velocity V, and so during the time interval t_1 that the light travels from O′ to the mirror, the mirror M_1 has moved away a distance Vt_1. Therefore, the distance travelled by the light waves in this reference frame on the outward journey is $d + Vt_1$. For the return trip, the point at O′ has moved a further distance Vt_2 and catches the returning light beam "early". The distance travelled by the light beam on the return path is $d - Vt_2$.

Since the velocity c is a constant, we can say that for the outward journey, from the point of view of the S reference frame:

$$\begin{aligned} c &= \frac{d + Vt_1}{t_1} = \frac{d}{t_1} + V \\ t_1 &= \frac{d}{c - V} \end{aligned} \tag{8.21}$$

and also, for the return journey:

$$\begin{aligned} c &= \frac{d - Vt_2}{t_2} = \frac{d}{t_2} - V \\ t_2 &= \frac{d}{c + V} \end{aligned} \tag{8.22}$$

The total time to make the round trip, as measured in the S reference frame, is thus:

$$t = t_1 + t_2 = \frac{d}{c + V} + \frac{d}{c - V} = \frac{2d}{c\left(1 - V^2/c^2\right)}$$

But, as we know from before, the time intervals in the two frames of reference are related by:

$$t = \frac{t'}{\sqrt{1 - V^2/c^2}} \tag{8.23}$$

and so:

$$t' = \frac{2d'}{c} = t\sqrt{1 - V^2/c^2} \tag{8.24}$$

thus:

$$\begin{aligned} t &= \frac{2d'}{c\sqrt{1 - V^2/c^2}} \\ \frac{2d}{c\left(1 - V^2/c^2\right)} &= \frac{2d'}{c\sqrt{1 - V^2/c^2}} \\ d &= d'\sqrt{1 - V^2/c^2} \end{aligned} \tag{8.25}$$

The length d' is called the proper length. The proper length of a body is that which is measured in the reference frame in which the body is at rest.

In the example above, the distance d between the point O′ and the mirror M_2, as measured from the S reference frame, is shorter than the distance d' as would be measured in the S′ reference frame.

Say a fishbowl, such as that shown in Figure 8.7, is moving with $V=0.5c$. Someone moving along with the fishbowl would take measurements x'_2 and x'_1 and might obtain $d'=1$ m. Someone in the S reference frame would observe the fishbowl to be at x_2 and x_1 and conclude the diameter to be $d=0.865$ m. As V approaches c, the diameter d approaches zero. As V approaches 0, the diameter d approaches d'.

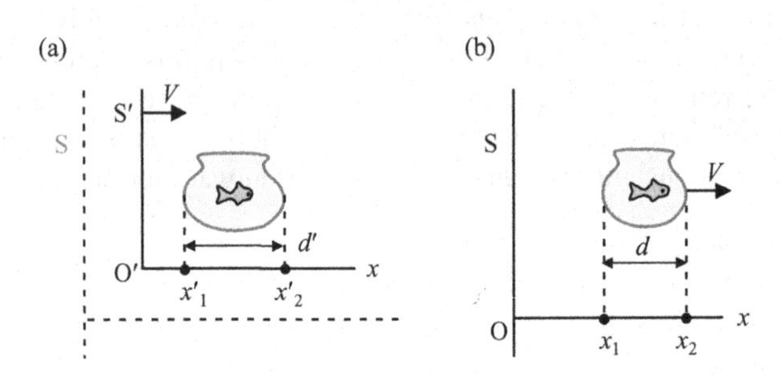

FIGURE 8.7 Comparison of the relativistic length of a body in the (a) moving and (b) stationary reference frames.

We showed earlier that from the point of view of the observer in the S reference frame, the time in the S′ reference frame in our hypothetical interferometer would be $t'=0.865t$. That is, the time for the light to travel the round trip in the parallel direction in the S′ reference frame is t' which is less than t by the same fraction as the shortening of the distance. That is, for the parallel path, a shorter distance over a shorter time (as measured by the stationary observer) gives the same value of c as the proper distance over the proper time measured by the moving observer. Both distance and time adjust to give the same value for c.

8.2.7 Lorentz Transformations

8.2.7.1 Lorentz Distance Transformation

An event, such as the reflection of a light beam from a mirror in the Michelson–Morley interferometer, or the sound of a bell, or the dropping of a ball, happens at a specific location and at a specific time. We might specify an event, in a S reference frame, using the quantities x and t. We would specify the same event in a S′ reference frame using the quantities x' and t'.

Our starting point is to say that at time $t=0$ and $t'=0$, the origin O of an S reference frame coincides with the origin O′ of an S′ reference frame. Let us say the event occurs at a point P at some later time t. During the time t, the S′ reference frame has moved, relative to the S reference frame, a distance $D=Vt$. At the time t, P has the coordinate x in the S reference frame. At the time t', P has the coordinate x' in the S′ reference frame. Our aim is to express the coordinate x in terms of the coordinate x'. The situation is shown in Figure 8.8(a).

The *Galilean* transformation (i.e. without any relativistic effects) is:

$$x = x' + Vt \tag{8.26}$$

However, according to Einstein, when we are measuring distances in the S′ reference frame from the point of view of the S reference frame, we now have to take into account our perceived contraction in x'.

As shown in Figure 8.8(b), from the point of view of someone in the S′ reference frame, x' remains unchanged, but from the point of view of someone in the S reference frame, the distance from O′ to x', and

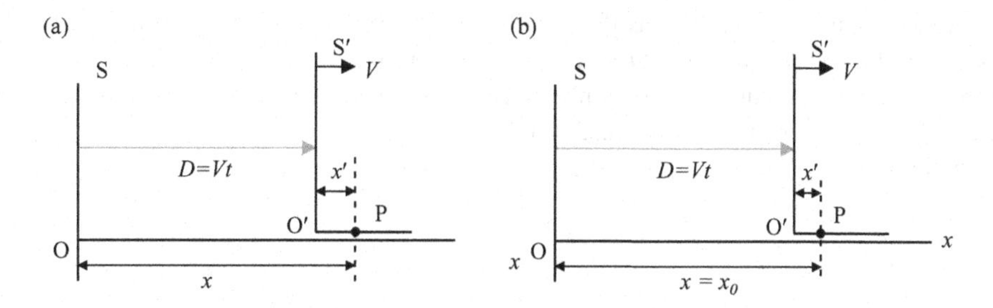

FIGURE 8.8 (a) Non-relativistic transformation of coordinate x in the S reference frame to the S′ reference frame. (b) Relativistic transformation of coordinate x in the S reference frame to the S′ reference frame.

hence the coordinate x, is now smaller than it would be compared to a non-relativistic case. To express the coordinate x in the S reference frame in terms of the coordinate x' in the S′ reference frame, we therefore have:

$$x = x'\sqrt{1 - V^2/c^2} + Vt \tag{8.27}$$

That is, this is the coordinate x, as would be seen at some time t, in an S reference frame, of a point moving with velocity V, and which has a coordinate x' in the S′ reference frame. In essence, we add on the foreshortened value of x' to the distance Vt to obtain the final coordinate position x. An observer in the S′ reference frame still sees the distance x' as marked off on the x axis, but the observer in the S reference frame sees a shortened value for x'.

Let's go back to our previous example for $V=0.5c$, and let the distance $D=1$. If x' also equals one in the S′ reference frame, then in the S reference frame, this distance would be seen as 0.865, and so the coordinate x is 1.865.

By a little rearrangement, we can specify the coordinate x' in the S′ reference frame with respect to the values of x and t that are seen in the S reference frame:

$$x' = \frac{x - Vt}{\sqrt{1 - V^2/c^2}} \tag{8.28}$$

This equation expresses the coordinate x', in the S′ reference frame, in terms of x and t in the S reference frame according to Einstein's postulates. Note that if the event occurs at $x'=0$, then $x=Vt$. If the event occurs at some place x, then at time $t=0$, the value of x' is that given by the length contraction equation we saw earlier.

8.2.7.2 Lorentz Time Transformation

Now, consider the time t for the point P to travel a distance D in the S reference frame.

$$D = Vt$$
$$t = \frac{D}{V} \tag{8.29}$$

Without accounting for any relativistic effects, then $D=x-x'$, and so:

$$t = \frac{x - x'}{V} \tag{8.30}$$

In the S′ reference frame, the point P has the coordinate x' at the time t'. Our aim now is to express the time t' in terms of the coordinate x and t in the S reference frame.

To do this, we need to focus on the distance D. In the S reference frame, this is equal to Vt.

In the S′ reference frame, we would see $D' = Vt'$.

Recognising that:

$$\begin{aligned} t' &= t\sqrt{1-V^2/c^2} \\ D &= Vt \end{aligned} \tag{8.31}$$

We have:

$$\begin{aligned} D' &= Vt' = Vt\sqrt{1-V^2/c^2} \\ D' &= D\sqrt{1-V^2/c^2} \end{aligned} \tag{8.32}$$

That is, from our perspective in the S′ reference frame, D' is somewhat *shorter* than that which would be measured by someone in the S reference frame. Note, this is somewhat different to our previous determination of length contraction where we found that the length *of a moving body* measured in the S reference frame would be perceived as being shorter than that measured in the moving frame of reference. That is, using our previous equation, we might expect:

$$D = D'\sqrt{1-V^2/c^2} \tag{8.33}$$

This predicts that if a body of length D' in the S′ reference frame is seen by someone from the S reference frame, then it would appear to have a shorter length D. Or:

$$D' = \frac{D}{\sqrt{1-V^2/c^2}} \tag{8.34}$$

That is, if we, in the S′ reference frame, are measuring the distance D', then it would seem to be *longer* than D that would be measured in the S reference frame.

The apparent contradiction is explained by realising that in the previous case, the body of length D' in the S′ reference frame is actually at rest in that frame of reference. If seen by someone in the S reference frame, it does appear shorter and has a length D which is less than D'. Conversely, our measurement of the length of the body, D', appears greater than that which would be computed by someone in the S reference frame *if the body is with us, in the moving frame of reference.*

However, if we are in the S′ reference frame, looking back at a body lying in the S reference frame, the roles are reversed. We see that body as moving away from us (to the left in Fig. 8.8), and so it has a shorter length than we would measure if it were with us at rest. The primed and unprimed quantities are swapped around. As we will see, this "looking back" means we must be very careful in interpreting the meaning of the transformations of physical quantities.

Thus, distances in the S reference frame which we, in the S′ reference frame, measure are shorter than those which would be measured by someone at rest to those distances. D' is therefore smaller than D in this case. This being the case, then the coordinate x is also smaller than that seen in the S reference frame. We might call our measurement of this coordinate x_0. This is shown in Figure 8.8 (b).

And so, we have:

$$\begin{aligned} D' &= Vt' \\ t' &= \frac{D'}{V} \end{aligned} \tag{8.35}$$

$$= \frac{x_0 - x'}{V}$$

$$= \frac{x\sqrt{1-V^2/c^2}}{V} - \frac{x - Vt}{V\sqrt{1-V^2/c^2}}$$

$$t' = \frac{t - xV/c^2}{\sqrt{1-V^2/c^2}}$$

This is the Lorentz transformation for time, where a time interval t' as measured in the S′ reference frame is expressed in terms of a time interval t and displacement x in the S reference frame. When $x'=0$, we have the event occurring at O′ so that $x=Vt$, and the Lorentz equation reduces to the previous time dilation equation.

We should also note that the relative velocity V between the two reference frames is the same in S as it is in S′.

$$V = \frac{d'}{t'} = \frac{d}{t} \tag{8.36}$$

8.2.7.3 Lorentz Velocity Transformation

Consider a body which has a velocity v in the S reference frame, and a velocity v' in the S′ reference frame, where the S′ reference frame has a different velocity V. This is shown in Figure 8.9.

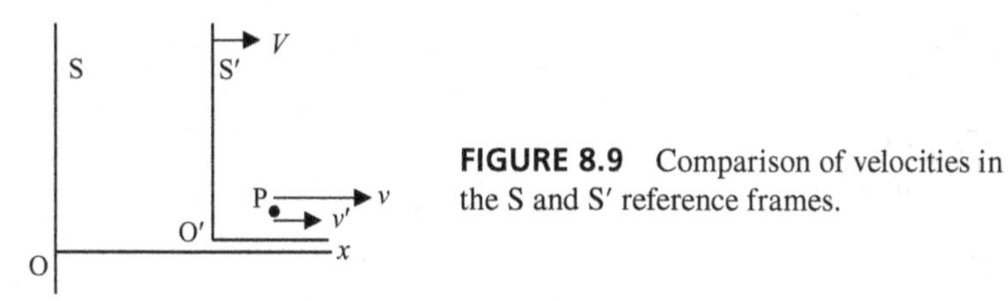

FIGURE 8.9 Comparison of velocities in the S and S′ reference frames.

The Galilean velocity transformation (in the x direction) was found to be:

$$v = \frac{dx'}{dt} + V$$
$$v' = v - V \tag{8.37}$$

- v is the velocity of a body as measured by an observer in the S reference frame.
- v' is the velocity of a body as measured by an observer in the S′ reference frame.
- V is the relative velocity in the x direction of the two reference frames.

To arrive at a relativistic velocity transformation between S and S′, we need to differentiate the relativistic displacement expressions:

$$v' = \frac{dx'}{dt'}$$
$$v = \frac{dx}{dt} \tag{8.38}$$

The relativistic transformation for x' to x involves the variable t, so a differential equation is the result.

$$\frac{dx'}{dt'} = \frac{dx'}{dt}\frac{dt}{dt'} \tag{8.39}$$

Now:

$$\begin{aligned} x' &= \frac{x - Vt}{\sqrt{1 - V^2/c^2}} \\ \frac{dx'}{dt} &= \frac{dx/dt - V}{\sqrt{1 - V^2/c^2}} \\ &= \frac{v - V}{\sqrt{1 - V^2/c^2}} \end{aligned} \tag{8.40}$$

and:

$$\begin{aligned} t' &= \frac{t - xV/c^2}{\sqrt{1 - V^2/c^2}} \\ t &= t'\sqrt{1 - V^2/c^2} + xV/c^2 \\ \frac{dt}{dt'} &= \sqrt{1 - V^2/c^2} + \frac{dx}{dt'}\frac{V}{c^2} \\ &= \sqrt{1 - V^2/c^2} + \frac{dx}{dt}\frac{dt}{dt'}\frac{V}{c^2} \\ &= \frac{\sqrt{1 - V^2/c^2}}{1 - dx/dt\left(V/c^2\right)} \\ &= \frac{\sqrt{1 - V^2/c^2}}{1 - vV/c^2} \end{aligned} \tag{8.41}$$

And so:

$$\begin{aligned} \frac{dx'}{dt'} &= \frac{dx'}{dt}\frac{dt}{dt'} \\ &= \left(\frac{v - V}{\sqrt{1 - V^2/c^2}}\right)\left(\frac{\sqrt{1 - V^2/c^2}}{1 - vV/c^2}\right) \\ v' &= \frac{dx'}{dt'} \\ &= \frac{v - V}{1 - vV/c^2} \end{aligned} \tag{8.42}$$

which is the relativistic velocity transformation we seek.

If we solve the above for v, we obtain:

$$v = \frac{v' + V}{1 + v'V/c^2} \tag{8.43}$$

Note, if the velocity of the body, say a light beam, is measured to be $v=c$ in the S reference frame, then we find that the velocity of the light beam v' in the S′ reference frame is:

$$\begin{aligned} v' &= \frac{c - V}{1 - cV/c^2} \\ &= \frac{c - V}{1 - V/c} \\ &= \frac{c(1 - V/c)}{1 - V/c} \\ &= c \end{aligned} \tag{8.44}$$

Indicating that the speed of light is measured to be the same in both frames of reference.

The above transformation is for a velocity in the x direction, v_x. If the velocity of the body has velocity components in the y and z directions, and the relative velocity V is still in the x direction, we obtain (by setting $V=0$ in the corresponding expressions for dy'/dt):

$$\begin{aligned} v'_y &= \frac{dy'}{dt'} = \frac{v_y\sqrt{1 - V^2/c^2}}{\left(1 - v_x V/c^2\right)} \\ v'_z &= \frac{dz'}{dt'} = \frac{v_z\sqrt{1 - V^2/c^2}}{\left(1 - v_x V/c^2\right)} \end{aligned} \tag{8.45}$$

The Lorentz transformations allow us to compute the distances, times and velocities in the S′ reference frame from the point of view of the S reference frame.

8.2.7.4 Momentum and Mass Transformations

If we have a moving body of mass m with velocity v (with respect to S), then the body has momentum $p = mv$. From the point of view of a reference frame S′ which has a velocity $V=v$, it has zero momentum. The momentum possessed by a body therefore depends on the frame of reference. However, during and after a collision with another body, the law of conservation of momentum applies for both reference frames.

Consider a body at point P in S′. The body and the reference frame are moving with velocity $v=V$ in the x direction relative to S. This is shown in Figure 8.10.

Now, let the body have a velocity v_y in the y direction. Due to the motion V (in the x direction), the velocities in the y direction in the two reference frames are not equal, but are related by:

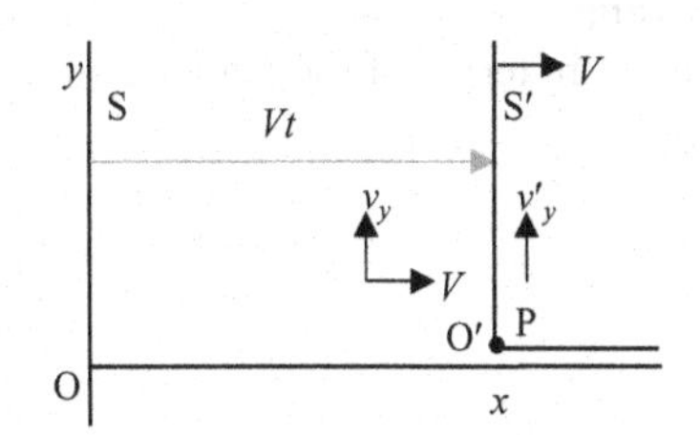

FIGURE 8.10 Velocities in the x and y directions in the S and S′ reference frames.

$$v_y' = \frac{dy'}{dt'} = \frac{v_y\sqrt{1-V^2/c^2}}{\left(1-v_x V/c^2\right)} \tag{8.46}$$

The momenta in the y direction for S and S′ are:

$$\begin{aligned} p &= mv_y \\ p' &= mv_y' \end{aligned} \tag{8.47}$$

In the y direction, we would expect these two momenta to be equal since there is no relative velocity between the two reference frames in this direction. In foreseeing that we will need a different mass in S and S′, we will write the expressions for momentum:

$$\begin{aligned} mv_y &= m'v_y' \\ &= m'\frac{v_y\sqrt{1-V^2/c^2}}{\left(1-V^2/c^2\right)} \end{aligned} \tag{8.48}$$

Therefore:

$$\begin{aligned} m &= \frac{m'}{\sqrt{1-v^2/c^2}} \\ m' &= m\sqrt{1-v^2/c^2} \end{aligned} \tag{8.49}$$

That is, the mass of the body in the S reference frame is m, but in the S′ reference frame, it is m'. For example, if $v=0.5c$, and the mass of the body is 1 kg in the S′ reference frame, then the body acts as if it has a mass of 1.155 kg in the S reference frame.

Since m' is stationary with respect to S′, we might call it the proper mass, but it is known instead as the rest mass m_0. It is the mass of a body as seen from the point of view of someone in the S′ reference frame who is at rest with respect to the body. In the above example, the rest mass of the body is 1 kg. The mass m is the relativistic mass measured by a person in the S reference frame. In the above example, the relativistic mass of the body is 1.155 kg.

The relativistic momentum is thus:

$$p = \frac{m_0}{\sqrt{1-v^2/c^2}}v \tag{8.50}$$

This is the relativistic momentum of a moving body when it is measured in the S reference frame, in which the body is moving with velocity v.

Thus far, we have considered the body in the S′ reference frame to be at rest with respect to that frame of reference, that is, $v'=0$ and $v=V$. Let us now consider the situation where the body has a velocity v' in the S′ reference frame as shown previously in Figure 8.9. That is, where v, from the S reference frame, is greater than V, the relative velocity between the two reference frames.

Now, the Lorentz velocity transformation gives the velocity v' in terms of the velocity as seen by an observer in the S reference frame:

$$v' = \frac{v-V}{1-vV/c^2} \tag{8.51}$$

$$v = \frac{v' + V}{1 + v'V/c^2} \tag{8.52}$$

- v is the velocity of the body as seen from the S reference frame.
- v' is the velocity of the body as seen from the S′ reference frame.
- V is the velocity of the S′ reference frame with respect to the S reference frame.

What might be the momentum p' of the body from the point of view of an observer in the S′ reference frame? Let us call this momentum p'.

$$p' = m'v' \tag{8.53}$$

m' is the mass as seen by an observer in the S′ reference frame – but this is not the rest mass. The rest mass is the mass seen by an observer *at rest* with respect to the moving body. That is, the rest mass would be the mass seen by an observer moving with the body.

- m is the mass of the body as seen by an observer in the S reference frame.
- m' is the mass of the body as seen by an observer in the S′ reference frame.
- m_0 is the rest mass, that seen by an observer moving with the body.

Now, the rest mass is:

$$m_0 = m\sqrt{1 - v^2/c^2} \tag{8.54}$$

We might ask the question, which m do we choose? m or m'? It doesn't matter. If we choose m, then we also use v as in the above. If we choose m', then we use v'.

Can we develop a relativistic transformation between p and p'?

$$p = \frac{m_0}{\sqrt{1 - v^2/c^2}} v'$$

$$= \frac{m_0}{\sqrt{1 - \left(\frac{v - V}{1 - vV/c^2}\right)^2 \Big/ c^2}} \frac{v' + V}{1 + v'V/c^2}$$

$$= \frac{m_0}{\sqrt{1 - \frac{(v' + V)^2}{(1 + v'V/c^2)^2} \frac{1}{c^2}}} \frac{v' + V}{1 + v'V/c^2}$$

$$= \frac{m_0 (1 + v'V/c^2)}{\sqrt{(1 + v'V/c^2)^2 - \left(\frac{v'}{c} + \frac{V}{c}\right)^2}} \frac{v' + V}{1 + v'V/c^2}$$

$$= \frac{m_0 (1 + v'V/c^2)}{\sqrt{1 + \frac{2v'V}{c^2} + \left(\frac{v'V}{c^2}\right)^2 - \frac{v'^2}{c^2} - \frac{2v'V}{c^2} - \frac{V^2}{c^2}}} \frac{v' + V}{1 + v'V/c^2} \tag{8.55}$$

$$= \frac{m_0(v'+V)}{\sqrt{1+\left(\frac{v'V}{c^2}\right)^2 - \frac{v'^2}{c^2} - \frac{V^2}{c^2}}}$$

$$= \frac{m_0(v'+V)}{\sqrt{(1-V^2/c^2)(1-v'^2/c^2)}}$$

But:

$$p' = m'v'$$

$$= \frac{m_0}{\sqrt{1-v'^2/c^2}}v' \tag{8.56}$$

Therefore:

$$p = \frac{m_0(v'+V)}{\sqrt{(1-V^2/c^2)(1-v'^2/c^2)}}$$

$$= \frac{p'(1+V/v')}{\sqrt{(1-V^2/c^2)}} \tag{8.57}$$

Solving for p' gives:

$$p' = \frac{p(1-V/v')}{\sqrt{(1-V^2/c^2)}} \tag{8.58}$$

These equations give the transformation between momentum from one reference frame to another where the body has a velocity v' in S' and S' has a velocity V with respect to S.

8.2.7.5 Mass and Energy Transformations

Newton's second law tells us that force F applied to a body of mass m is the time rate of change of momentum p. This results in the mass receiving an acceleration a:

$$F = \frac{dp}{dt} = m\frac{dv}{dt} = ma \tag{8.59}$$

In Newtonian mechanics, the kinetic energy of a body is given by:

$$\text{KE} = \int_0^x F dx$$

$$= \int_0^x m\frac{dv}{dt}dx \tag{8.60}$$

$$= \int_0^v mv dv$$

$$= \frac{1}{2}mv^2$$

Where m is the mass of the body assumed to be a constant. This is the case for v very much less than c.

Consider the case of a body of mass m being given a velocity v in the x direction. Since $p = mv$, we can express the integral above in terms of momentum p where we use the relativistic momentum with $V = v$ to obtain a more precise expression for the kinetic energy of the body. In terms of the rest mass m_0, we can write the kinetic energy of the body, from the point of view of an S reference frame, as:

$$\begin{aligned} \text{KE} &= \int_0^v \frac{m_0 v}{\sqrt{1 - v^2/c^2}} dv \\ &= m_0 c \int_0^v \frac{v}{\sqrt{c^2 - v^2}} dv \end{aligned} \tag{8.61}$$

m_0 the rest mass, is the mass the body would have if it were measured in the S′ reference frame.

When solved by parts (see Appendix 8a), the above integral gives:

$$\text{KE} = \frac{m_0 c^2}{\sqrt{1 - v^2/c^2}} - m_o c^2 \tag{8.62}$$

But the relativistic mass is:

$$m = \frac{m_0}{\sqrt{1 - v^2/c^2}} \tag{8.63}$$

Thus:

$$\text{KE} = mc^2 - m_0 c^2 \tag{8.64}$$

This is the kinetic energy of the moving body of relativistic mass m and rest mass m_0. If the rest mass is equal to the relativistic mass, then the velocity is zero and there is zero kinetic energy. In essence, the kinetic energy of a moving body depends upon the difference between its relativistic mass and its rest mass multiplied by c^2. It is the relativistic mass that increases with v and hence leads to an increase in kinetic energy.

The term $m_0 c^2$ is a new kind of energy associated with the rest mass of the body; it is the rest mass energy. That is, total energy E of a moving body is thus made up of two components, its kinetic energy and its rest mass energy:

$$E = \text{KE} + m_0 c^2 \tag{8.65}$$

Expressed in terms of the relativistic mass m, the total energy of a body is:

$$E = mc^2 \tag{8.66}$$

where:

$$m = \frac{m_0}{\sqrt{1 - v^2/c^2}} \tag{8.67}$$

The kinetic energy arises due to the velocity v of the moving body of relativistic mass m. The total energy E is the sum of the kinetic energy of motion and the energy bound up in the rest mass of the body. We see that mass is energy. Even if a body has no velocity, it has rest energy.

The total energy of a body depends upon the frame of reference. If the body has a velocity v' in the S′ reference frame, then the total energy would be computed from:

$$E' = m'c^2 \tag{8.68}$$

where:

$$m' = \frac{m_0}{\sqrt{1 - v'^2/c^2}} \tag{8.69}$$

The law of conservation of energy still applies, but it only applies within the reference frame under consideration.

The total energy of a particle in a reference frame S is:

$$E = \frac{m_0}{\sqrt{1 - v^2/c^2}} c^2 \tag{8.70}$$

where v is the velocity of the particle as measured in the S reference frame.

Now, say reference frame S′ moves with velocity V and in which the particle has a velocity v'. We've seen how the velocity transforms from one reference frame to another:

$$v' = \frac{v - V}{1 - vV/c^2} \tag{8.71}$$

What might be the energy of the particle as seen by an observer in the S′ reference frame?

$$E' = \frac{m_0}{\sqrt{1 - v'^2/c^2}} c^2 \tag{8.72}$$

We now need to express E' in terms of E. That is, we need a transformation equation. Let's begin with the velocity. Squaring both sides of the above, we have:

$$v'^2 = \frac{v^2 - 2vV + V^2}{\left(1 - vV/c^2\right)^2} \tag{8.73}$$

$$\frac{v'^2}{c^2} = \frac{v^2 - 2vV + V^2}{c^2\left(1 - vV/c^2\right)^2}$$

$$1 - \frac{v'^2}{c^2} = 1 - \frac{v^2 - 2vV + V^2}{c^2\left(1 - vV/c^2\right)^2}$$

$$= \frac{c^2\left(1 - vV/c^2\right)^2 - \left(v^2 - 2vV + V^2\right)}{c^2\left(1 - vV/c^2\right)^2}$$

$$= \frac{c^2\left(1 - 2vV/c^2 + v^2V^2/c^4\right) - \left(v^2 - 2vV + V^2\right)}{c^2\left(1 - vV/c^2\right)^2}$$

$$= \frac{c^2 - 2vV + v^2V^2/c^2 - v^2 + 2vV - V^2}{c^2\left(1 - vV/c^2\right)^2}$$

$$= \frac{c^2 + v^2V^2/c^2 - v^2 - V^2}{c^2\left(1 - vV/c^2\right)^2}$$

$$= \frac{1 + v^2V^2/c^4 - v^2/c^2 - V^2/c^2}{\left(1 - vV/c^2\right)^2}$$

$$= \frac{\left(1 - v^2/c^2\right)\left(1 - V^2/c^2\right)}{\left(1 - vV/c^2\right)^2}$$

$$\frac{1}{1 - v'^2/c^2} = \frac{\left(1 - vV/c^2\right)^2}{\left(1 - v^2/c^2\right)\left(1 - V^2/c^2\right)}$$

$$\frac{1}{\sqrt{1 - v'^2/c^2}} = \frac{\left(1 - vV/c^2\right)}{\sqrt{\left(1 - v^2/c^2\right)}\sqrt{\left(1 - V^2/c^2\right)}}$$

Therefore:

$$E' = \frac{m_0c^2}{\sqrt{1 - v'^2/c^2}} = \frac{m_0c^2\left(1 - vV/c^2\right)}{\sqrt{\left(1 - v^2/c^2\right)}\sqrt{\left(1 - V^2/c^2\right)}} \tag{8.74}$$

But:

$$E = \frac{m_0c^2}{\sqrt{1 - v^2/c^2}} \tag{8.75}$$

And so:

$$E' = E\frac{\left(1 - vV/c^2\right)}{\sqrt{\left(1 - V^2/c^2\right)}} \tag{8.76}$$

This is the transformation of the total energy from one reference frame to another.

We can express this transformation in terms of momentum.

$$E' = \frac{m_0c^2\left(1 - vV/c^2\right)}{\sqrt{\left(1 - v^2/c^2\right)}\sqrt{\left(1 - V^2/c^2\right)}} \tag{8.77}$$

where V is the relative velocity between the two reference frames. Working in one dimension, what might this be in terms of the momentum p_x?

$$E' = \frac{m_0c^2 - m_0c^2\left(vV/c^2\right)}{\sqrt{\left(1 - v^2/c^2\right)}\sqrt{\left(1 - V^2/c^2\right)}} \tag{8.78}$$

$$= \left(\frac{m_0c^2}{\sqrt{\left(1 - v^2/c^2\right)}} - \frac{vm_0V}{\sqrt{\left(1 - v^2/c^2\right)}}\right)\left(\frac{1}{\sqrt{\left(1 - V^2/c^2\right)}}\right)$$

$$= \frac{E - Vp_x}{\sqrt{\left(1 - V^2/c^2\right)}}$$

Remember that in this part of the book, m is the rest mass.

The transformation between momentum in the S and S′ frames can thus be expressed in terms of the energy E and momentum p:

$$E' = \frac{m_0 c^2 \left(1 - vV/c^2\right)}{\sqrt{\left(1 - v^2/c^2\right)}\sqrt{\left(1 - V^2/c^2\right)}}$$

$$p'_x c^2 = E'v'$$

$$= \frac{m_0 c^2 \left(1 - vV/c^2\right)}{\sqrt{\left(1 - v^2/c^2\right)}\sqrt{\left(1 - V^2/c^2\right)}} \frac{v - V}{1 - vV/c^2}$$

$$= \frac{m_0 c^2 \left(v - V\right)}{\sqrt{\left(1 - v^2/c^2\right)}\sqrt{\left(1 - V^2/c^2\right)}}$$

$$= \frac{m_0 v c^2 - m_0 V c^2}{\sqrt{\left(1 - v^2/c^2\right)}\sqrt{\left(1 - V^2/c^2\right)}} \tag{8.79}$$

$$= \frac{m_0 v c^2 - m_0 V c^2}{\sqrt{\left(1 - v^2/c^2\right)}} \frac{1}{\sqrt{\left(1 - V^2/c^2\right)}}$$

$$= \left(\frac{m_0 v c^2}{\sqrt{\left(1 - v^2/c^2\right)}} - \frac{m_0 V c^2}{\sqrt{\left(1 - v^2/c^2\right)}} \right) \frac{1}{\sqrt{\left(1 - V^2/c^2\right)}}$$

$$p'_x = \frac{p_x - VE/c^2}{\sqrt{\left(1 - V^2/c^2\right)}}$$

This gives the transformation of momentum from one reference frame to another in terms of the momentum and energy.

8.2.8 Consequences of Special Relativity

From this point on, we will use the symbol v for the velocity of the S′ reference frame where the velocity of the body in that reference frame is zero. That is, $V = v$ and $v' = 0$. When the body has a velocity v' in the S′ reference frame, then we will make the distinction between v (the velocity of the body from the point of view of the S reference frame) and V, the velocity of the S′ reference frame.

As well, in anticipation of what will follow in later chapters, we will hereafter refer to an object as a particle. This gives the impression that we mean a small particle, and for the most part, in mechanics, we mean the centre of mass of an object at a point.

8.2.8.1 Energy and Momentum

The total energy of a body can be expressed in terms of its relativistic momentum.

The relativistic momentum p is expressed:

$$\begin{aligned} p &= mv \\ &= \frac{m_0}{\sqrt{1-v^2/c^2}}v \end{aligned} \tag{8.80}$$

This is the momentum of a body as seen from the S reference frame. That is, m is the relativistic mass, and v is the velocity of the particle as seen from the S reference frame, and m_0 is the rest mass.

Now:

$$\begin{aligned} p &= \frac{m_0}{\sqrt{1-v^2/c^2}}v \\ p^2 &= \frac{m_0{}^2}{1-v^2/c^2}v^2 \\ \frac{p^2}{m_0{}^2c^2} &= \frac{v^2/c^2}{1-v^2/c^2} \end{aligned} \tag{8.81}$$

The total energy of the body can be written:

$$\begin{aligned} E &= \frac{m_0}{\sqrt{1-v^2/c^2}}c^2 \\ \left(\frac{E}{m_0c^2}\right)^2 &= \frac{1}{1-v^2/c^2} \end{aligned} \tag{8.82}$$

To eliminate v, we subtract as follows:

$$\begin{aligned} \left(\frac{E}{m_0c^2}\right)^2 - \frac{p^2}{m_0^2c^2} &= \frac{1}{1-v^2/c^2} - \frac{v^2/c^2}{1-v^2/c^2} \\ &= 1 \\ \left(\frac{E}{m_0c^2}\right)^2 &= 1 + \frac{p^2}{m_0^2c^2} \\ E^2 &= m_0^2c^4 + \frac{m_0^2c^4p^2}{m_0^2c^2} \\ &= m_0^2c^4 + p^2c^2 \end{aligned} \tag{8.83}$$

We might regard the rest mass as something inherent to the body. The rest mass does not change depending on the frame of reference. It is invariant. Rearranging the above we have:

$$m_0^2c^4 = E^2 - p^2c^2 \tag{8.84}$$

The left-hand side of this equation involving m_0 and c is invariant. We know that the relativistic momentum depends on v, and the total energy E possessed by the body also depends also on v. It is the quantity $E^2 - p^2c^2$ that is invariant.

That is, if the body has a velocity v which is different to V, then the energy E' from the point of view of an observer in the S′ reference frame would be:

$$\begin{aligned} E' &= m'c^2 \\ &= \frac{m_0c^2}{\sqrt{1-v'^2/c^2}} \end{aligned} \tag{8.85}$$

And the momentum:

$$p' = m'v' \tag{8.86}$$

And so:

$$\begin{aligned} E^2 &= m_0^2c^4 + p^2c^2 \\ E'^2 &= m_0^2c^4 + p'^2c^2 \\ m_0^2c^4 &= E^2 - p^2c^2 = E'^2 - p'^2c^2 \end{aligned} \tag{8.87}$$

Returning for a moment to our relativistic momentum equations, a little algebraic substitution gives:

$$p = \frac{p' + VE'/c^{2\prime}}{\sqrt{\left(1-V^2/c^2\right)}} \tag{8.88}$$

8.2.8.2 Kinetic Energy

Special relativity deals with the mechanics of motion where the motion of one reference frame is a constant with respect to another reference frame. Einstein's theory of general relativity deals with the case where the velocity between the two reference frames is not constant – that is, where there is a relative acceleration. Special relativity is a special case of general relativity.

The kinetic energy of a body is given by:

$$\begin{aligned} KE &= mc^2 - m_oc^2 \\ &= \frac{m_0c^2}{\sqrt{1-v^2/c^2}} - m_0c^2 \\ &= m_0c^2\left(1 - \frac{1}{\sqrt{1-v^2/c^2}}\right) \end{aligned} \tag{8.89}$$

This square root term may be expanded into a binomial series (see Appendix 8b), and when $v << c$, reduces to the classical form:

$$KE = \frac{1}{2}m_0v^2 \quad v << c \tag{8.90}$$

It should be noted that the kinetic energy of a body, that is, the relativistic kinetic energy, is *not* given by $1/2mc^2$ where m is the relativistic mass. The kinetic energy of a body, as seen from the S reference frame, can really only be determined from:

$$KE = mc^2 - m_0c^2 = \frac{m_0c^2}{\sqrt{1-v^2/c^2}} - m_0c^2 \tag{8.91}$$

Kinetic energy is not an invariant quantity. Its value depends upon the frame of reference used to measure it. From the point of view of a frame of reference moving with the particle, the kinetic energy of a body would be zero.

When a body of rest mass m_0 is supplied with an increment of energy ΔE (for example, by the application of a force applied by an electric field) work is done by a force acting through a distance. The velocity of the body changes to a new value, v. Because the velocity has increased, its relativistic mass has increased. Because the relativistic mass m has increased, the kinetic energy of the body has increased. The increment of energy supplied has gone into the increase in kinetic energy of the body.

When $v << c$, we compute the new kinetic energy of the body to be $\frac{1}{2}m_0v^2$. When v is of the same order as c, every increment in velocity results in a much larger change in relativistic mass. A small change in velocity results in a very large change in m. The kinetic energy increases as before, but the velocity increase is no longer found from $\frac{1}{2}m_0v^2$; the velocity increase becomes smaller, and the relativistic mass becomes larger. As we approach the speed of light, the rate of increase in velocity decreases, and the velocity approaches an asymptotic value of c, and the relativistic mass approaches infinity. From the point of view of the applied force, the acceleration (and hence change of velocity) is smaller, and the mass against which the force acts becomes larger. The body appears to have more inertia as its velocity increases. This concept was the subject of Einstein's second paper on special relativity published in 1905 entitled "Does the Inertia of a Body Depend upon its Energy Content?"

The concept of relativistic mass does lead to some confusion. For example, as we've seen, the relativistic kinetic energy of a body is not given by $\frac{1}{2}mv^2$ where m is the relativistic mass. On the other hand, when we look at momentum and inertia, it does appear as though the mass of a body increases with its velocity. The word "mass" is usually taken to mean the rest mass, something inherent to the body, and any apparent increase in mass is more of a relativistic correction applied to the quantity being measured rather than interpreted as an increase in mass.

8.2.8.3 Photons

Photons travel at velocity c. Their total energy is given by $E=hf$. Photons have no rest mass. A photon (in free space) can never be slowed down. If they are absorbed or stopped, then their energy goes into other forms, and not into kinetic energy of a rest mass – since they have none. However, due to their relativistic mass, they are affected by gravity.

A photon has zero rest mass, but it does have a relativistic mass and thus momentum.

$$\begin{aligned} E &= hf = mc^2 \\ \frac{h}{\lambda} &= mc = p \end{aligned} \tag{8.92}$$

The expression for time dilation:

$$t = \frac{t'}{\sqrt{1-v^2/c^2}} \tag{8.93}$$

shows that as v approaches c, the time in the S′ reference frame t' (attached to the photons) slows down compared to that in our observer's reference frame to the point that t approaches infinity. In a sense, when we see photons travelling past us, we judge *their* time to have stopped. That is, while we see photons take time to travel from place to place, a photon itself has no sense of time.

8.2.9 Summary of Special Relativity

Einstein found that classical concepts of length, time and mass required modification to satisfy the requirement for the general applicability of physical laws and the observed constant speed of light. We will need to have some appreciation of these issues in our study of quantum physics, not the least of reasons being that one of the chief players, the photon, has no rest mass and moves with velocity c, yet carries a certain momentum and kinetic energy.

In this section, the body moves with a velocity v, with respect to a frame of reference S. The S′ reference frame has a velocity V with respect to the S reference frame. In most cases, $V=v$, but in instances where the body has a velocity v' in the S′ reference frame, then we will make the distinction between v and V.

8.2.9.1 Length

The x coordinate in two different reference frames after time t:

$$x = x'\sqrt{1-V^2/c^2} + Vt \tag{8.94}$$

$$x' = \frac{x - Vt}{\sqrt{1-V^2/c^2}} \tag{8.95}$$

This is the Lorentz transformation for coordinates.

Or, for a body of length l moving with velocity v with respect to an observer in an S reference frame:

$$\begin{aligned} l &= l'\sqrt{1-v^2/c^2} \\ l' &= \frac{l}{\sqrt{1-v^2/c^2}} \end{aligned} \tag{8.96}$$

- l is the relativistic length of the body measured by the observer in the S reference frame.
- l' is the length of the body measured by an observer in the S′ reference frame.
- v is the velocity of the body measured by the observer in the S reference frame.

Note: as v approaches c, the length l of the body, as it appears to the observer in the S reference frame, shrinks to zero. $l \le l'$.

8.2.9.2 Time

For a time period Δt between two events where the events occur at a point which is in a reference frame S′ with velocity v with respect to a reference frame S:

$$t = \frac{t'}{\sqrt{1-v^2/c^2}} \tag{8.97}$$

This is a time dilation equation.

Or:

$$t' = \frac{t - xv/c^2}{\sqrt{1-v^2/c^2}} \tag{8.98}$$

This is the Lorentz transformation for time.

This applies for the case where a time interval t' as measured in the S′ reference frame is expressed in terms of a time interval t and displacement x in the S reference frame. When $x'=0$, we have the event occurring at O′ so that $x=Vt$, and this reduces to the former time dilation equation.

Combining the two above equations, we obtain:

$$t' = \sqrt{t^2 - x^2/c^2} \tag{8.99}$$

- Δt is the relativistic time interval measured by the observer in the S reference frame.
- $\Delta t'$ is the time interval measured by an observer in the S′ reference frame, where the events occur in the same place, the proper time.
- x is the coordinate of the body in the S reference frame.
- V is the velocity of the S′ reference frame measured by the observer in the S reference frame.

Note: as V approaches c, the time t as measured by an observer in S goes to infinity. $\Delta t \geq \Delta t'$.

Note, if the body is at $x=0$, then we obtain:

$$t' = \frac{t}{\sqrt{1 - v^2/c^2}} \tag{8.100}$$

This is in effect "looking back" from the S′ reference frame, and so the primed and unprimed quantities are reversed.

8.2.9.3 Velocity

For a body moving with velocity v with respect to an observer in a reference frame S:

$$v' = \frac{v - V}{1 - vV/c^2} \tag{8.101}$$

$$v = \frac{v' + V}{1 + v'V/c^2} \tag{8.102}$$

- v is the relativistic velocity measured by the observer in the S reference frame.
- v' is the velocity measured by an observer in an S′ reference frame which itself has velocity V.
- V is the velocity of the S′ reference frame measured by the observer in the S reference frame.

8.2.9.4 Mass

For a body of mass m moving with velocity v with respect to an observer in a reference frame S:

$$m = \frac{m_0}{\sqrt{1 - v^2/c^2}} \tag{8.103}$$

- m is the relativistic mass of the body as measured by the observer in the S reference frame.
- m_0 is the rest mass of the body as would be measured in the S′ reference frame where the observer is at rest with respect to that mass. $m \geq m_0$.
- v is the velocity of the body measured by the observer in the S reference frame.

Note: as v approaches c, the relativistic mass m goes to infinity. At $v=0$, $m=m_0$.

8.2.9.5 Momentum

For a body of mass m moving with velocity v with respect to an observer in a reference frame S:

$$\begin{aligned} p &= mv \\ &= \frac{m_0}{\sqrt{1-v^2/c^2}}v \end{aligned} \tag{8.104}$$

- p is the relativistic momentum of the body as measured by the observer in the S reference frame.
- m is the relativistic mass of the body as measured by the observer in the S reference frame.
- m_0 is the rest mass of the body as would be measured in the S′ reference frame. $m \geq m_0$.
- v is the velocity of the body measured by the observer in the S reference frame.

Note: as v approaches c, the relativistic momentum p goes to infinity.

Also, in the case where the body has a velocity v' in a reference frame S′, which itself is moving with velocity V, we have:

$$p' = m'v'$$

and the transformation between S and S′ is:

$$p = \frac{p'(1+V/v')}{\sqrt{(1-V^2/c^2)}} \tag{8.105}$$

$$p' = \frac{p(1-V/v')}{\sqrt{(1-V^2/c^2)}} \tag{8.106}$$

- p' is the momentum as measured by the observer in the S′ reference frame.
- m' is the relativistic mass of the body as measured by an observer in the S′ reference frame (note, in this case, an observer in the S′ reference frame sees the body moving with a velocity v' and so to that observer, the body is not at rest and the mass m' of the body is not the rest mass).

8.2.9.6 Energy

The total energy of a body is:

$$E = mc^2 = KE + m_o c^2 \tag{8.107}$$

where:

$$\begin{aligned} KE &= mc^2 - m_o c^2 \\ &= \frac{m_0}{\sqrt{1-v^2/c^2}}c^2 - m_o c^2 \\ &\approx \frac{1}{2}m_0 v^2 \quad \text{when } v << c \end{aligned} \tag{8.108}$$

The transformation of E from one frame to another is given by:

$$E' = E\frac{\left(1 - vV/c^2\right)}{\sqrt{\left(1 - V^2/c^2\right)}} \tag{8.109}$$

- m is the relativistic mass of the body as measured by the observer in the S reference frame.
- m_0 is the rest mass of the body as would be measured in the S′ reference frame. $m \geq m_0$.
- The body has a velocity v' in a reference frame S′, which itself is moving with velocity V with respect to S.

8.3 GENERAL RELATIVITY

8.3.1 Introduction

In special relativity, we saw that there is no distinction to be made between being at rest and travelling with a constant velocity. It was found that "without looking outside", it was not possible to determine whether we, performing experiments inside, were moving (ignoring rotations) or not. That is, there is no preferred or stationary frame of reference required for physical laws. As a consequence of this, the velocity of light (in free space) had the same value no matter whether the source or the observer was moving.

The theory of general relativity is concerned with reference frames that are accelerating with respect to one another. Imagine you are in an elevator in a very tall building, about to descend. The situation is shown in Figure 8.11. Unfortunately for you, the elevator cable snaps, and the elevator falls towards the ground. After a short period, you will find that you, inside the elevator, can no longer feel the force of gravity. You can float around the inside of the elevator just as if you were in outer space.

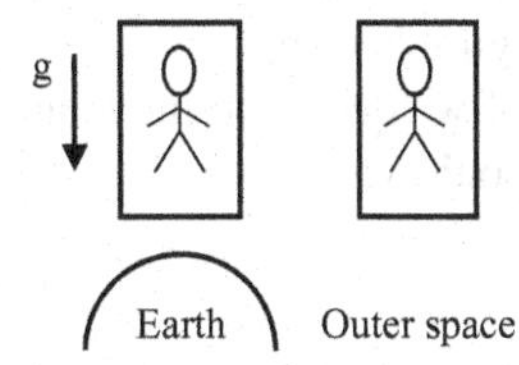

FIGURE 8.11 Body in an elevator in a gravitational field and in free space.

Imagine it is a very long trip to the ground floor. You have time to perform some physics experiments. You find that the laws of motion are valid, but the only thing missing is the force of gravity that you are so used to dealing with. Indeed, after a while, you come to accept your lot and remain weightless in perfect harmony with nature.

So harmonious do you feel that you are completely unaware of a very large mass right below you imposing its gravitational force upon you. Without looking outside, you would not know if you were falling in the gravitational field of this large mass, or just drifting through space without a care in the world. From your point of view, the force of gravity has been switched off.

Einstein's theory of general relativity deals with the concepts of acceleration, gravity and the nature of space and time. It is said that Einstein was impressed by the shout of a man who was falling from a roof of a house in Berlin, whereupon he exclaimed that for a short period, he did not feel the force of gravity. That gravity and acceleration are equivalent is one of the main conclusions of general relativity.

Our interest is mainly concerned with the nature of space and time rather than the effects of gravity, and so we need to become familiar with the concept of space-time.

8.3.2 Space-Time

A particular feature of relativity is the combination of space and time which, for the want of a better name, is called space-time. Space-time has four dimensions. A particle's position can be specified in space-time by the coordinates (x, y, z, ct). We can recognise that the product ct has units of distance so all three axes are ones of distance. The distance on the vertical axis would thus be that travelled by a photon (with velocity c) over a time t. Even though the vertical axis is one of distance, it is in effect a time dimension.

We actually are very familiar with this concept when we measure large distances, such as that to the stars, in units of light years. That is, a light year is the distance travelled by a photon in one year. If our distances on our axes in space time are expressed in metres, each division on our vertical axis represents the time needed for a photon to travel one metre – that is, ct has the units of light metres. One light metre is about 3.3×10^{-9} seconds.

To make our understanding of space-time easier, we need to work with three of these coordinates – and suppress one of the familiar spatial coordinates. To do this, we draw our three coordinate axes as shown in Figure 8.12.

Each point in space-time, such as point P shown in Figure 8.12, is an event. The continued existence of the point P in time is represented by a line in space-time – that is, a series of events. Without a line, a point is just an instantaneous event. The point may represent a physical body, or particle, and if so, the line is called the world-line of the particle. If the particle is not moving with respect to the x and y axes, then the world-line is vertical.

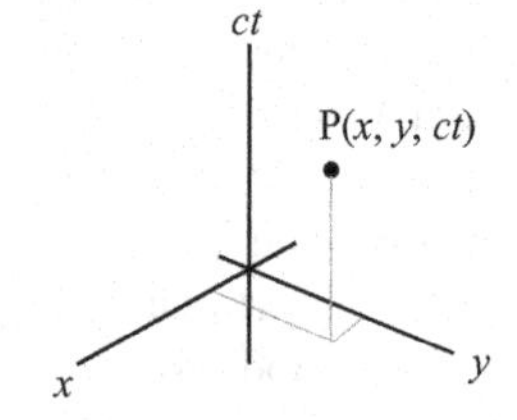

FIGURE 8.12 Space-time diagram of time and two-dimensional space.

We shall begin our existence as a stationary body at coordinates (0,0,0) as shown in Figure 8.13(a). As time proceeds, our world-line follows the ct axis as shown in Figure 8.13(b).

We might now ask ourselves where our world-line will go if we decide to move at constant velocity, say $0.3c$, in the x direction. In our space-time coordinates, we would write:

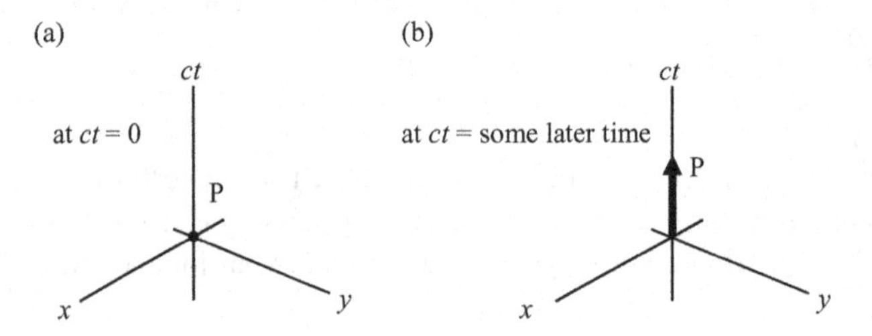

FIGURE 8.13 Space-time diagram showing position of particle at (a) $t=0$ and (b) some later time t.

$$\frac{x}{t} = 0.3c$$
$$x = 0.3ct \tag{8.110}$$

Note that this velocity is the slope of the world-line (as seen from the ct axis) as shown in Figure 8.14.

But, if we are a photon, then our velocity is limited to c in any direction, and so the surface of a cone, drawn at 45°, represents the only possible coordinates in space-time that we can have.

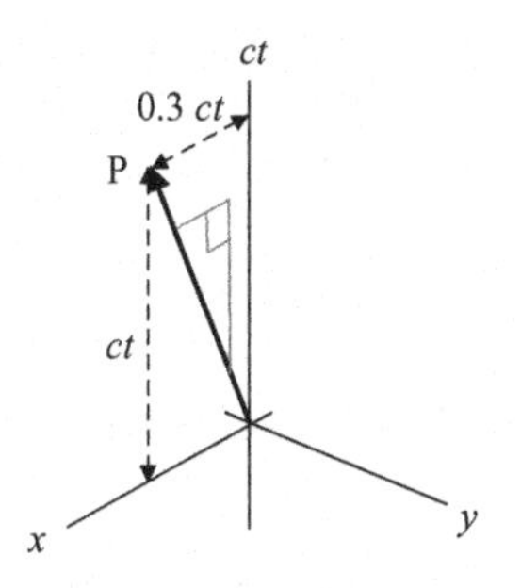

FIGURE 8.14 Space-time diagram for constant velocity of particle.

This cone is called a light-cone. The interior of the light-cone represents the future available coordinates in space-time of any physical particle P.

The origin of the light-cone can be translated to any convenient point in space-time. There is a light-cone at every point P(x, y, z, ct). We only need to draw the one that is convenient for our purposes (usually that for our present position). If we ignore the effects of gravity, then all light-cones point in the same direction (i.e. the axes are aligned in the same way no matter where the origin of the light-cone is). This is illustrated in Figure 8.15.

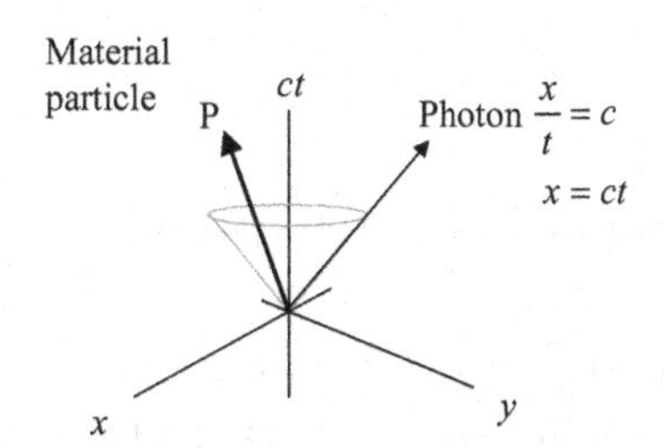

FIGURE 8.15 Light-cone and particle in space-time.

Particles that move with a constant velocity are represented by world-lines that are straight, but lie within the light-cone. Particles that accelerate result in a curvature of their world-line. For any particle with rest mass, the world-line falls within the bounds of the light-cone at any point. Only particles with no rest mass, such as photons, can move along the 45° world-lines.

As mentioned above, we have expressed time in terms of distance by multiplying t by c. This gives us a convenient 45° world-line for massless particles such as photons.

In space-time, the concept of distance in the coordinate system has important physical meaning. In Euclidean geometry, the distance d from O to P would be expressed:

$$d^2 = x^2 + y^2 \tag{8.111}$$

That is, the distance d between two points (O and P) in Euclidean space, as shown in Figure 8.16(a), is a linear spatial distance.

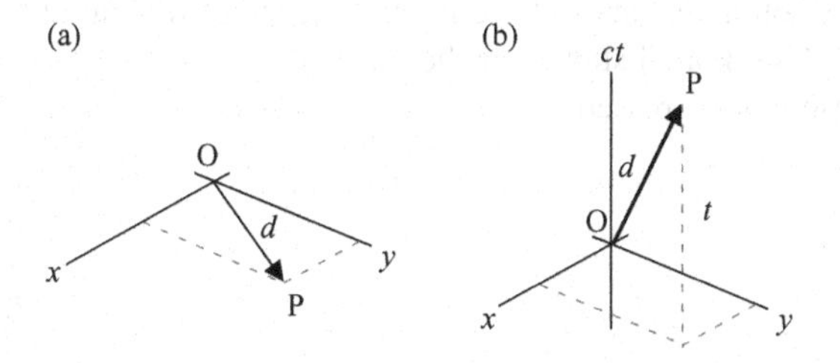

FIGURE 8.16 Distance between two points in (a) Euclidean space and (b) space-time.

In space-time, we might ask what the distance between two events might actually represent? We might be tempted to calculate the distance from:

$$d^2 = x^2 + y^2 + c^2t^2 \tag{8.112}$$

This is shown in Figure 8.16(b).

Before we do that, let's take a simpler situation in two dimensions. The distance d is calculated from

$$d^2 = x^2 + c^2t^2 \tag{8.113}$$

as shown in Figure 8.17(a).

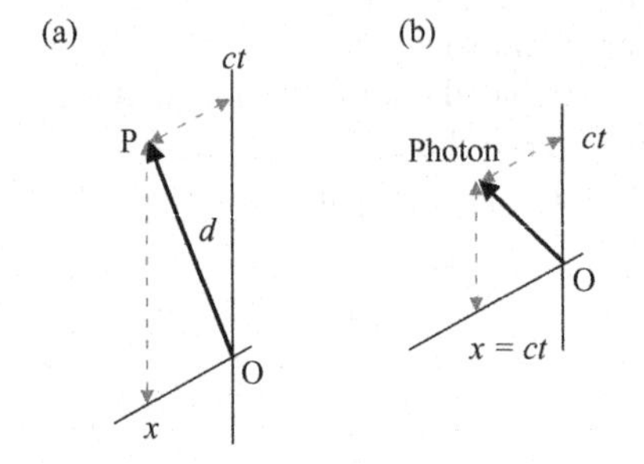

FIGURE 8.17 Distance between two points in two-dimensional space-time for (a) a particle and (b) a photon.

This distance d is the interval between two events. But an interval of what? Space? Time?

If our interval is directly along the ct axis (i.e. the particle is at rest), the distance d would be a time interval (although expressed in terms of the distance ct that a photon would travel in time t). The more the interval d aligns with the x axis, the more space-like it is. The more it is aligned with the ct axis, the more the interval is time-like. In the general case, the distance d is an interval in space-time. d only represents time for a particle at rest where d lies along the ct axis. That is, for a general interval O to P, the time coordinate ct does not represent the time experienced by the particle unless the particle is at rest. How then can we extract the time information from the space-time interval d?

The time experienced by the particle is given, not by the distance d (converted from light metres into seconds), but by the quantity s (in light metres) defined by:

$$s^2 = x^2 - c^2t^2 \tag{8.114}$$

different from the usual formula for three-dimensional space by the – sign.

In ordinary experience, the distance x is very much less than the product ct (i.e. most particles move at velocities less than c), and so for slowly moving bodies, the distance s is only a little less than ct. For a photon, $x = ct$ and so $s = 0$. That is, when a photon travels from O to P, from our perspective in the stationary coordinate frame, the photon appears to have no experience of time at all. Its time has stopped. The quantity s is called the Minkowskian distance, or the interval.

An interval in space-time is expressed:

$$s^2 = x^2 + y^2 + z^2 - c^2t^2 \tag{8.115}$$

which can be also written:

$$s^2 = ct^2 - x^2 - y^2 - z^2 \tag{8.116}$$

One important significance of this measure of distance, the interval between two events, is that it will be the same value for every observer. Unlike ordinary (Euclidean) spatial distance – which depends upon the velocity of the observer – the interval is invariant. The interval is the "space-time distance" between two space-time points.

The interval for two events involving a photon is zero. That is, when the x coordinate is $x=ct$ (the photon is travelling at c) $s=0$. This important result means that no matter which observer is observing the photon, it must appear to travel at velocity c because it is only when the velocity of the photon, for any observer, is equal to c does the interval equal zero. On a space-time diagram, we might show the path of a photon as a line at 45° as shown in Figure 8.17(b).

The interval is not shown on the space-time diagram as a distance. The linear distance d between O and P is not the interval. The linear distance O to P represents the space-time interval but not its magnitude. Space-time is not linear distance, or linear time. The term "distance" means distance only along the distance axes, while the term "time" only means time along the ct axis. Although the distance in space-time is a calculated quantity, and it is not shown on a space-time diagram, the path taken by a particle may be shown on a space-time diagram to indicate the object going from one place to another.

When a particle moves in a stationary reference frame (say in two dimensions), with velocity v, then $x=vt$, and so the distance travelled in space-time from $t=0$ to some time t is:

$$s^2 = v^2t^2 - c^2t^2 \tag{8.117}$$

But, for an observer moving along with the particle at velocity v (with respect to the stationary reference frame), the same space-time interval must be observed (since s is invariant). But, in this case, the particle is at rest, and the world line of the particle is vertical. The coordinate of the particle at some time t is the distance ct'.

The space-time distance is the same in each reference frame, and so:

$$\begin{aligned} s^2 &= -c^2t'^2 \\ &= v^2t^2 - c^2t^2 \\ c^2t'^2 &= t^2\left(c^2 - v^2\right) \end{aligned} \tag{8.118}$$

and so, the time that it takes for the particle to move from place to place is:

$$\begin{aligned} t'^2 &= t^2\left(1 - \frac{v^2}{c^2}\right) \\ t' &= \frac{t}{\sqrt{1 - v^2/c^2}} \end{aligned} \tag{8.119}$$

But, as we saw for the relativistic transformation of velocity, when we are moving along with the particle, and "looking back" at the stationary reference frame, the primed and unprimed quantities change around, and from the point of view of the stationary reference frame:

$$t = \frac{t'}{\sqrt{1 - v^2/c^2}} \tag{8.120}$$

This applies for the case where a time interval t' as measured in the S′ reference frame is expressed in terms of a time interval t and displacement x in the S reference frame.

When a particle moves, its world-line (in the stationary coordinate system) is tilted over towards the length axis.

The concept of a light-cone can be useful in picturing the relationship between the past, present and the future.

As shown in Figure 8.18, the past events involving the particle are contained within the past light-cone, while the future events must exist within the future light-cone.

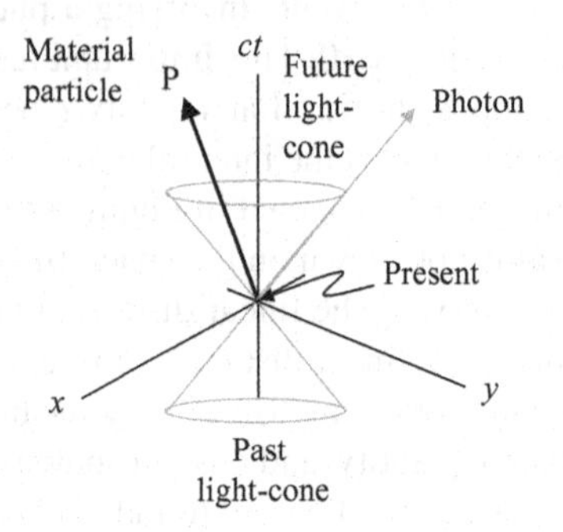

FIGURE 8.18 Past and future light-cones in space time and world-lines of particle and photon.

Imagine a flash of light is created at some particular point in space (event #1) and is then seen by an observer located at another particular point in space (event #2). The light-cones attached to the two events indicate what and when this happens. This is shown in Figure 8.19.

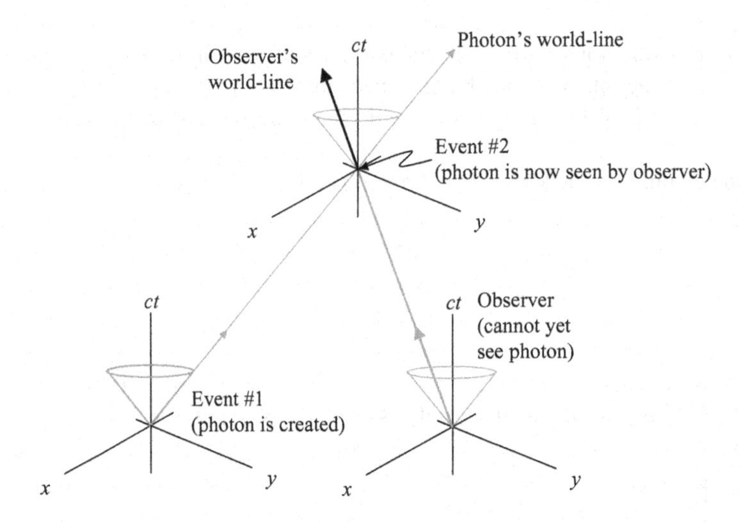

FIGURE 8.19 How and when two events are observed by an observer.

8.4 CONCLUSION

It is not often in science that our beliefs undergo a profound shift. The Copernican revolution removed us from the centre of the universe to the backwaters of an out-of-the-way galaxy. Newton's laws overturned Aristotle's spiritual explanations of physics and provided a clockwork mechanism for the universe with immutable matter. Einstein set all this adrift in space–time and energy, but at least things were still deterministic in different frames of reference. Quantum mechanics threw reality up in the air and made everything probabilistic.

By the 1920s, things had calmed down to an uneasy equilibrium with many, including Einstein, refusing to believe in the probabilities of quantum mechanics. As we shall see, by the end of the 1920s, we were all at sea again, literally in a sea of negative energy, when Paul Dirac proposed the existence of antimatter. This then led to a completely new formulation of quantum physics where particles were given second billing to quantum fields – the subject of Part 2 of this book.

Advanced Mathematics 9

9.1 VECTOR CALCULUS

9.1.1 Vector Differential Operator

Consider a scalar function $\phi = \phi(x,y,z)$. It is often needed to determine the change in ϕ with respect to the x, y or z directions. That is, the spatial rate of change (as distinct from the time rate of change we might be more used to with derivatives). These rates of change are found by taking the partial derivatives with respect to x, y and z. The resulting derivatives form the basis of a vector where the partial derivatives are the magnitudes of the components. To this end, the three-dimensional differential operator is defined as:

$$\nabla = \frac{\partial}{\partial x}\mathbf{i} + \frac{\partial}{\partial y}\mathbf{j} + \frac{\partial}{\partial z}\mathbf{k} \tag{9.1}$$

$\nabla\varphi$ is called the gradient of ϕ and is written:

$$\nabla\varphi = \frac{\partial\varphi}{\partial x}\mathbf{i} + \frac{\partial\varphi}{\partial y}\mathbf{j} + \frac{\partial\varphi}{\partial z}\mathbf{k} \tag{9.2}$$

The dot product of the differential operator and a vector is called the divergence. For example, if **A** is a vector:

$$\mathbf{A}(x,y,z) = f(x,y,z)\mathbf{i} + g(x,y,z)\mathbf{j} + h(x,y,z)\mathbf{k} \tag{9.3}$$

then:

$$\begin{aligned}\nabla\cdot\mathbf{A} &= \left(\frac{\partial}{\partial x}\mathbf{i} + \frac{\partial}{\partial y}j + \frac{\partial}{\partial z}\mathbf{k}\right)\cdot\left(f\mathbf{i} + g\mathbf{j} + h\mathbf{k}\right)\\ &= \frac{\partial f}{\partial x} + \frac{\partial g}{\partial y} + \frac{\partial h}{\partial z}\end{aligned} \tag{9.4}$$

The divergence is a scalar quantity.

The curl is given by the cross product of the differential operator and a vector:

$$\nabla\times\mathbf{A} = \begin{vmatrix}\mathbf{i} & \mathbf{j} & \mathbf{k}\\ \frac{\partial}{\partial x} & \frac{\partial}{\partial y} & \frac{\partial}{\partial z}\\ f & g & h\end{vmatrix} \tag{9.5}$$

$$= \mathbf{i}\begin{vmatrix} \frac{\partial}{\partial y} & \frac{\partial}{\partial z} \\ g & h \end{vmatrix} - \mathbf{j}\begin{vmatrix} \frac{\partial}{\partial x} & \frac{\partial}{\partial z} \\ f & h \end{vmatrix} + \mathbf{k}\begin{vmatrix} \frac{\partial}{\partial x} & \frac{\partial}{\partial y} \\ f & g \end{vmatrix}$$

$$= \left(\frac{\partial h}{\partial y} - \frac{\partial g}{\partial z}\right)\mathbf{i} + \left(\frac{\partial f}{\partial z} - \frac{\partial h}{\partial x}\right)\mathbf{j} + \left(\frac{\partial g}{\partial x} - \frac{\partial f}{\partial y}\right)\mathbf{k}$$

where:

$$\mathbf{A}(x,y,z) = f(x,y,z)\mathbf{i} + g(x,y,z)\mathbf{j} + h(x,y,z)\mathbf{k} \tag{9.6}$$

The curl is a vector quantity.

In summary, the properties of the differential operator are that when this operator operates directly on a scalar function, we get a vector, the gradient:

$$\nabla\varphi = \frac{\partial\varphi}{\partial x}\mathbf{i} + \frac{\partial\varphi}{\partial y}\mathbf{j} + \frac{\partial\varphi}{\partial z}\mathbf{k} \tag{9.7}$$

The dot product of the differential operator with a vector function gives a scalar. The divergence:

$$\nabla\cdot\mathbf{A} = \frac{\partial f}{\partial x} + \frac{\partial g}{\partial y} + \frac{\partial h}{\partial z} \tag{9.8}$$

The cross product of the differential operator with a vector gives a vector, the curl:

$$\nabla\times\mathbf{A} = \left(\frac{\partial h}{\partial y} - \frac{\partial g}{\partial z}\right)\mathbf{i} + \left(\frac{\partial f}{\partial z} - \frac{\partial h}{\partial x}\right)\mathbf{j} + \left(\frac{\partial g}{\partial x} - \frac{\partial f}{\partial y}\right)\mathbf{k} \tag{9.9}$$

The product of the differential operator with itself is a dot product and is, in general, a scalar. This is called the Laplacian:

$$\begin{aligned} \nabla\cdot\nabla\varphi &= \frac{\partial^2\varphi}{\partial x^2} + \frac{\partial^2\varphi}{\partial y^2} + \frac{\partial^2\varphi}{\partial^2 z} \\ &= \nabla^2\varphi \end{aligned} \tag{9.10}$$

When the Laplacian operator acts on a vector, the result is a new vector. In this case, the Laplacian is called the vector Laplacian:

$$\nabla\cdot\nabla\mathbf{A} = \nabla^2\mathbf{A} = \left[\frac{\partial^2 A}{\partial x^2}; \frac{\partial^2 A}{\partial y^2}; \frac{\partial^2 A}{\partial^2 z}\right] \tag{9.11}$$

Note that $\nabla^2\varphi$ is not the same as $(\nabla\varphi)^2$. We cannot write $\nabla\nabla\varphi$. The differential operator only acts directly on a scalar to produce a vector. The differential operator does not act directly on a vector. The correct expression is $\nabla\cdot\nabla\varphi$. This is the dot product between two vectors which results in a scalar.

9.1.2 Line Integral

Consider a surface $z = F(x,y)$. Let a line increment on the curve C in the xy plane be given by Δs as shown in Figure 9.1(a).

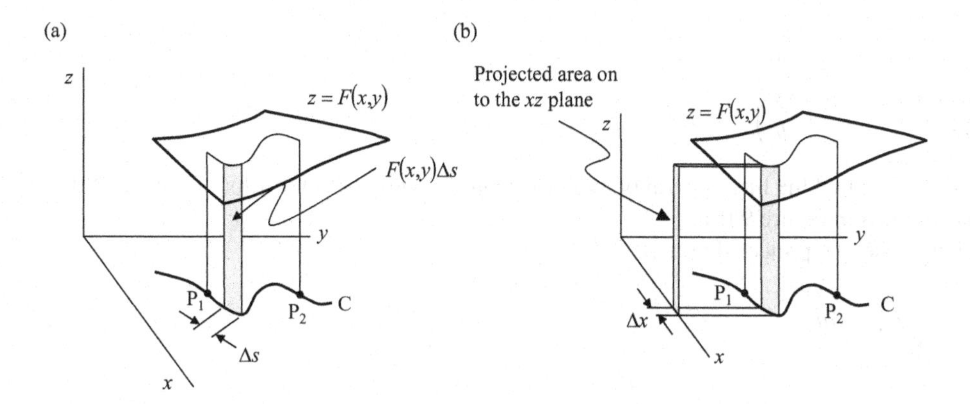

FIGURE 9.1 (a) Surface represented by the function $z=F(x,y)$ and a curve C over which an integral is to be summed. (b) Projected area on to the xz plane.

The area of the band formed from the line increment to the surface is given by the product:

$$F(x,y)\Delta s \tag{9.12}$$

The total area under the surface along the curve from P_1 to P_2 is thus the limit of the sum as Δs goes to zero.

$$\lim_{\substack{n\to\infty \\ \Delta s\to 0}} \sum F(x,y)\Delta s \tag{9.13}$$

This limit is called the line integral of $F(x,y)$ along the curve C from P_1 to P_2 and is written:

$$\int_{P_1}^{P_2} F(x,y)\,ds = \int_{C} F(x,y)\,ds \tag{9.14}$$

If the curve C is given as a function $y=f(x)$, then the line integral is:

$$\int_{P_1}^{P_2} F(x,y)\,ds = \int_{C} F(x,f(x))\,ds \tag{9.15}$$

To evaluate line integrals, it is often necessary to connect the function $F(x,y)$ and the curve C through a parameter t, and so:

$$\int_{C} F(x,y)\,ds = \int_{t_1}^{t_2} F(x(t),y(t))\frac{ds}{dt}\,dt \tag{9.16}$$

where:

$$\frac{ds}{dt} = \sqrt{\left(\frac{dx}{dt}\right)^2 + \left(\frac{dy}{dt}\right)^2} \tag{9.17}$$

Extended to three dimensions, we have:

$$\int_{C} F(x,y,z)\,ds = \int_{t_1}^{t_2} F(x(t),y(t),z(t))\frac{ds}{dt}\,dt \tag{9.18}$$

with:

$$\frac{ds}{dt} = \sqrt{\left(\frac{dx}{dt}\right)^2 + \left(\frac{dy}{dt}\right)^2 + \left(\frac{dz}{dt}\right)^2} \tag{9.19}$$

We may wish to find the line integral of $z = F(x,y)$ along C with respect to x (or y) instead of the arc length s. This is shown in Figure 9.1(b).

In this case, the projected area is:

$$\begin{aligned} A_x &= \int_C F(x,y)\,dx \\ &= \int_{t_1}^{t_2} F(x(t),y(t))\frac{dx}{dt}dt \end{aligned} \tag{9.20}$$

In some circumstances, a line integral is computed from the addition of the projected line integrals of two different functions:

$$\int_C F(x,y)\,ds = \int_C f(x,y)\,dx + \int_C g(x,y)\,dy \tag{9.21}$$

For example, the functions f, g may be the components of a vector function $\mathbf{F}$:

$$\mathbf{F} = f(x,y)\mathbf{i} + g(x,y)\mathbf{j} \tag{9.22}$$

The position vector $\mathbf{r}$ as we travel a distance s along the curve C is:

$$\mathbf{r} = x\mathbf{i} + y\mathbf{j} \tag{9.23}$$

An increment of $\mathbf{r}$ is thus:

$$d\mathbf{r} = dx\mathbf{i} + dy\mathbf{j} \tag{9.24}$$

The dot product of $\mathbf{F}$ and $d\mathbf{r}$ is thus:

$$\mathbf{F}\cdot d\mathbf{r} = f(x,y)\,dx + g(x,y)\,dy \tag{9.25}$$

And summing over the distance s along C, we obtain:

$$\int_C \mathbf{F}\cdot d\mathbf{r} = \int_C f(x,y)\,dx + \int_C g(x,y)\,dy \tag{9.26}$$

If $\mathbf{F}$ represents a force vector, then the line integral gives the work done in traversing a distance s along the path C. In terms of the arc length s:

$$\int_C \mathbf{F}\cdot d\mathbf{r} = \int_C \mathbf{F}\cdot\frac{d\mathbf{r}}{ds}ds \tag{9.27}$$

where:

$$\frac{d\mathbf{r}}{ds} = \frac{dx}{ds}\mathbf{i} + \frac{dy}{ds}\mathbf{j} \tag{9.28}$$

If the integral is taken around a closed curve, then we write:

$$\oint_C \mathbf{F} \cdot d\mathbf{r} \tag{9.29}$$

This integral is sometimes referred to as the circulation of **F** around C.

In general, the value of the line integral depends upon the path of integration. If, however, the vector field **F** can be represented by:

$$\mathbf{F} = \nabla\varphi \tag{9.30}$$

then the line integral is independent of the path. In this case, ϕ is called the scalar potential. The field **F** is said to be a conservative field, and $d\phi$ is called an exact differential.

Let:

$$\mathbf{F} = \nabla\varphi \tag{9.31}$$

Then, in three dimensions:

$$\begin{aligned}
\int \mathbf{F} \cdot d\mathbf{r} &= \int_{P_1}^{P_2} \nabla\varphi \cdot d\mathbf{r} \\
&= \int_{P_1}^{P_2} \left(\frac{\partial\varphi}{\partial x}\mathbf{i} + \frac{\partial\varphi}{\partial y}\mathbf{j} + \frac{\partial\varphi}{\partial z}\mathbf{k} \right) \cdot (dx\mathbf{i} + dy\mathbf{j} + dz\mathbf{k}) \\
&= \int_{P_1}^{P_2} \frac{\partial\varphi}{\partial x}dx + \frac{\partial\varphi}{\partial y}dy + \frac{\partial\varphi}{\partial z}dz \\
&= \int_{P_1}^{P_2} d\varphi \\
&= \varphi(x_2, y_2, z_2) - \varphi(x_1, y_1, z_1)
\end{aligned} \tag{9.32}$$

The integral depends only on the end points P_1 and P_2 and not on the path joining them (since the form or equation of the curve C is not given in the above).

If the path of integration C is a closed curve, and if **F** is conservative, then:

$$\oint_C \mathbf{F} \cdot d\mathbf{r} = 0 \tag{9.33}$$

where the circle in the integral sign denotes integration around a closed curve.

Therefore, if $\mathbf{F} = \nabla\varphi$ then **F** is called a conservative field and is the gradient of a scalar potential. We will later show that the curl of **F** is equal to zero.

$$\nabla \times \mathbf{F} = 0 \tag{9.34}$$

Now, consider the curl of a field **F**.

$$\nabla\times\mathbf{F}=\begin{vmatrix}\mathbf{i} & \mathbf{j} & \mathbf{k}\\ \frac{\partial}{\partial x} & \frac{\partial}{\partial y} & \frac{\partial}{\partial z}\\ f & g & h\end{vmatrix}=\mathbf{i}\begin{vmatrix}\frac{\partial}{\partial y} & \frac{\partial}{\partial z}\\ g & h\end{vmatrix}-\mathbf{j}\begin{vmatrix}\frac{\partial}{\partial x} & \frac{\partial}{\partial z}\\ f & h\end{vmatrix}+\mathbf{k}\begin{vmatrix}\frac{\partial}{\partial x} & \frac{\partial}{\partial y}\\ f & g\end{vmatrix}$$

$$=\left(\frac{\partial h}{\partial y}-\frac{\partial g}{\partial z}\right)\mathbf{i}+\left(\frac{\partial f}{\partial z}-\frac{\partial h}{\partial x}\right)\mathbf{j}+\left(\frac{\partial h}{\partial x}-\frac{\partial f}{\partial y}\right)\mathbf{k} \tag{9.35}$$

The divergence of this result is:

$$\nabla\cdot(\nabla\times\mathbf{F})=\left(\frac{\partial}{\partial x}\mathbf{i}+\frac{\partial}{\partial y}\mathbf{j}+\frac{\partial}{\partial z}\mathbf{k}\right)\cdot\left(\left(\frac{\partial h}{\partial y}-\frac{\partial g}{\partial z}\right)\mathbf{i}+\left(\frac{\partial f}{\partial z}-\frac{\partial h}{\partial x}\right)\mathbf{j}+\left(\frac{\partial g}{\partial x}-\frac{\partial f}{\partial y}\right)\right)\mathbf{k}$$

$$=\frac{\partial}{\partial x}\left(\frac{\partial h}{\partial y}-\frac{\partial g}{\partial z}\right)+\frac{\partial}{\partial y}\left(\frac{\partial f}{\partial z}-\frac{\partial h}{\partial x}\right)+\frac{\partial}{\partial z}\left(\frac{\partial g}{\partial x}-\frac{\partial f}{\partial y}\right)$$

$$=\frac{\partial}{\partial x}\frac{\partial h}{\partial y}-\frac{\partial}{\partial x}\frac{\partial g}{\partial z}+\frac{\partial}{\partial y}\frac{\partial f}{\partial z}-\frac{\partial}{\partial y}\frac{\partial h}{\partial x}+\frac{\partial}{\partial z}\frac{\partial g}{\partial x}-\frac{\partial}{\partial z}\frac{\partial f}{\partial y} \tag{9.36}$$

$$=\frac{\partial}{\partial x}\frac{\partial h}{\partial y}-\frac{\partial}{\partial y}\frac{\partial h}{\partial x}+\frac{\partial}{\partial z}\frac{\partial g}{\partial x}-\frac{\partial}{\partial x}\frac{\partial g}{\partial z}+\frac{\partial}{\partial y}\frac{\partial f}{\partial z}-\frac{\partial}{\partial z}\frac{\partial f}{\partial y}$$

$$=0$$

since by the ordinary rules of calculus:

$$\frac{d}{dx}\frac{ds}{dy}=\frac{d}{dy}\frac{ds}{ds} \tag{9.37}$$

This important result:

$$\nabla\cdot(\nabla\times\mathbf{F})=0 \tag{9.38}$$

says that the divergence of the curl of a conservative vector field is zero. It will be used to study the magnetic field **B**.

9.1.3 Multiple Integrals

Consider a surface $z=f(x,y)$ as shown in Figure 9.2. Let an incremental area of a region R in the xy plane be given by ΔA.

The volume of the column from the surface z over the area ΔA is given by the product:

$$f(x,y)\Delta A \tag{9.39}$$

The total volume under the surface is thus the limit of the sum as ΔA goes to zero.

$$\lim_{\substack{n\to\infty\\ \Delta A\to 0}}\sum f(x,y)\Delta A \tag{9.40}$$

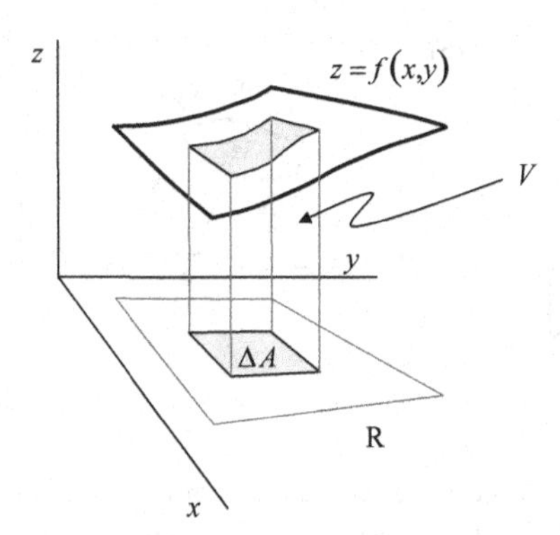

FIGURE 9.2 Surface represented by $z=f(x,y)$ and a region R in the xy plane.

This limit is called the double integral of $f(x,y)$ over the region R.

$$\iint_R f(x,y)\,dA \tag{9.41}$$

Expressed in terms of increments in the x and y axes directions, we obtain:

$$\iint_R f(x,y)\,dA = \int_{x_1}^{x_2}\int_{y_1}^{y_2} f(x,y)\,dy\,dx \tag{9.42}$$

This integral is evaluated by holding x fixed and performing the inside sum with respect to dy, and then integrating that expression with respect to x. In double integrals, the integration is performed over a region R.

For functions of three variables, the region of integration is a volume V so that the triple integral is given by:

$$\iiint_V f(x,y,z)\,dV = \int_{x_1}^{x_2}\int_{y_1}^{y_2}\int_{z_1}^{z_2} f(x,y,z)\,dz\,dy\,dx \tag{9.43}$$

For example, if $\rho(x,y,z)$ is a function giving the density of a solid at any point, then the mass of a volume element is given by:

$$dM = \rho(x,y,z)\,dV \tag{9.44}$$

The total mass is:

$$M = \iiint_V dM = \iiint_V \rho(x,y,z)\,dz\,dy\,dx \tag{9.45}$$

9.1.4 Surface and Volume Integrals

A surface integral is a generalisation of the double integral. Double integrals are integrals over a flat surface (a region R of integration in the xy plane). Surface integrals are double integrals over a curved surface S.

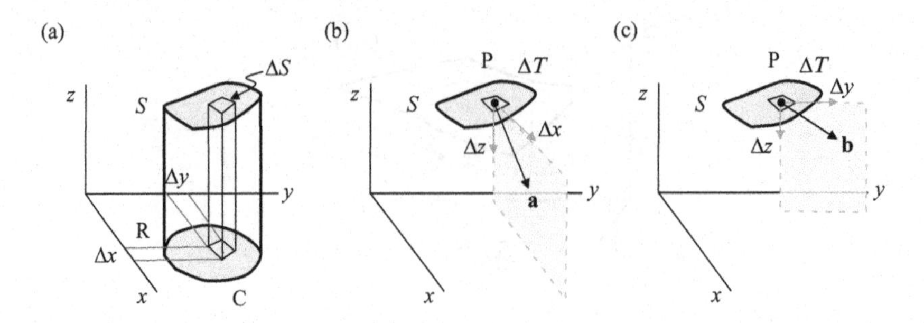

FIGURE 9.3 (a) Surface S over which the integral is to be summed. (b) Tangent plane at point P with vector **a** in the *xz* plane. (c) Tangent plane at point P with vector **b** in the *yz* plane.

Let the equation of the surface S in Figure 9.3(a) be:

$$z = f(x,y) \tag{9.46}$$

Let $\phi(x,y,z)$ be the function to be integrated (not shown in the figure) over the surface. We divide S into increments ΔS and let ΔT be a tangent plane to ΔS at P(x, y, z) as shown in Figure 9.3(b). The curved surface S has a projection R onto the xy plane.

Consider a point P and let **a** be a vector tangent to the surface S parallel to the xz plane as shown in Figure 9.3(b). Let **b** be a vector tangent to the surface S parallel to the yz plane as shown in Figure 9.3(c).

Thus:

$$\begin{aligned} \mathbf{a} &= \Delta x\mathbf{i} + \Delta z\mathbf{k} \\ \frac{\Delta z}{\Delta x} &= \frac{\partial z}{\partial x} \\ \Delta z &= \frac{\partial z}{\partial x}\Delta x \\ \mathbf{a} &= \Delta x\mathbf{i} + \frac{\partial z}{\partial x}\Delta x\mathbf{k} \end{aligned} \tag{9.47}$$

and:

$$\mathbf{b} = \Delta y\mathbf{j} + \frac{\partial z}{\partial y}\Delta y\mathbf{k} \tag{9.48}$$

The area of $\Delta T = |\mathbf{a} \times \mathbf{b}|$

$$\mathbf{a} \times \mathbf{b} = \begin{vmatrix} \mathbf{i} & \mathbf{j} & \mathbf{k} \\ \Delta x & 0 & \frac{\partial z}{\partial x}\Delta x \\ 0 & \Delta y & \frac{\partial z}{\partial y}\Delta y \end{vmatrix} \tag{9.49}$$

which expands to:

$$\mathbf{a}\times\mathbf{b}=\left(-\frac{\partial z}{\partial x}\Delta x\Delta y\right)\mathbf{i}+\left(\frac{\partial z}{\partial y}\Delta x\Delta y\right)\mathbf{j}+\left(\Delta x\Delta y\right)\mathbf{k}$$

$$=\sqrt{\left(\frac{\partial z}{\partial x}\right)^2\Delta x^2\Delta y^2+\left(\frac{\partial z}{\partial y}\right)^2\Delta x^2\Delta y^2+\Delta x^2\Delta y^2} \tag{9.50}$$

$$=\ \Delta x\Delta y\sqrt{\left(\frac{\partial z}{\partial x}\right)^2+\left(\frac{\partial z}{\partial y}\right)^2+1}$$

$$=\Delta \mathrm{T}$$

$\Delta T = \Delta S$ if ΔT is small. And so:

$$dS=\left(\sqrt{\left(\frac{\partial z}{\partial x}\right)^2+\left(\frac{\partial z}{\partial y}\right)^2+1}\right)dxdy \tag{9.51}$$

This gives us dS expressed in terms of dx and dy.

The surface integral of $f(x,y,z)$ over S can thus be written:

$$\oiint_S \varphi(x,y,z)dS=\iint_R \varphi(x,y,z)\sqrt{\left(\frac{\partial z}{\partial x}\right)^2+\left(\frac{\partial z}{\partial y}\right)^2+1}\ dxdy \tag{9.52}$$

where R is the projection of S on the xy plane.

Now, the vector result of $\mathbf{a}\times\mathbf{b}$ is a vector $\perp$ to ΔT, the magnitude of which is the area of ΔT. Let $\mathbf{n}$ be an outward normal unit vector in this same direction as shown in Figure 9.4(a).

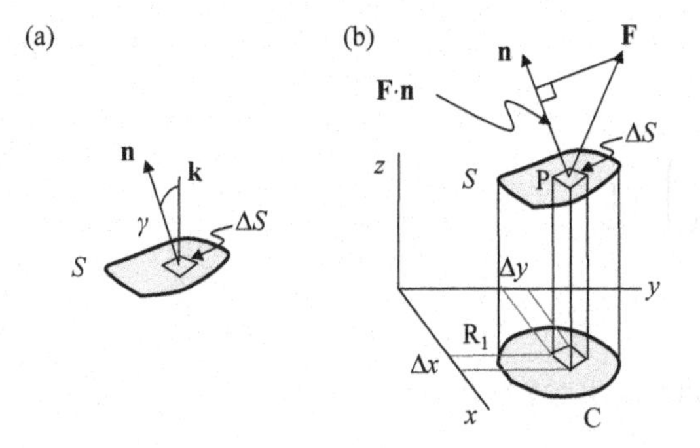

FIGURE 9.4 (a) Outward normal n to the surface at ΔS. (b) Outward normal unit vector at $P(x,y,z)$ and integrating the normal component of **F** over the surface S at $P(x,y,z)$.

$$\mathbf{n}=\frac{\mathbf{a}\times\mathbf{b}}{|\mathbf{a}\times\mathbf{b}|}=\frac{-\dfrac{\partial z}{\partial x}\Delta x\Delta y\mathbf{i}-\dfrac{\partial z}{\partial y}\Delta x\Delta y\mathbf{j}+\Delta x\Delta y\mathbf{k}}{\Delta x\Delta y\sqrt{\left(\dfrac{\partial z}{\partial x}\right)^2+\left(\dfrac{\partial z}{\partial y}\right)^2+1}} \tag{9.53}$$

But:

$$dS = \sqrt{\left(\frac{\partial z}{\partial x}\right)^2 + \left(\frac{\partial z}{\partial y}\right)^2 + 1}\, dxdy \tag{9.54}$$

and also:

$$\mathbf{n}\cdot\mathbf{k} = \frac{1}{\sqrt{\left(\frac{\partial z}{\partial x}\right)^2 + \left(\frac{\partial z}{\partial y}\right)^2 + 1}} \tag{9.55}$$

Thus:

$$\begin{aligned} dS &= \frac{1}{\mathbf{n}\cdot\mathbf{k}} dxdy \\ \iint_S \varphi(x,y,z)\, dS &= \iint_R \varphi(x,y,z) \frac{1}{\mathbf{n}\cdot\mathbf{k}} dxdy \end{aligned} \tag{9.56}$$

The normal unit vector **n** may also be represented by:

$$\mathbf{n} = l\mathbf{i} + m\mathbf{j} + n\mathbf{k} \tag{9.57}$$

and so:

$$\begin{aligned} |\mathbf{n}\cdot\mathbf{k}| &= |\mathbf{n}||\mathbf{k}|\cos\gamma = \cos\gamma \\ \frac{1}{|\mathbf{n}\cdot\mathbf{k}|} &= \sec\gamma \end{aligned} \tag{9.58}$$

since $|\mathbf{n}|, |\mathbf{k}| = 1$ and where γ is the angle between the outward normal vector **n** and **k** (which is in the direction of the z axis).

So:

$$\cos\gamma = \frac{1}{\sqrt{\left(\frac{\partial z}{\partial x}\right)^2 + \left(\frac{\partial z}{\partial y}\right)^2 + 1}} \tag{9.59}$$

Thus:

$$\iint_S \varphi(x,y,z)\, dS = \iint_R \varphi(x,y,z) \sec\gamma\, dxdy \tag{9.60}$$

In the case of vector functions that require integrating over a surface:

$$\mathbf{F} = f(x,y,z)\mathbf{i} + g(x,y,z)\mathbf{j} + h(x,y,z)\mathbf{k} \tag{9.61}$$

If **n** is an outward normal unit vector at $P(x,y,z)$ as shown in Figure 9.4(b), then for the special case of integrating the normal component of **F** over the surface S, we have:

$$\mathbf{F}\cdot\mathbf{n} \tag{9.62}$$

as the normal component of **F** at $P(x,y,z)$.

The surface integral of the normal component of **F** over *S* is:

$$\iint_S \mathbf{F}\cdot\mathbf{n}\,dS = \iint_S (fl + gm + hn)\,dS$$

$$= \iint_S (f\cos\alpha + g\cos\beta + h\cos\gamma)\,dS \quad (9.63)$$

$$= \iint_{R_3} f dydx + \iint_{R_2} g dxdz + \iint_{R_1} h dxdy$$

Noting that $\cos\gamma\sec\gamma = 1$, and where R_3 is the projection on the *yz* plane, R_2 is the projection on the *xz* plane and R_1 is the projection on the *xy* plane.

This integral is sometimes referred to as the flux of **F** over *S*.

The physical interpretation of surface integrals depends upon the nature of the function being integrated. For example:

Area of the surface:

$$A = \iint_S dS \quad (9.64)$$

Mass of the surface:

$$M = \iint_S \rho(x,y,z)\,dS \quad (9.65)$$

Moment of inertia:

$$I = \iint_S (x^2 + y^2)\rho\,dS \quad (9.66)$$

Surface integrals are generally found by solving double integrals of parametric equations of the surface.

9.1.5 Stokes' Theorem

Let *S* be an open, two-sided surface bounded by a closed curve C with a normal unit vector **n**. Stokes' theorem says that:

$$\oint_C \mathbf{F}\cdot d\mathbf{r} = \iint_S (\nabla\times\mathbf{F})\cdot\mathbf{n}\,dS \quad (9.67)$$

where C is traversed in an anti-clockwise direction. We won't prove Stokes' theorem here, but take it on trust.

Now, if the line integral of **F** around C is zero, then **F** must be a conservative field. That is, if:

$$\oint_C \mathbf{F}\cdot d\mathbf{r} = 0 \quad (9.68)$$

then:

$$\nabla\times\mathbf{F} = 0 \quad (9.69)$$

A conservative vector field has no curl.

9.2 GAUSS' LAW

Gauss' law is a classical field theory and is concerned with the calculation of the magnitude of the electric field around a point charge as shown in Figure 9.5(a).

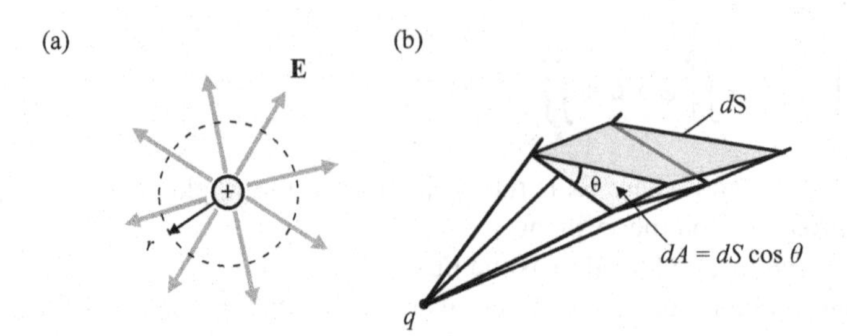

FIGURE 9.5 (a) Electric field lines around a point charge. (b) Area dA over which the flux density is computed.

We consider a spherical surface at radius r from the charge. The surface area of the sphere is of course:

$$A = 4\pi r^2 \tag{9.70}$$

Faraday envisaged that the electric field could be represented by lines of force. The strength of the electric field is defined as being proportional to the number of lines of force per unit area. The number of lines is called the electric flux. The number of lines per unit area is called the electric flux density, or more simply, the electric field **E**. **E** is a vector field.*

$$\begin{aligned} E &\propto \frac{N}{A} \\ EA &\propto N \end{aligned} \tag{9.71}$$

We know from Coulomb's law that the magnitude of the electric field E at some distance r from a single point charge can be calculated from the magnitude of the charge q and, for a charge in free space, is expressed:

$$E = \frac{1}{4\pi\varepsilon_o}\frac{q}{r^2} \tag{9.72}$$

That is:

$$E = \frac{1}{A}\frac{q}{\varepsilon_o} \tag{9.73}$$

where A is the surface area of a sphere of radius r surrounding the charge. The product of EA is therefore proportional to the electric flux ϕ, and so:

$$EA = \frac{q}{\varepsilon_o} = \varphi \tag{9.74}$$

* In this part of the book we have to be careful about notation for scalars and vectors. A vector is written in bold face where the vector characteristic of the quantity is important. Otherwise, the magnitude of a vector, and scalars, are written in italics.

The constant of proportionality is $1/\varepsilon_0$ where ε_0 is the permittivity of free space – often taken to be the same as that of air.

While the above is straight forward for a spherical surface, we wish to make it more general. Consider now the portion dS of non-spherical surface of area S as shown in Figure 9.5(b).

As shown in Fig. 9.5(b), the elemental area on an underlying spherical surface surrounding the charge is given by:

$$dA = dS\cos\theta \tag{9.75}$$

The total flux (i.e. number of flux lines) over the whole surface S is thus the same as that over an underlying spherical surface A as shown in Figure 9.5(b). Thus:

$$\begin{aligned}\varphi &= \oiint_S \frac{1}{4\pi\varepsilon_o}\frac{q}{r^2}\cos\theta dS \\ &= \oiint_S E\cos\theta dS\end{aligned} \tag{9.76}$$

Since the dot product is defined as:

$$\mathbf{a}\cdot\mathbf{b} = ab\cos\theta \tag{9.77}$$

Then the flux is expressed as a dot product of E over dS:

$$\begin{aligned}\varphi &= \oiint_S \mathbf{E}\cdot d\mathbf{S} = \frac{q}{\varepsilon_o} \\ q &= \varepsilon_o \oiint_S \mathbf{E}\cdot d\mathbf{S}\end{aligned} \tag{9.78}$$

q is the net charge enclosed by the surface. This is the integral form of Gauss' law and one of Maxwell's equations.

Let an elemental volume ΔV have a charge Δq distributed uniformly within it. The total charge within the volume is given by Gauss' law and can be expressed in terms of the charge density ρ within the volume:

$$\Delta q = \varepsilon_o \iint_S \mathbf{E}\cdot d\mathbf{S} = \rho\Delta V \tag{9.79}$$

The charge density has units of Coulombs per unit volume.

The charge density at a single point in space is expressed as the limit as $\Delta V \to 0$:

$$\begin{aligned}\rho &= \lim_{\Delta V\to 0}\frac{\Delta q}{\Delta V} \\ &= \lim_{\Delta V\to 0}\frac{\varepsilon_o \iint_S \mathbf{E}\cdot d\mathbf{S}}{\Delta V}\end{aligned} \tag{9.80}$$

This limit is called the divergence of the vector field **E**. If a field has a divergence at a particular point, then this signifies the existence of a charge at that point.

Written in the form of a differential operator, in three dimensions:*

$$\rho = \nabla \cdot \varepsilon_o \mathbf{E} \tag{9.81}$$

which is called the differential form of Gauss' law.

The total charge q within a total volume V can also be expressed in terms of the summation of the charge densities over all the volume elements within it.

$$q = \iiint_V \rho dV \tag{9.82}$$

And so:

$$q = \iint_S \varepsilon_o \mathbf{E} \cdot d\mathbf{S} = \iiint_V \nabla \cdot \varepsilon_o \mathbf{E}\, dV \tag{9.83}$$

More generally, we can say: if $\mathbf{F} = M\mathbf{i} + N\mathbf{j} + O\mathbf{k}$, represents a vector quantity, then:

$$\oiint_S \mathbf{F} \cdot \mathbf{n} dS = \iiint_V \nabla \cdot \mathbf{F}\, dV \tag{9.84}$$

where $\mathbf{n}$ is a positive unit outer normal vector, and S is a closed surface.

Or:

$$\iint_S \left(M\cos\alpha + N\cos\beta + O\cos\gamma\right) dS = \iiint_V \left(\frac{\partial M}{\partial x} + \frac{\partial N}{\partial y} + \frac{\partial O}{\partial z}\right) dV \tag{9.85}$$

The cos terms are the direction cosines of the vector $\mathbf{F}$. The flux of a vector $\mathbf{F}$ over S is equal to the triple integral of the divergence of $\mathbf{F}$ over V. This is referred to as the divergence theorem or Gauss' theorem.

9.3 CONTINUITY EQUATION

The rate of flow of electric charge per unit time is called the electric current i. In general:

$$i = \frac{dq}{dt} \tag{9.86}$$

Where i is measured in Coulombs/sec which is usually expressed in the unit of Amps. The current density $\mathbf{J}$ is the amount of charge crossing a unit cross-sectional area per unit time and has the units A/m^2. $\mathbf{J}$ is a vector quantity since its value depends on the direction of current with respect to the orientation of the area under consideration.

$$\mathbf{J} = \frac{i}{\mathbf{A}} \tag{9.87}$$

$$i = \mathbf{J} \cdot \mathbf{A}$$

* When the differential operator acts on a vector, the dot between the operator and the vector function is usually omitted if it is clear that a vector function is involved.

Now, take a volume V which contains a certain amount of charge. If there is an additional amount of charge entering the volume across its surface from the outside, then this constitutes an ingoing current across the surface. If there is an outflow of charge from the interior to the outside, then there is an outgoing current. The time rate of increase or decrease of charge (Coulombs/sec) within the volume must equal the net flow of current (Coulombs/sec) across the surface of the volume.

The net current i is found from the summation of the current density over all the elemental surface areas dA:

$$i = \oiint_S \mathbf{J} \cdot d\mathbf{A} \tag{9.88}$$

Within the volume, we might have a net increase or decrease of charge. That is, a change in current density ρ. The time rate of change of charge within the volume can be expressed as:

$$\frac{di}{dt} = \iiint_V \rho \, dV \tag{9.89}$$

Since the net current (Coulombs/sec) must equal the time rate of change of charge (also in Coulombs/sec), we have:

$$\oiint_S \mathbf{J} \cdot d\mathbf{A} = -\frac{\partial}{\partial t} \iiint_V \rho \, dV \tag{9.90}$$

The negative sign means that a decrease in the time rate of change of charge within the volume results in an outward positive current through the surface – assuming of course that we have a conservation of charge.

By the divergence theorem:

$$\oiint_S \mathbf{J} \cdot dA = \iiint_V \nabla \cdot \mathbf{J} \, dV \tag{9.91}$$

and so:

$$\iiint_V \nabla \cdot \mathbf{J} dV = -\frac{\partial}{\partial t} \iiint_V \rho \, dV \tag{9.92}$$

The terms inside the integral must therefore be equal, and so:

$$\nabla \cdot \mathbf{J} = -\frac{\partial \rho}{\partial t} \tag{9.93}$$

That is, the current diverging from the volume is equal to the time rate of decrease of charge within it. This is known as a continuity equation and has a wide range of applicability besides that in electrical theory.

9.4 FOUR-VECTORS

9.4.1 Four-Position

In relativity equations and also in particle physics, many of the equations are written in four-dimensional space-time with the time dimension represented by a distance ct. An object, or a point, can be specified in space-time by the coordinates (ct, x, y, z). It is customary in particle physics to write the time-like dimension first.

The distance between two points in space-time, that is, two events, is given by what is called the interval s:

$$s^2 = c^2t^2 - x^2 - y^2 - z^2 \tag{9.94}$$

One important significance of this measure of space-time distance, the interval between two events, is that it will be the same value for every observer. The interval in space-time is Lorentz invariant.

In many of the equations we will encounter, the functions are written in terms of (ct,x,y,z), and to avoid repetition, a general variable x is used with a subscript μ, so that x_μ means the set:

$$\begin{aligned} x_0 &= ct \\ x_1 &= x \\ x_2 &= y \\ x_3 &= z \end{aligned} \tag{9.95}$$

Often, the time-like quantities need to be separated out from the three space-like quantities. Three-dimensional vector quantities can be written with an arrow or in bold face.

$$x_\mu = \left(ct, \vec{x}\right) \tag{9.96}$$

In this notation, μ ranges from 0 to 3. For a vector, $\mu = 1, 2, 3$ refers to the directions of the x, y, z axes. The arrow above the space components signifies a three-dimensional vector, or a three-vector (x, y, z). If we wish to refer to the space components separately, we use the subscript i instead of μ. So, x_i refers to the three space components x, y and z, where i goes from 1 to 3.*

Now, a distance Δs in space-time is:

$$\begin{aligned} \Delta s^2 &= \Delta c^2t^2 - \Delta\vec{x}^2 \\ \Delta s &= \sqrt{\Delta c^2t^2 - \Delta\vec{x}^2} \end{aligned} \tag{9.97}$$

Dividing through by c, we obtain:

$$\frac{s}{c} = \sqrt{\frac{c^2t^2}{c^2} - \frac{\vec{x}^2}{c^2}} = \sqrt{t^2 - \frac{\vec{x}^2}{c^2}} \tag{9.98}$$

This has the units of time.

In a coordinate system S′ moving with the particle, the particle is at rest, and so any space-time distance is seen directly on the ct axis. It is completely time-like. It is a proper time. From the point of view of the S reference frame, the proper time is:

$$\begin{aligned} t' &= \sqrt{t^2 - \frac{\vec{x}^2}{c^2}} \\ &= t\sqrt{1 - \frac{\vec{x}^2}{t^2c^2}} \\ &= t\sqrt{1 - \vec{v}^2/c^2} \end{aligned} \tag{9.99}$$

* If there is a subscript, μ or i written with a vector quantity, we don't have to write the vector in bold or with an arrow.

Note that:

$$\vec{x}^2 = x^2 + y^2 + z^2 \tag{9.100}$$

and:

$$\vec{v}^2 = {v_x}^2 + {v_y}^2 + {v_z}^2 \tag{9.101}$$

In any particular reference frame, the more the interval is aligned with the *ct* axis, the more the interval is time-like. The more the interval aligns with the *x* axis, the more the interval is space-like. At 45°, the particle is moving with velocity *c* and so must have no rest mass.

9.4.2 Four-Velocity

In three dimensions, the velocity **v** (or $\vec{v}$) has components in the *x*, *y* and *z* directions:

$$\vec{v} = \left(v_x, v_y, v_z\right) \tag{9.102}$$

and:

$$\vec{v}^2 = v_x^2 + v_y^2 + v_z^2 \tag{9.103}$$

The velocity components are the space derivatives with respect to time:

$$v_x = \frac{dx}{dt};\ v_y = \frac{dy}{dt};\ v_z = \frac{dz}{dt} \tag{9.104}$$

Written as a three-vector:

$$\vec{v} = \frac{d\vec{x}}{dt} \tag{9.105}$$

We need to be able to express velocity in an invariant way. In a reference frame that is moving with the particle, the displacement of the particle is on the time axis (the world line is vertical). In this case, intervals in space-time are intervals in time, and time between two events in the same place is a proper time.

$$t' = t\sqrt{1 - v^2/c^2} \tag{9.106}$$

The four-velocity vector $u\mu$ in four-dimensional space-time is defined as:

$$u_\mu = \frac{dx_\mu}{dt'} \tag{9.107}$$

This is a set of four equations, one for each value of μ:

$$\begin{aligned} u_0 &= \frac{\partial ct}{\partial t'} = \frac{c}{\sqrt{1-\vec{v}^2/c^2}} \\ u_1 &= \frac{\partial x}{\partial t'} = \frac{\partial x}{\partial t\sqrt{1-\vec{v}^2/c^2}} = \frac{v_x}{\sqrt{1-\vec{v}^2/c^2}} \end{aligned} \tag{9.108}$$

$$u_2 = \frac{\partial y}{\partial t'} = \frac{\partial y}{\partial t\sqrt{1-\vec{v}^2/c^2}} = \frac{v_y}{\sqrt{1-\vec{v}^2/c^2}}$$

$$u_3 = \frac{\partial z}{\partial t'} = \frac{\partial z}{\partial t\sqrt{1-\vec{v}^2/c^2}} = \frac{v_z}{\sqrt{1-\vec{v}^2/c^2}}$$

The four-velocity can be written:

$$u_\mu = (u_0, \vec{u}) \tag{9.109}$$

Now, since:

$$s^2 = c^2t^2 - x^2 - y^2 - z^2 \tag{9.110}$$

Then, we can expect that:

$$u_\mu^2 = u_0^2 - u_1^2 - u_2^2 - u_3^2 \tag{9.111}$$

And so:

$$\begin{aligned} u_0^2 - u_1^2 - u_2^2 - u_3^2 &= \frac{c^2}{1-\vec{v}^2/c^2} - \frac{v_x^2}{1-\vec{v}^2/c^2} - \frac{v_y^2}{1-\vec{v}^2/c^2} - \frac{v_z^2}{1-\vec{v}^2/c^2} \\ &= \frac{c^2 - \left(v_x^2 + v_y^2 + v_z^2\right)}{1-\vec{v}^2/c^2} \\ &= \frac{c^2 - \vec{v}^2}{1-\vec{v}^2/c^2} \\ &= c^2 \end{aligned} \tag{9.112}$$

That is, the magnitude of the four-velocity is always a constant c.

The four-velocity is the rate of change of four-position with respect to the proper time.

9.4.3 Four-Momentum

The four-momentum is written:

$$p_\mu = mu_\mu \tag{9.113}$$

In this part of the book, the mass m is the rest mass (i.e. as measured in the reference frame where the particle is seen to be at rest). For the x axis, $\mu = 1$. This would be:

$$p_x = m\frac{v_x}{\sqrt{1-\vec{v}^2/c^2}} \tag{9.114}$$

The four-momentum has a connection with energy. The time component of the four-momentum is an energy component.

$$p_o = mu_o = \frac{mc}{\sqrt{1-\vec{v}^2/c^2}} = \frac{E}{c} \tag{9.115}$$

In short:

$$p_\mu = (E, \vec{p}) \tag{9.116}$$

and:

$$\begin{aligned} p_0^2 - p_1^2 - p_2^2 - p_3^2 &= \frac{m^2c^4}{1-\vec{v}^2/c^2} - \frac{m^2v_x^2}{1-\vec{v}^2/c^2} - \frac{m^2v_y^2}{1-\vec{v}^2/c^2} - \frac{m^2v_z^2}{1-\vec{v}^2/c^2} \\ &= m^2\frac{c^2-\vec{v}^2}{1-\vec{v}^2/c^2} \\ &= m^2c^4 \\ |p_\mu| &= mc^2 \end{aligned} \tag{9.117}$$

where m is the rest mass.

9.4.4 Dot Product of Four-Vectors

We saw that an interval in space-time is defined by:

$$s^2 = c^2t^2 - x^2 - y^2 - z^2 \tag{9.118}$$

This is the dot product between the four-vector s and itself.

Consider two four-vectors A_μ and B_μ.

$$\begin{aligned} A_\mu &= (A_0, A_x, A_y, A_z) \\ B_\mu &= (B_0, B_x, B_y, B_z) \end{aligned} \tag{9.119}$$

Then, the dot product between the two is:

$$A_\mu \cdot B_\mu = A_0B_0 - A_xB_x - A_yB_y - A_zB_z \tag{9.120}$$

This can also be written:*

$$A_\mu \cdot B_\mu = A_0B_0 - \vec{A}\cdot\vec{B} \tag{9.121}$$

* Usually, and somewhat unfortunately, the dot product is written without the dot: $A_\mu B_\mu$. The context (that is, the terms involved have to be recognised as vectors) determines how the product is interpreted.

9.4.5 Four-Differential Operator, and the d'Alembertian

The three-dimensional differential operator is defined as:

$$\vec{\nabla} = \left[\frac{\partial}{\partial x}, \frac{\partial}{\partial y}, \frac{\partial}{\partial z} \right] \tag{9.122}$$

- When the differential operator operates directly on a scalar function, we get a vector – the gradient of the vector.
- The dot product of the differential operator with a vector function gives a scalar – the divergence of the vector.
- The cross product of the differential operator with a vector gives a vector – the curl of the vector.

A small arrow above the operator symbol signifies that this is a three-vector space operator.

We might then enquire about the four-dimensional form of the differential operator.

The four-dimensional differential operator is defined as:

$$\begin{aligned} \nabla_\mu &= \left[\frac{\partial}{\partial ct}, -\frac{\partial}{\partial x}, -\frac{\partial}{\partial y}, -\frac{\partial}{\partial z} \right] \\ &= \left[\frac{\partial}{\partial ct}, -\vec{\nabla} \right] \end{aligned} \tag{9.123}$$

The unexpected minus signs arise from the definition of Minkowskian space in relativity. The subscript μ signifies the four-dimensional operator.

If we have a scalar given by:

$$\varphi = \varphi(ct, x, y, z) \tag{9.124}$$

then the four-gradient of that scalar is the four-vector:

$$\begin{aligned} \nabla_\mu \varphi &= \left[\frac{\partial \varphi}{\partial ct}, -\frac{\partial \varphi}{\partial x}, -\frac{\partial \varphi}{\partial y}, -\frac{\partial \varphi}{\partial z} \right] \\ &= \left[\frac{\partial \varphi}{\partial ct}, -\vec{\nabla} \varphi \right] \end{aligned} \tag{9.125}$$

When we take the dot product of the differential operator with a vector, we get the divergence – which is a scalar. For example, if we have a four-vector given by:

$$A_\mu = \left(A_{ct}, \vec{A} \right) \tag{9.126}$$

then the four-divergence of A_μ is:

$$\begin{aligned} \nabla_\mu \cdot A_\mu &= \frac{\partial}{\partial x_\mu} A_\mu \\ &= \frac{\partial}{\partial ct} A_{ct} - \frac{\partial}{\partial x}(-A_x) - \frac{\partial}{\partial y}(-A_y) - \frac{\partial}{\partial z}(-A_z) \end{aligned} \tag{9.127}$$

$$= \frac{\partial A_{ct}}{\partial ct} + \frac{\partial A_x}{\partial x} + \frac{\partial A_y}{\partial y} + \frac{\partial A_z}{\partial z}$$

$$= \frac{\partial A_{ct}}{\partial ct} + \vec{\nabla} \cdot \vec{A}$$

The dot product between two four-vectors on its own carries minus signs for the space components, and the differential operator also carries minus signs, hence the + signs appearing here in the space components of the divergence. The divergence of a four-vector is a scalar.

The dot product is implied when the differential operator is shown acting on a vector.

The four-dimensional version of the Laplacian is the d'Alembertian and is defined to be:

$$\begin{aligned} \nabla^2 = \nabla \cdot \nabla &= \frac{\partial^2}{\partial ct^2} - \vec{\nabla}^2 \\ &= \frac{\partial^2}{\partial ct^2} - \frac{\partial^2}{\partial x^2} - \frac{\partial^2}{\partial y^2} - \frac{\partial^2}{\partial z^2} \end{aligned} \tag{9.128}$$

The d'Alembertain is a scalar differential operator. If it operates on a scalar, the result is a scalar.

However, like the vector Laplacian, if the d'Alembertian operates on a four-vector, the result is a new four-vector.

For a scalar, we have:

$$\nabla^2 \varphi = \frac{\partial^2 \varphi}{\partial ct^2} - \frac{\partial^2 \varphi}{\partial x^2} - \frac{\partial^2 \varphi}{\partial y^2} - \frac{\partial^2 \varphi}{\partial z^2} \tag{9.129}$$

The result is a scalar.

For a four-vector A_μ, we obtain:

$$\begin{aligned} \nabla^2 A_\mu &= \frac{\partial^2 A_\mu}{\partial {x_\mu}^2} \\ &= \left[\frac{\partial^2 A_{ct}}{\partial ct^2}; \frac{\partial^2 A_x}{\partial x^2}; \frac{\partial^2 A_y}{\partial y^2}; \frac{\partial^2 A_z}{\partial z^2} \right] \end{aligned} \tag{9.130}$$

which is a vector.

The d'Alembertian is also given the symbol $\Box$ so that:

$$\Box = \nabla^2 = \frac{\partial^2}{\partial ct^2} - \vec{\nabla}^2 \tag{9.131}$$

From here on, the symbol ∇ will mean the four-dimensional differential operator, and $\vec{\nabla}$ will mean the three-dimensional differential operator.

9.5 THE HAMILTONIAN

Expressed in terms of the vector differential operator, the Schrödinger equation is:

$$-\frac{\hbar^2}{2m} \nabla^2 \Psi + V\Psi = i\hbar \frac{\partial \Psi}{\partial t} \tag{9.132}$$

From this, we can form a new operator $\hat{H}$, so that:

$$\hat{H} = -\frac{\hbar^2}{2m}\nabla^2 + V \tag{9.133}$$

$\hat{H}$ is called the energy operator (since it is now equivalent to the operator on the right-hand side of the Schrödinger equation, which is also the energy operator acting on the wave function). It is more familiarly known as the Hamiltonian operator. The Schrödinger equation is thus written:

$$\hat{H}\Psi(x,t) = i\hbar\frac{\partial\Psi(x,t)}{\partial t} \tag{9.134}$$

The time-independent Schrödinger equation, for say the x direction, becomes:

$$\hat{H}\psi(x) = E\psi(x) \tag{9.135}$$

We say that $\psi(x)$ are eigenfunctions of the operator $\hat{H}$ with eigenvalues E. Other quantities (like position, momentum, etc.) can also be represented by the eigenvalues of operators.

In many cases of practical interest, the ideal quantum system is perturbed by some small effect – for example, the application of an electric field from a nearby particle or some external source. How may the eigenfunctions and eigenvalues of the perturbed system be represented? The Hamiltonian is written as the sum of the original, plus the perturbation.

For example, say we have an electron in the harmonic oscillator potential. An electric field is applied in the x direction, and this perturbs the system. For the unperturbed case:

$$V(x) = \frac{1}{2}Cx^2 \tag{9.136}$$

and the Hamiltonian takes the form:

$$\hat{H}_0 = -\frac{\hbar^2}{2m}\nabla^2 + \frac{1}{2}Cx^2 \tag{9.137}$$

Now, the perturbation is an electric field E acting on the charge q. The potential energy associated with this depends on the position x and is given by the product $U = qEx = \lambda x$ since E and q are assumed to be constant. The product $\lambda = qE$ thus quantifies the strength of the perturbation.

The Hamiltonian for the perturbed system becomes:

$$\begin{aligned}\hat{H} &= \hat{H}_0 + \lambda\hat{H}_1 \\ &= -\frac{\hbar^2}{2m}\nabla^2 + \frac{1}{2}Cx^2 + \lambda x\end{aligned} \tag{9.138}$$

that is, $\hat{H}_1 = x$ in this case.

When the perturbation λ is small, the eigenfunctions for the system as a whole can be represented by a convergent series:

$$\psi_n = \psi_n^{(0)} + \lambda\psi_n^{(1)} + \lambda^2\psi_n^{(2)}\ldots \tag{9.139}$$

and the eigenvalues:

$$E_n = E_n^{(0)} + \lambda E_n^{(1)} + \lambda^2 E_n^{(2)}\ldots \tag{9.140}$$

An approximation for the expectation value of $E^{(1)}$ for the first term of the perturbation is:

$$E^1 = \int_{-\infty}^{\infty} \psi_n^* \hat{H}_1 \psi_n dx \tag{9.141}$$

Often, but not always, a first order approximation is reasonable for our purposes, and the effect on the energies of the perturbation can be obtained analytically.

The expression for the next term, the second order perturbation is:

$$E_n^{(2)} = \sum_{p \neq n} \frac{\left| \left\langle \psi_p | \hat{H}_1 | \psi_n \right\rangle \right|^2}{E_n^0 - E_p^0} \tag{9.142}$$

Here, n is the final state and p is the initial state.

We might call a first order perturbation being the result of a shift in the potential V, or a small Coulomb force $\lambda = qE$ acting through a distance x. $\hat{H}_1$ is the perturbing Hamiltonian (i.e. V or qE).

9.6 THE LAGRANGIAN

9.6.1 Action

In many physical situations, a particle might move from one place to another in the presence of a field. For example, an electron may move from A to B in the presence of an electric field. A golf ball might move from one place to another through three-dimensional space within a gravitational field. The motion of the particle is not haphazard. When plotted against time, the motion (say the height above the ground, or the distance travelled) is a smooth function. We might ask, what is the function that describes this path?

We can use equations of motion to determine the path taken by the particle if those equations are known. For example, say a ball is dropped from a height $x = h$. We know that the motion of the ball will be straight down, with a constant acceleration g, until it hits the ground. The equations of motion are:

$$\begin{aligned} v &= gx \\ v^2 &= 2gx \\ x &= \frac{1}{2}gt^2 \end{aligned} \tag{9.143}$$

In terms of energy, the ball has an initial potential energy $V = mgh$, and at the end of its travel, the potential energy will have been transferred into kinetic energy $E_k = ½mv^2$.

Although the motion of the ball can be calculated from the equations of motion above, we might not always be able to do this, especially if the equations of motion are not known in advance. However, it is found that as a general principle, the path taken by a ball is such that the differences in kinetic and potential energies, integrated over the total time, are a minimum. This sum is called the action and given the symbol S.

The function which gives the difference in energies is called the Lagrangian and is given the symbol L. In terms of the above example, action is defined as the sum of the Lagrangian, over time.

$$S = \int_{t_1}^{t_2} L(x, v) dt \tag{9.144}$$

v is the velocity, and x is the distance above the ground. Both are functions of t. $L(x,v)$ is an energy term, so the units of S are energy × time. In mechanics, L is taken as the difference between the kinetic and potential energies for each increment of time along the path.

$$E_{KE} = \frac{1}{2}mv^2$$
$$E_{PE} = V(x) = mgx \tag{9.145}$$

For the case of a falling ball, the Lagrangian is:

$$L = \frac{1}{2}mv^2 - mgx \tag{9.146}$$

and the action would be calculated as:

$$S = \int_{t_1}^{t_2}\left(\frac{1}{2}mv^2 - mgx\right)dt \tag{9.147}$$

And so, since $v=gt$, and $x=\frac{1}{2}gt^2$, then:

$$S = \int_{t_1}^{t_2}\frac{1}{2}m(gt)^2 - mg\frac{1}{2}gt^2dt \tag{9.148}$$

whereupon we have:

$$S = 0 \tag{9.149}$$

Now, this is the action S for the straight down path. The action is 0, indicating that any increase in kinetic energy is matched by a decrease in potential energy. The collection of h values as the ball falls straight down could be thought of as the path which gives the true minimum action. We will call this path x_0.

What about the action for other possible paths between the starting point for the ball and the same ending point? What if the ball, for some reason, were acted upon by an upwards force at some point during its journey and forced to slow up, and then later a downwards force acts to speed it up, so that it reaches the ground at the same place in the same time? Is the action the same as before? Calculations show that the action will always be greater than the minimum calculated above for these other paths. The unrestrained free motion of the ball is such that the action, the sum of the differences between the kinetic and potential energies is a minimum. The minimum in S may not always be zero, but any other path will result in a greater value for S. If we didn't know the equations of motion, we would have to describe the path by the one which gave the minimum action.

How do we find this minimum in S? Trial and error would be very inefficient, but we may proceed using what is called variational calculus.

9.6.2 Variational Calculus

Say we have a function $y=6(x-2)^2+3$ as shown in Figure 9.6(a). The minimum might be at say $x_0=+2$ which gives $y_0=3$. This is shown as point A in Figure 9.6(a). This is the true minimum value. If we select an x coordinate a small distance away from $x_0=2$, say $x=2.1$, the new value of y would be 3.06. Not much of a difference from the value at the true minimum.

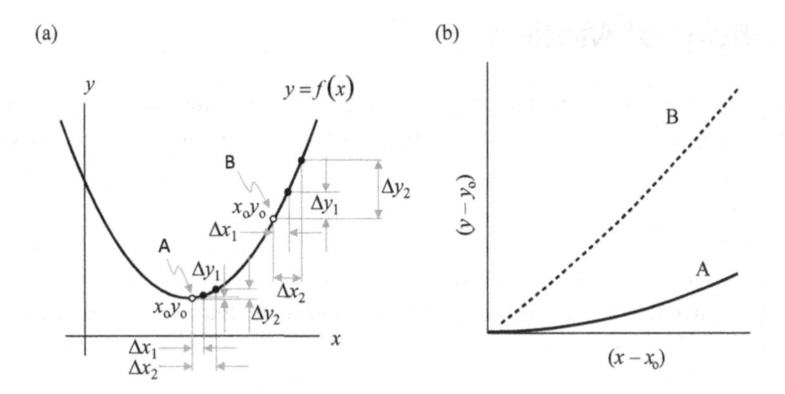

FIGURE 9.6 (a) Hypothetical function with a minimum at (x_0,y_0). (b) Difference in y vs difference in x for two points on the curve of (a), A being near the minimum, and B being elsewhere.

If we go a series of small steps away from the minimum value, and plot the difference $\Delta y = y - y_0$ as a function of $\Delta x = x - x_0$, we find that the relationship between the two is like a parabola, and this is shown as curve A in Figure 9.6(b).

But say we didn't know where the minimum was, and made a guess – say at $x_0' = 5$ as shown as point B in Figure 9.6(a). If we then plot the differences, $(y - y_0')$ as a function of $(x - x_0')$ for this new value x_0', we find that the relationship between the two is nearly linear. This is shown in Figure 9.6(b) as curve B. The greater distance away from the true minimum we are, the more linear the relationship between the difference between $(y - y_0)$ and the difference $(x - x_0)$. Note, we are not saying these increments are Δx and ΔS as if we were calculating the slope of the tangent to the function, but the difference between the values of x and S compared to their values at our selected values of x_0 and y_0. That is, the slope of the secant.

So, by examining a plot of $(y - y_0)$ against $(x - x_0)$ we can tell if we are at or near the true minimum by the linearity of the plot. If the plot is almost linear, that is, first order in $(x - x_0)$, we are not near the minimum. If the plot is parabolic, or second order, we are very close to the minimum.

Let the vertical axis in Figure 9.6(b) be the change in the value of the action: ΔS. The horizontal axis is the difference in the time-dependent path variable from a selected value, say x_0, i.e. $\gamma = x - x_0$. Say we said that the plot of ΔS against γ was characterised by an equation:

$$\Delta S = A\gamma^2 + B\gamma \tag{9.150}$$

The closer we select x_0 to be near that which gives the true minimum of S the greater the value of the factor A and the smaller the value of the factor B. In fact, if we had selected x_0 as being the one which gives the true minimum of S, then the factor B would be exactly zero, and the *change* in S, i.e. ΔS, with the variation of paths (i.e. $\gamma = x - x_0$) would be the second order function $A\gamma^2$. The further away we take x_0 from the true minimum, the smaller the factor A and the greater the factor B. What we are saying is that with x_0 at the value which gives the true minimum in S, any variations in paths from that path show up as a second order change in the action ΔS, and there would be no first order change at all if ΔS were to be plotted against γ. At x_0 at a minimum other than the true minimum, any change in action from that reference point would show up as a first order variation. That is, at x_0 at the minimum, there is no first order change in S for a change γ.

We can thus identify the true minimum when we plot ΔS against γ so that there is only a parabolic relationship between them. Or, put another way, we can identify the true minimum value of x_0 when we plot ΔS against γ so that any first order dependence of ΔS on γ equals zero. That is, with reference to the above equation:

$$\begin{gathered} \Delta S = B\gamma = 0 \\ B = 0 \end{gathered} \tag{9.151}$$

9.6.3 Equations of Motion

We return to the example of the ball falling from a height. The time-dependent path variable is x. Expressing the velocity in terms of the derivative of x with respect to t, the action becomes:

$$S = \int_{t_1}^{t_2} \left(\frac{1}{2} m \left(\frac{dx}{dt} \right)^2 - mgx \right) dt \tag{9.152}$$

Let's consider a path such that the x coordinates for the motion are changed by a small amount γ. That is:

$$\gamma = x - x_0 \tag{9.153}$$

which is a function of t. Let's say the minimum action is S_0. The new action S′ for the varied path is written:

$$S' = \int_{t_1}^{t_2} \left(\frac{1}{2} m \left(\frac{d}{dt}(x_0 + \gamma) \right)^2 - mg(x_0 + \gamma) \right) dt \tag{9.154}$$

Because there has been a change in the path, there has also been a change in the action $\delta S = S' - S_0$ as a result of the change in x, i.e. $x = x_0 + \gamma$.

If x_0 is on the true minimum path, then any deviation from that path will be second order only. That is, there will no linear dependence between δS and γ. We do not know what x_0 is yet, but we do know that any variation in S will have no first order dependence on the variation in x. This is, in effect, a boundary condition. We need therefore to examine the first order terms in the expression for the new action. This is because if there are first order terms present, then δS will be zero.

Focussing on the kinetic energy term, we have:

$$\begin{aligned} S' &= \int_{t_1}^{t_2} \left(\frac{1}{2} m \left(\frac{dx_0}{dt} + \frac{d\gamma}{dt} \right)^2 - mg(x_0 + \gamma) \right) dt \\ &= \int_{t_1}^{t_2} \left(\frac{1}{2} m \left(\left(\frac{dx_0}{dt} \right)^2 + 2 \frac{dx_0}{dt} \frac{d\gamma}{dt} + \left(\frac{d\gamma}{dt} \right)^2 \right) - mg(x_0 + \gamma) \right) dt \end{aligned} \tag{9.155}$$

Since we are looking for an expression for a first order relationship between ΔS and Δx, we can eliminate any second order terms in γ:

$$S = \int_{t_1}^{t_2} \left(\frac{1}{2} m \left(\left(\frac{dx_0}{dt} \right)^2 + 2 \frac{dx_0}{dt} \frac{d\gamma}{dt} \right) - mg(x_0 + \gamma) \right) dt \tag{9.156}$$

Now, for the potential energy term, we have:

$$V(x_0 + \gamma) = mg(x_0 + \gamma) \tag{9.157}$$

Because γ is small compared to x_0 this can be expressed in terms of a Taylor series:

$$f(x) = \sum_{n=0}^{\infty} a_n (x - c)^n = a_0 + a_1(x - c) + a_2(x - c)^2 + \ldots \tag{9.158}$$

which is written:

$$f(x)=f(c)+\frac{(x-c)}{1!}f'(c)+\frac{(x-c)^2}{2!}f''(c)+\ldots+\frac{(x-c)^n}{n!}f^n(c)+\ldots \tag{9.159}$$

In this case, c in the above becomes x_0 since $x-x_0=\gamma$, then:

$$V(x_0+\gamma)=V(x_0)+\frac{\gamma}{1!}V'(x_0)+\frac{\gamma^2}{2!}V''(x_0)+\ldots+\frac{\gamma^n}{n!}V^n(x_0)+\ldots \tag{9.160}$$

where:

$$V'(x_0)=\frac{\partial V}{\partial x_0};\ V''(x_0)=\frac{\partial^2 V}{\partial {x_0}^2}\ldots \tag{9.161}$$

Because we are only interested in the first order terms, then the potential term we want is:

$$\begin{aligned}V(x_0+\gamma)&=V(x_0)+\gamma V'(x_0)\\&=mgx_0+\gamma mg\end{aligned} \tag{9.162}$$

In fact, in this case, the potential energy is linear with respect to x_0, there are no higher terms anyway, but it is instructive to show the procedure.

The full expression for the action is thus:

$$\begin{aligned}S&=\int_{t_1}^{t_2}\left(\frac{1}{2}m\left(\left(\frac{dx_0}{dt}\right)^2+2\frac{dx_0}{dt}\frac{d\gamma}{dt}\right)-mgx_0-\gamma mg\right)dt\\&=\int_{t_1}^{t_2}\frac{1}{2}m\left(\frac{dx_0}{dt}\right)^2+m\frac{dx_0}{dt}\frac{d\gamma}{dt}-mgx_0-\gamma mg\,dt\end{aligned} \tag{9.163}$$

With some rearrangement, this becomes:

$$\begin{aligned}S&=\int_{t_1}^{t_2}\frac{1}{2}m\left(\frac{dx_0}{dt}\right)^2-mgx_0+m\frac{dx_0}{dt}\frac{d\gamma}{dt}-\gamma mg\,dt\\&=\int_{t_1}^{t_2}\frac{1}{2}m\left(\frac{dx_0}{dt}\right)^2-mgx_0\,dt+\int_{t_1}^{t_2}m\frac{dx_0}{dt}\frac{d\gamma}{dt}-\gamma mg\,dt\end{aligned} \tag{9.164}$$

The first term in the above is the action that would have been calculated as the true minimum:

$$S_0=\int_{t_1}^{t_2}\left(\frac{1}{2}m\left(\frac{dx_0}{dt}\right)^2-mgx_0\right)dt \tag{9.165}$$

The difference in action δS between S' and S_0 is thus the second term:

$$\begin{aligned}\delta S&=\int_{t_1}^{t_2}m\frac{dx_0}{dt}\frac{d\gamma}{dt}-\gamma mg\,dt\\&=0\end{aligned} \tag{9.166}$$

Because this expression came about from a consideration of the *first order terms only*, we know that at x_0 which gives the minimum value of S, then there will be no *first order* change in action, and so the difference δS in the above will be zero.

We are looking for the value of x_0, that which gives the true minimum in S. The derivative of x_0 is contained within the above expression, so we need to extract x_0 somehow from this. This is done by integrating by parts.

$$\int u dv = uv - \int v du \tag{9.167}$$

$$\begin{aligned} u &= f(t); du = f'(t)dt \\ v &= g(t); dv = g'(t)dt \end{aligned} \tag{9.168}$$

Thus:

$$\begin{aligned} &u = m\frac{dx_0}{dt};\ v = \gamma;\ du = \frac{d}{dt}\left(m\frac{dx_0}{dt}\right)dt;\ dv = \frac{d\gamma}{dt}dt \\ &\delta S = \left[m\gamma\frac{dx_0}{dt}\right]_{t_1}^{t_2} - \int_{t_1}^{t_2}\gamma\frac{d}{dt}\left(m\frac{dx_0}{dt}\right)dt - \int_{t_1}^{t_2}\gamma mg\, dt \end{aligned} \tag{9.169}$$

Now, the first term in the above contains the quantity γ which is a function of t. But we know that at the start and end of the path, there is no difference between the path under consideration and the minimum path so that at t_1 and t_2, $\gamma = 0$. So, the difference in action becomes:

$$\begin{aligned} \delta S &= -\int_{t_1}^{t_2}\gamma\,\frac{d}{dt}\left(m\frac{dx_0}{dt}\right)dt - \int_{t_1}^{t_2}\gamma mg\, dt \\ &= -\int_{t_1}^{t_2}\gamma m\,\frac{d^2x_0}{dt^2}dt - \int_{t_1}^{t_2}\gamma mg\, dt \\ &= -\left(m\frac{d^2x_0}{dt^2}\right)\int_{t_1}^{t_2}\gamma\, dt - mg\int_{t_1}^{t_2}\gamma\, dt \\ &= \left(-m\frac{d^2x_0}{dt^2} - mg\right)\int_{t_1}^{t_2}\gamma\, dt \\ &= 0 \end{aligned} \tag{9.170}$$

That is, the equation of motion is:

$$0 = m\frac{d^2x_0}{dt^2} + mg \tag{9.171}$$

At the beginning of this problem we defined x as the distance above the ground, but from the point of view of the accelerations, the displacement was in the downward, or negative direction, and so:

$$0 = -m\frac{d^2x_0}{dt^2} + mg \tag{9.172}$$

Note that the first term is the gravitational force, $F = mg$, and the second term came from the derivative of the potential energy with respect to distance, which is also a force. This second force is something that manifests itself as an inertia force if our point of view was that travelling with the falling ball. From that point of view, the ball is at static equilibrium, the weight force pulling down and the inertia force pulling up. This method of analysis is called d'Alembert's principle and turns a dynamic problem into one expressed in terms of statics.

The variational calculus computation has shown that, for the example of the falling ball, the path followed by the ball is not $(x_0 + \gamma)$, but x_0 since this is the path that gives the minimum S.

When more than one particle is involved, the kinetic and potential energies for all of them are included in the integral. That is, the Lagrangians for all particles are just added together within the action integral.

This principle of least action is also known as Hamilton's principle.

Note that although it appears long-winded for this familiar physical example, the principle of least action allows us to determine the equations of motion for a particle.

An important point to note is that when the ball is falling, or when a particle is going from place to place, it does not know in advance where it will finish its journey. The particle does not perform the integration over time taken to get from its origin to its destination. A particle moves in a series of infinitesimal steps, and so the integration, or the travel, proceeds incrementally. It is a local phenomenon. The only way a particle knows which way to make the next step is to sample nearby steps and determine if the change in the Lagrangian is first or second order, and to pick the direction which gives a second order change, that is, a zero first order change. The spatial scale of this searching for the minimum path is the wavelength of the particle – the de Broglie wavelength.

The principle of least action is a more general statement of Fermat's principle of least time. In the above, we talked about the motion of a particle travelling from place to place in which the journey took the same time, but the paths taken were different. According to the boundary conditions of our integration above, this meant that the particle would have to change velocity from path to path to keep the travel time constant. If it is specified that the velocity is to be kept constant from path to path over the distance travelled from source to destination, then the particle takes the path of least time. The principle is the same; some quantity, when integrated (or counted) over time, is to be a minimum.

Relativistic Quantum Mechanics

10

10.1 THE DIRAC EQUATION

In 1928, Dirac applied Einstein's relativistic energy equation for a complete description of the motion of a particle, such as an electron, at any speed. To see how this is done, we begin with the non-relativistic energy equation for a particle of mass m moving with velocity v:

$$\begin{aligned} E &= \frac{1}{2}mv^2 + V \\ &= \frac{p^2}{2m} + V \end{aligned} \tag{10.1}$$

V is the potential energy, m is the rest mass of the particle and p is the momentum.

The Schrödinger equation builds upon this by postulating a wave equation of the form (in one dimension):

$$-\frac{\hbar^2}{2m}\frac{\partial^2\Psi(x,t)}{\partial x^2} + V(x,t)\Psi(x,t) = i\hbar\frac{\partial\Psi(x,t)}{\partial t} \tag{10.2}$$

where Ψ is the wave function for the matter wave. The $-i\hbar\partial/\partial x$ and $i\hbar\partial/\partial t$ terms are the momentum and energy quantum mechanical operators respectively.

In three spatial dimensions, with $\Psi = \Psi(x,y,z,t)$, this becomes:

$$-\frac{\hbar^2}{2m}\left(\frac{\partial^2\Psi}{\partial x^2} + \frac{\partial^2\Psi}{\partial y^2} + \frac{\partial^2\Psi}{\partial z^2}\right) + V(x,y,z,t)\Psi = i\hbar\frac{\partial\Psi}{\partial t} \tag{10.3}$$

In three-dimensional operator form, this is expressed:

$$-\frac{\hbar^2}{2m}\nabla^2\Psi + V\Psi = i\hbar\frac{\partial\Psi}{\partial t} \tag{10.4}$$

where $\vec{\nabla}$ is the three-vector differential operator:

$$\vec{\nabla} = \frac{\partial}{\partial x}\mathbf{i} + \frac{\partial}{\partial y}\mathbf{j} + \frac{\partial}{\partial z}\mathbf{k} \tag{10.5}$$

When relativity is taken into account, the total energy possessed by a body is given by Einstein's relativistic energy equation:

$$\begin{aligned} E^2 &= p^2c^2 + m^2c^4 \\ E^2 - p^2c^2 - m^2c^4 &= 0 \end{aligned} \tag{10.6}$$

In this equation, m is the rest mass of the body and p is the magnitude of the momentum vector.

In three dimensions, the quantum mechanical momentum operator is:

$$\hat{p} = -i\hbar\left(\frac{\partial}{\partial x}+\frac{\partial}{\partial y}+\frac{\partial}{\partial z}\right) = -i\hbar\nabla \tag{10.7}$$

The quantum mechanical energy operator is as before:

$$\hat{E} = i\hbar\frac{\partial}{\partial t} \tag{10.8}$$

The relativistic form of the energy equation in three dimensions is therefore:

$$\hbar^2\frac{\partial^2\Psi}{\partial t^2} = \hbar^2c^2\nabla^2\Psi - m^2c^4\Psi \tag{10.9}$$

This is known as the Klein–Gordon equation. With a little rearrangement, this becomes:

$$-\frac{1}{c^2}\frac{\partial^2\Psi}{\partial t^2}+\nabla^2\Psi-\frac{m^2c^2}{\hbar^2}\Psi = 0 \tag{10.10}$$

Note the differences with the Schrödinger equation. The Klein–Gordon equation involves an energy squared term, implying the existence of both positive and negative values of the energy E. As well, we do not have the first time derivative of the wave function but the second time derivative. To solve this mathematically, we will need the first time derivative of the wave function as a boundary condition. That is, the wave function itself may not be sufficient to specify the matter wave, and so the probability amplitude calculated from it may no longer be sufficient to determine the probability of finding the electron at a specific point at a specific time.

The Klein–Gordon equation was given some attention by Schrödinger, but Schrödinger was unsure of the interpretation of negative energies that arose as a result. Since the probability density P depends linearly on the energy operator, the square of the energy term in the Klein–Gordon equation implies the existence of negative probabilities which was unphysical. Dirac sought a relativistic equation which was first-order in E.

Dirac attempted to factorise the energy equation into the product of two first order equations. One way of doing this is by letting:

$$E = \alpha_x p_x c + \alpha_y p_y c + \alpha_z p_z c + \beta mc^2 \tag{10.11}$$

The relativistic energy equation then becomes the product:

$$\begin{aligned} E^2 &= \left(\alpha_x p_x c + \alpha_y p_y c + \alpha_z p_z c + \beta mc^2\right)\left(\alpha_x p_x c + \alpha_y p_y c + \alpha_z p_z c + \beta mc^2\right) \\ &= \alpha_x^2 p_x^2 c^2 + \alpha_y^2 p_y^2 c^2 + \alpha_z^2 p_z^2 c^2 + \beta^2 m^2 c^4 + 12 \text{ cross terms} \end{aligned} \tag{10.12}$$

which must eventually agree with:

$$E^2 = p_x^2c^2 + p_y^2c^2 + p_z^2c^2 + m^2c^4 \tag{10.13}$$

Note that the magnitude of the components of momentum add, in vector form, to give the total magnitude of the momentum p:

$$|\vec{p}|^2 = p_x^2 + p_y^2 + p_z^2 \tag{10.14}$$

To maintain compatibility with the relativistic energy equation, all the 12 cross terms must vanish, and the square of the multiplicative constants must all be equal to 1. However, no combination of values for the constant coefficients α_x, α_y, α_z and β will meet these requirements.

In anticipation of the final form of these equations, we will rearrange the proposed factors in a slightly different way and with different notation so that:

$$\gamma_o E - \gamma_1 p_x c - \gamma_1 p_2 c - \gamma_3 p_z c - mc^2 = 0 \tag{10.15}$$

Dirac found that a solution could be obtained if the coefficients γ_0, γ_1, γ_2, γ_3 were represented by a set of 4×4 matrices:*

$$\gamma_0 = \begin{bmatrix} 1 & 0 & 0 & 0 \\ 0 & 1 & 0 & 0 \\ 0 & 0 & -1 & 0 \\ 0 & 0 & 0 & -1 \end{bmatrix} \quad \gamma_1 = \begin{bmatrix} 0 & 0 & 0 & 1 \\ 0 & 0 & 1 & 0 \\ 0 & -1 & 0 & 0 \\ -1 & 0 & 0 & 0 \end{bmatrix}$$
$$\gamma_2 = \begin{bmatrix} 0 & 0 & 0 & -i \\ 0 & 0 & i & 0 \\ 0 & i & 0 & 0 \\ -i & 0 & 0 & 0 \end{bmatrix} \quad \gamma_3 = \begin{bmatrix} 0 & 0 & 1 & 0 \\ 0 & 0 & 0 & -1 \\ -1 & 0 & 0 & 0 \\ 0 & 1 & 0 & 0 \end{bmatrix} \tag{10.16}$$

Note the similarity between the γ_1, γ_2 and γ_3 matrices and the Pauli matrices which describe electron spin. It is as if the Pauli matrices $\hat{\sigma}_x, \hat{\sigma}_y, \hat{\sigma}_z$ are a subset of the Dirac matrices γ_1, γ_2, γ_3. It is no surprise therefore to learn that spin is an integral part of the Dirac formulation.

By squaring the coefficient gamma matrices, we can see they all reduce to a unit matrix. For example, for γ_1 we have:

$$\begin{bmatrix} 0 & 0 & 0 & 1 \\ 0 & 0 & 1 & 0 \\ 0 & -1 & 0 & 0 \\ -1 & 0 & 0 & 0 \end{bmatrix}^2 = \begin{bmatrix} 0 & 0 & 0 & 1 \\ 0 & 0 & 1 & 0 \\ 0 & -1 & 0 & 0 \\ -1 & 0 & 0 & 0 \end{bmatrix}\begin{bmatrix} 0 & 0 & 0 & 1 \\ 0 & 0 & 1 & 0 \\ 0 & -1 & 0 & 0 \\ -1 & 0 & 0 & 0 \end{bmatrix}$$
$$= \begin{bmatrix} -1 & & & \\ & -1 & & \\ & & -1 & \\ & & & -1 \end{bmatrix} \tag{10.17}$$

Similarly, for the other matrices.

Before we proceed further, it is convenient to express the time variable in terms of a distance dimension *ct*. We've seen that the product *ct* is a distance dimension in our study of relativity, and so this is not such an unreasonable substitution. As we will see later, this leads to a very elegant form of these equations.

The quantum mechanical energy operator becomes:

$$\hat{E} = i\hbar\frac{\partial}{\partial t} = i\hbar c\frac{\partial}{\partial ct} \tag{10.18}$$

* There are various forms of the γ matrices possible.

and so, with the momentum operator as before, $\hat{p} = -i\hbar\dfrac{\partial}{\partial x}$, the complete operator equation is:

$$i\gamma_0\hbar c\frac{\partial}{\partial ct} + i\gamma_1\hbar c\frac{\partial}{\partial x} + i\gamma_2\hbar c\frac{\partial}{\partial y} + i\gamma_3\hbar c\frac{\partial}{\partial z} - mc^2 = 0 \tag{10.19}$$

When applied to the wave function Ψ, say for an electron, the Dirac relativistic wave equation becomes:

$$i\gamma_0\hbar c\frac{\partial\Psi}{\partial ct} + i\gamma_1\hbar c\frac{\partial\Psi}{\partial x} + i\gamma_2\hbar c\frac{\partial\Psi}{\partial y} + i\gamma_3\hbar c\frac{\partial\Psi}{\partial z} - mc^2\Psi = 0 \tag{10.20}$$

Dividing through by $\hbar c$ we obtain:

$$i\gamma_0\frac{\partial\Psi}{\partial ct} + i\gamma_1\frac{\partial\Psi}{\partial x} + i\gamma_2\frac{\partial\Psi}{\partial y} + i\gamma_3\frac{\partial\Psi}{\partial z} - \frac{mc}{\hbar}\Psi = 0 \tag{10.21}$$

The notable feature of the equation is the presence of Planck's constant (quantum physics) and the speed of light (relativity).

For the matrix formulation of the equation to make sense, we need to express the wave function in terms of four components Ψ_1, Ψ_2, Ψ_3, Ψ_4:

$$\Psi = \begin{bmatrix}\Psi_1\\\Psi_2\\\Psi_3\\\Psi_4\end{bmatrix} \tag{10.22}$$

Thus:

$$\begin{aligned}
\gamma_0\frac{\partial\Psi}{\partial ct} &= \frac{\partial}{\partial ct}\begin{bmatrix}1&0&0&0\\0&1&0&0\\0&0&-1&0\\0&0&0&-1\end{bmatrix}\begin{bmatrix}\Psi_1\\\Psi_2\\\Psi_3\\\Psi_4\end{bmatrix} = \frac{\partial}{\partial ct}\begin{bmatrix}\Psi_1\\\Psi_2\\-\Psi_3\\-\Psi_4\end{bmatrix}\\
\gamma_1\frac{\partial\Psi}{\partial x} &= \frac{\partial}{\partial x}\begin{bmatrix}0&0&0&1\\0&0&1&0\\0&-1&0&0\\-1&0&0&0\end{bmatrix}\begin{bmatrix}\Psi_1\\\Psi_2\\\Psi_3\\\Psi_4\end{bmatrix} = \frac{\partial}{\partial x}\begin{bmatrix}\Psi_4\\\Psi_3\\-\Psi_2\\-\Psi_1\end{bmatrix}\\
\gamma_2\frac{\partial\Psi}{\partial y} &= \frac{\partial}{\partial y}\begin{bmatrix}0&0&0&-i\\0&0&i&0\\0&i&0&0\\-i&0&0&0\end{bmatrix}\begin{bmatrix}\Psi_1\\\Psi_2\\\Psi_3\\\Psi_4\end{bmatrix} = \frac{\partial}{\partial y}\begin{bmatrix}-i\Psi_4\\i\Psi_3\\i\Psi_2\\-i\Psi_1\end{bmatrix}\\
\gamma_3\frac{\partial\Psi}{\partial z} &= \frac{\partial}{\partial z}\begin{bmatrix}0&0&1&0\\0&0&0&-1\\-1&0&0&0\\0&1&0&0\end{bmatrix}\begin{bmatrix}\Psi_1\\\Psi_2\\\Psi_3\\\Psi_4\end{bmatrix} = \frac{\partial}{\partial z}\begin{bmatrix}\Psi_3\\-\Psi_4\\-\Psi_1\\\Psi_2\end{bmatrix}
\end{aligned} \tag{10.23}$$

The complete Dirac equation is therefore:

$$i\frac{\partial}{\partial ct}\begin{bmatrix}\Psi_1\\ \Psi_2\\ -\Psi_3\\ -\Psi_4\end{bmatrix}+i\frac{\partial}{\partial x}\begin{bmatrix}\Psi_4\\ \Psi_3\\ -\Psi_2\\ -\Psi_1\end{bmatrix}+i\frac{\partial}{\partial y}\begin{bmatrix}-i\Psi_4\\ i\Psi_3\\ i\Psi_2\\ -i\Psi_1\end{bmatrix}+i\frac{\partial}{\partial z}\begin{bmatrix}\Psi_3\\ -\Psi_4\\ -\Psi_1\\ \Psi_2\end{bmatrix}-\frac{m}{\hbar}c\begin{bmatrix}\Psi_1\\ \Psi_2\\ \Psi_3\\ \Psi_4\end{bmatrix}=0 \tag{10.24}$$

These wave equations can thus be written as a set of four simultaneous linear partial differential equations:

$$\begin{aligned}
&i\frac{\partial}{\partial ct}\Psi_1+i\frac{\partial}{\partial x}\Psi_4-\frac{\partial}{\partial y}\Psi_4+i\frac{\partial}{\partial z}\Psi_3-\frac{mc}{\hbar}\Psi_1=0\\
&i\frac{\partial}{\partial ct}\Psi_2+i\frac{\partial}{\partial x}\Psi_3+\frac{\partial}{\partial y}\Psi_3-i\frac{\partial}{\partial z}\Psi_4-\frac{mc}{\hbar}\Psi_2=0\\
&-i\frac{\partial}{\partial ct}\Psi_3-i\frac{\partial}{\partial x}\Psi_2+\frac{\partial}{\partial y}\Psi_2-i\frac{\partial}{\partial z}\Psi-\frac{mc}{\hbar}\Psi_3=0\\
&-i\frac{\partial}{\partial ct}\Psi_4-i\frac{\partial}{\partial x}\Psi_1-\frac{\partial}{\partial y}\Psi_1+i\frac{\partial}{\partial z}\Psi_2-\frac{mc}{\hbar}\Psi_4=0
\end{aligned} \tag{10.25}$$

When spin was expressed in matrix terms by Pauli and applied to the Schrödinger equation, we had a two-component wave function representing spin-up and spin-down. With the four components required by relativistic considerations, there must be something more than just spin that is of significance here, and just what this is will become evident when we come to solve this system of equations.

The four-component matrix form of the wave function has a corresponding complex conjugate, but the conjugate is a row matrix. That is:

$$\begin{aligned}
\Psi^*\Psi&=\begin{bmatrix}\Psi_1^*, & \Psi_2^*, & \Psi_3^*, & \Psi_4^*\end{bmatrix}\begin{bmatrix}\Psi_1\\ \Psi_2\\ \Psi_3\\ \Psi_4\end{bmatrix}\\
&=\left[\Psi_1^*\Psi_1+\Psi_2^*\Psi_2+\Psi_3^*\Psi_3+\Psi_4^*\Psi_4\right]
\end{aligned} \tag{10.26}$$

That is, $\Psi^*\Psi$ gives the total probability P since:

$$P(x,t)=|\Psi|^2=\Psi^*\Psi \tag{10.27}$$

10.2 SOLUTIONS TO THE DIRAC EQUATION

10.2.1 At Rest

The first problem to be solved is the case for the particle, say an electron, at rest. In this case, the momentum terms are zero and so are their spatial derivatives. The wave function Ψ is a function of time only.

The position is not known, but wherever the electron is, we know it is at rest. All the momentum terms drop out, and so the four equations to be solved are:

$$\begin{aligned}
i\frac{\partial}{\partial ct}\Psi_1 - \frac{m}{\hbar}c\Psi_1 &= 0 \\
i\frac{\partial}{\partial ct}\Psi_2 - \frac{m}{\hbar}c\Psi_2 &= 0 \\
-i\frac{\partial}{\partial ct}\Psi_3 - \frac{m}{\hbar}c\Psi_3 &= 0 \\
-i\frac{\partial}{\partial ct}\Psi_4 - \frac{m}{\hbar}c\Psi_4 &= 0
\end{aligned} \tag{10.28}$$

In this case, each of the four equations involves only one of the component wave functions.

Consider the first wave equation:

$$i\frac{\partial}{\partial ct}\Psi_1 - \frac{m}{\hbar}c\Psi_1 = 0 \tag{10.29}$$

In this equation, we have the energy operator applied to a wave function. The wave function must be of the form of a travelling wave of frequency ω and with $E = \hbar\omega$ and the exponential can be expressed in terms of E. The amplitude is set to be equal to constant A. This means that the position of the electron is unknown – it has a constant probability (A^2) of being anywhere at any time t. But, in this section, we are expressing time as a function of the product ct, and so the wave function thus becomes:

$$\Psi_1(ct) = Ae^{-i\frac{E}{\hbar c}ct} \tag{10.30}$$

Therefore:

$$\begin{aligned}
\frac{\partial}{\partial ct}\Psi_1 &= -i\frac{E}{\hbar c}Ae^{-i\frac{E}{\hbar c}ct} \\
i\frac{\partial}{\partial ct}\Psi_1 &= \frac{E}{\hbar c}Ae^{-i\frac{E}{\hbar c}ct} \\
\frac{mc}{\hbar}\Psi &= \frac{mc}{\hbar}Ae^{-i\frac{E}{\hbar c}ct}
\end{aligned} \tag{10.31}$$

and so, by the first wave equation:

$$\begin{aligned}
i\frac{\partial}{\partial ct}\Psi_1 - \frac{mc}{\hbar}\Psi_1 &= 0 \\
\frac{E}{\hbar c}Ae^{-i\frac{E}{\hbar c}ct} - \frac{mc}{\hbar}Ae^{-i\frac{E}{\hbar c}ct} &= 0 \\
EAe^{-i\frac{E}{\hbar c}ct} &= mc^2Ae^{-i\frac{E}{\hbar c}ct} \\
E &= mc^2
\end{aligned} \tag{10.32}$$

This is as expected since m is the rest mass, and the kinetic energy term is zero for this case. The energy is a constant.

For the second equation, this is the same as the first with just a change of subscript.

$$\Psi_2(ct) = E(ct) = Ae^{-i\frac{E}{\hbar c}ct} \tag{10.33}$$

For the third and fourth equations, we obtain:

$$-i\frac{\partial}{\partial ct}\Psi_{3,4} - \frac{m}{\hbar}c\Psi_{3,4} = 0 \tag{10.34}$$

and so:

$$\begin{aligned}
\Psi_3 &= Ae^{-i\frac{E}{\hbar c}ct} \\
\frac{\partial}{\partial ct}\Psi_3 &= -i\frac{E}{\hbar c}Ae^{-i\frac{E}{\hbar c}ct} \\
-i\frac{\partial}{\partial ct}\Psi_3 &= -\frac{E}{\hbar c}Ae^{-i\frac{E}{\hbar c}ct} \\
\frac{mc}{\hbar}\Psi &= \frac{mc}{\hbar}Ae^{-i\frac{E}{\hbar c}ct}
\end{aligned} \tag{10.35}$$

And so, by the third and fourth equations, we have:

$$\begin{gathered}
-\frac{E}{\hbar c}Ae^{-i\frac{E}{\hbar}ct} - \frac{mc}{\hbar}Ae^{-i\frac{E}{\hbar c}ct} = 0 \\
-EAe^{-\frac{E}{\hbar}ict} = mc^2Ae^{-i\frac{E}{\hbar c}ct} \\
E = -mc^2
\end{gathered} \tag{10.36}$$

Either we have negative energy, or negative mass. Having a negative energy in itself is not such a problem. We often speak of negative energies, particularly in the case of potential energies, where, for example, the first energy level of the hydrogen atom is taken to be −13.6 eV. However, here, we do not have a potential energy or even a kinetic energy, we have mass-energy. It is difficult to conceive a physical notion for negative mass.

We could have arrived at the expression $E = mc^2$ by using $-E$ instead of E in the third and fourth equations above.

The wave function is expressed in terms of four components Ψ_1, Ψ_2, Ψ_3, Ψ_4. In this case, the first two solutions (which in this case comprise only Ψ_1 and Ψ_2) are associated with an electron with a positive energy, and last two solutions (which in this case comprise only Ψ_3 and Ψ_4) are associated with an electron with a negative energy. By the Pauli exclusion principle, we cannot have electrons occupying the same energy state at the same time unless they have opposite spins. As with the case of the two-component Schrödinger wave function where spin had to be added afterwards, here we have two spin states, up and down, as well as two energy states, positive and negative. In this case, the spin properties are part of the overall solution – they didn't need to have them added afterwards.

For the spin-up and spin-down states, we can write:

$$\Psi\uparrow = \begin{bmatrix} \Psi_1 \\ 0 \\ \Psi_3 \\ 0 \end{bmatrix}; \ \Psi\downarrow = \begin{bmatrix} 0 \\ \Psi_2 \\ 0 \\ \Psi_4 \end{bmatrix} \tag{10.37}$$

Unlike the non-relativistic Schrödinger equation, intrinsic spin is a consequence of the relativistic Dirac equation – but so are negative energies.

What this tells us is that there is a probability that the electron at some particular time t will be at rest (we don't know where), and the electron may be in either a spin-up or a spin-down configuration. Also, there is a probability that the electron may be in either the spin-up or spin-down configuration if the electron has a negative energy (the consequence of which we will discuss soon).

The four solutions Ψ satisfy the four simultaneous equations that comprise the matrix form of the Dirac equation. Each solution has the form:

$$\Psi(ct) = Ae^{-i\frac{E}{\hbar c}ct} \tag{10.38}$$

And so, the four solutions for the electron-at-rest case are:

$$\Psi = A\begin{bmatrix} \uparrow 1 \\ 0 \\ 0 \\ 0 \end{bmatrix} e^{-i\frac{E}{\hbar}t}; \quad \Psi = A\begin{bmatrix} 0 \\ \downarrow 1 \\ 0 \\ 0 \end{bmatrix} e^{-i\frac{E}{\hbar}t}; \quad \Psi = A\begin{bmatrix} 0 \\ 0 \\ \uparrow 1 \\ 0 \end{bmatrix} e^{-i\frac{E}{\hbar}t}; \quad \Psi = A\begin{bmatrix} 0 \\ 0 \\ 0 \\ \downarrow 1 \end{bmatrix} e^{-i\frac{E}{\hbar}t} \tag{10.39}$$

Note the last two solutions represent the negative energy states.

The square-bracketed terms, the spin matrices, signify the weight contribution of $\Psi_{1,2,3,4}$ to each of the solutions. In this case, there is a weight of 1 for Ψ_1 for the first solution and so on, but, as we shall see, other components of the wave function may also contribute to each solution. The term A is a normalisation factor which has the same significance as in the Schrödinger equation.

10.2.2 Constant Velocity

Now let's consider an electron moving in one direction, the x direction, with a constant velocity and zero potential energy term.

The equations are:

$$\begin{aligned}
&i\frac{\partial}{\partial ct}\Psi_1 + i\frac{\partial}{\partial x}\Psi_4 - \frac{m}{\hbar}c\Psi_1 = 0 \\
&i\frac{\partial}{\partial ct}\Psi_2 + i\frac{\partial}{\partial x}\Psi_3 - \frac{m}{\hbar}c\Psi_2 = 0 \\
&-i\frac{\partial}{\partial ct}\Psi_3 - i\frac{\partial}{\partial x}\Psi_2 - \frac{m}{\hbar}c\Psi_3 = 0 \\
&-i\frac{\partial}{\partial ct}\Psi_4 - i\frac{\partial}{\partial x}\Psi_1 - \frac{m}{\hbar}c\Psi_4 = 0
\end{aligned} \tag{10.40}$$

In this case, the wave function Ψ is a function of both x and t. Using the de Broglie relationships, it was shown in Section 7.3.3 that the solution to the wave equation will be of the general form:

$$\Psi(x, ct) = Ae^{i\left(\frac{p}{\hbar}x - \frac{E}{\hbar c}ct\right)} \tag{10.41}$$

where p is the momentum in the x direction and E is the energy. Let's examine the first of the Dirac equations. It can be seen that the equation depends on a mixture of Ψ_1 and Ψ_4. We don't know in advance precisely what contribution the Ψ_1 component might make to the total solution here, so we will give it a coefficient, α. Thus:

$$i\alpha\frac{\partial}{\partial ct}\Psi_1 + i\frac{\partial}{\partial x}\Psi_4 - \alpha\frac{mc}{\hbar}\Psi_1 = 0 \tag{10.42}$$

So, we can write:

$$\begin{aligned}
\alpha\frac{\partial}{\partial ct}\Psi_1 &= -i\alpha\frac{E}{\hbar c}Ae^{i\left(\frac{p_x}{\hbar}x - \frac{E}{\hbar c}ct\right)} \\
i\alpha\frac{\partial}{\partial ct}\Psi_1 &= \alpha\frac{E}{\hbar c}Ae^{i\left(\frac{p_x}{\hbar}x - \frac{E}{\hbar c}ct\right)} \\
\frac{\partial}{\partial x}\Psi_4 &= i\frac{p_x}{\hbar}Ae^{i\left(\frac{p_x}{\hbar}x - \frac{E}{\hbar c}ct\right)} \\
i\frac{\partial}{\partial x}\Psi_4 &= -\frac{p_x}{\hbar}Ae^{i\left(\frac{p_x}{\hbar}x - \frac{E}{\hbar c}ct\right)} \\
\alpha\frac{mc}{\hbar}\Psi_1 &= \frac{mc}{\hbar}Ae^{i\left(\frac{p_x}{\hbar}x - \frac{E}{\hbar c}ct\right)}
\end{aligned} \tag{10.43}$$

and thus, from the first equation, we have:

$$\begin{aligned}
\alpha\frac{E}{\hbar c}Ae^{i\left(\frac{p_x}{\hbar}x - \frac{E}{\hbar c}ct\right)} - \frac{p_x}{\hbar}Ae^{i\left(\frac{p_x}{\hbar}x - \frac{E}{\hbar c}ct\right)} - \alpha\frac{mc}{\hbar}Ae^{i\left(\frac{p_x}{\hbar}x - \frac{E}{\hbar c}ct\right)} &= 0 \\
\alpha EAe^{i\left(\frac{p_x}{\hbar}x - \frac{E}{\hbar c}ct\right)} - p_xcAe^{i\left(\frac{p_x}{\hbar}x - \frac{E}{\hbar c}ct\right)} - \alpha mc^2Ae^{i\left(\frac{p_x}{\hbar}x - \frac{E}{\hbar c}ct\right)} &= 0 \\
\alpha E - \alpha mc^2 &= p_xc \\
\alpha &= \frac{p_xc}{E - mc^2}
\end{aligned} \tag{10.44}$$

For the first equation, the solution (i.e. the wave function) is a combination of the Ψ_1 and Ψ_4 wave functions. In matrix form, we can write that a solution (i.e. a wave function) that satisfies the first equation is:

$$\Psi = \begin{bmatrix} \alpha\Psi_1 \\ 0 \\ 0 \\ \Psi_4 \end{bmatrix} \quad \text{or,} \quad \Psi = A\begin{bmatrix} \uparrow\alpha \\ 0 \\ 0 \\ \downarrow 1 \end{bmatrix}e^{i\left(\frac{p_x}{\hbar}x - \frac{E}{\hbar}t\right)} \tag{10.45}$$

Let us now consider the fourth equation. This time, the factor α is applied to the Ψ_4 component.

$$-i\alpha\frac{\partial}{\partial ct}\Psi_4 - i\frac{\partial}{\partial x}\Psi_1 - \alpha\frac{m}{\hbar}c\Psi_4 = 0 \tag{10.46}$$

Thus:

$$\frac{\partial}{\partial ct}\Psi_4 = -i\frac{E}{\hbar c}Ae^{i\left(\frac{p_x}{\hbar}x-\frac{E}{\hbar c}ct\right)}$$

$$-i\alpha\frac{\partial}{\partial ct}\Psi_4 = -\alpha\frac{E}{\hbar c}Ae^{i\left(\frac{p_x}{\hbar}x-\frac{E}{\hbar c}ct\right)}$$

$$\frac{\partial}{\partial x}\Psi_1 = i\frac{p_x}{\hbar}Ae^{i\left(\frac{p_x}{\hbar}x-\frac{E}{\hbar c}ct\right)} \qquad (10.47)$$

$$-i\frac{\partial}{\partial x}\Psi_1 = \frac{p_x}{\hbar}Ae^{i\left(\frac{p_x}{\hbar}x-\frac{E}{\hbar c}ct\right)}$$

$$\alpha\frac{mc}{\hbar}\Psi_4 = \alpha\frac{mc}{\hbar}Ae^{i\left(\frac{p}{\hbar_x}x-\frac{E}{\hbar c}ct\right)}$$

And so:

$$-\alpha\frac{E}{\hbar c}Ae^{i\left(\frac{p_x}{\hbar}x-\frac{E}{\hbar c}ct\right)}+\frac{p}{\hbar}Ae^{i\left(\frac{p_x}{\hbar}x-\frac{E}{\hbar c}ct\right)}-\alpha\frac{mc}{\hbar}Ae^{i\left(\frac{p_x}{\hbar}x-\frac{E}{\hbar c}ct\right)}=0$$

$$-\alpha EAe^{i\left(\frac{p_x}{\hbar}x-\frac{E}{\hbar c}ct\right)}+p_xcAe^{i\left(\frac{p_x}{\hbar}x-\frac{E}{\hbar c}ct\right)}-\alpha mc^2Ae^{i\left(\frac{p_x}{\hbar}x-\frac{E}{\hbar c}ct\right)}=0 \qquad (10.48)$$

$$p_xc = \alpha E + \alpha mc^2$$

$$\alpha = \frac{p_xc}{E+mc^2}$$

That is, a wave function which satisfies the fourth wave equation is:

$$\Psi = \begin{bmatrix} \uparrow 1 \\ 0 \\ 0 \\ \downarrow \alpha \end{bmatrix} Ae^{i\left(\frac{p_x}{\hbar}x-\frac{E}{\hbar}t\right)} \qquad (10.49)$$

One of these must correspond to the negative energy solution and one to the positive energy solution. If p is set to zero, and the equations compared to the at-rest case, it can easily be seen that the solution above for the first wave equation corresponds to that obtained earlier for the negative energy solution.

Unlike the "at-rest" case, as well as the spin associated with the positive energy spin-up configuration, there is a contribution to the wave function from what must be the negative energy spin-down state. For what must be the negative energy state, in the spin-down configuration, there is a contribution from the positive energy spin-up state.

The second and third equations can be analysed in the same manner as above, and for the second equation:

$$\Psi = A\begin{bmatrix} 0 \\ \downarrow \alpha \\ \uparrow 1 \\ 0 \end{bmatrix} e^{i\left(\frac{p_x}{\hbar}x-\frac{E}{\hbar}t\right)} \text{ where } \alpha = \frac{p_xc}{E-mc^2} \qquad (10.50)$$

For the third equation:

$$\Psi = A\begin{bmatrix} 0 \\ \downarrow 1 \\ \uparrow \alpha \\ 0 \end{bmatrix} e^{i\left(\frac{p_x}{\hbar}x - \frac{E}{\hbar}t\right)} \text{ where } \alpha = \frac{p_x c}{E + mc^2} \tag{10.51}$$

In each case, there is a contribution to the wave function from the negative energy state of opposite spin.

The four solutions show spin-up and spin-down states for both positive and negative energy states, but more than that, it shows that there is a contribution to both states of spin from both positive and negative energies. That is, the negative energy states cannot be written off as being mathematical artefacts; they are an integral part of the overall solution. Further, it appears that the wave function for an electron shows that there is a weighted possibility of finding the electron in either spin state when it is moving.

We might therefore state the free-particle wave function solutions as:

$$\Psi = A\begin{bmatrix} \uparrow 1 \\ 0 \\ 0 \\ \downarrow \dfrac{p_x c}{E^+ + mc^2} \end{bmatrix} e^{i\left(\frac{p_x}{\hbar}x - \frac{E}{\hbar}t\right)}; \quad \Psi = A\begin{bmatrix} 0 \\ \downarrow 1 \\ \uparrow \dfrac{p_x c}{E^+ + mc^2} \\ 0 \end{bmatrix} e^{i\left(\frac{p_x}{\hbar}x - \frac{E}{\hbar}t\right)}$$

$$\Psi = A\begin{bmatrix} 0 \\ \downarrow \dfrac{p_x c}{E^- - mc^2} \\ \uparrow 1 \\ 0 \end{bmatrix} e^{i\left(\frac{p_x}{\hbar}x - \frac{E}{\hbar}t\right)}; \quad \Psi = A\begin{bmatrix} \uparrow \dfrac{p_x c}{E^- - mc^2} \\ 0 \\ 0 \\ \downarrow 1 \end{bmatrix} e^{i\left(\frac{p_x}{\hbar}x - \frac{E}{\hbar}t\right)} \tag{10.52}$$

The term A is a normalisation factor. Unlike the at-rest solution, these four solutions are not just the original component wave functions $\Psi_{1,2,3,4}$, they are a weighted mixture of these component wave functions.

These solutions were obtained for the case of a velocity in the x direction.

When these analyses are repeated for velocities in the y and z directions, the complete solutions are written:

$$\Psi = A\begin{bmatrix} \uparrow 1 \\ \downarrow 0 \\ \uparrow \dfrac{p_z c}{E^+ + mc^2} \\ \downarrow \dfrac{p_x c + ip_y c}{E^+ + mc^2} \end{bmatrix} e^{i\left(\frac{p}{\hbar}x - \frac{E}{\hbar}t\right)}; \quad \Psi = A\begin{bmatrix} \uparrow 0 \\ \downarrow 1 \\ \uparrow \dfrac{p_x c - ip_y c}{E^+ + mc^2} \\ \downarrow \dfrac{-p_z c}{E^+ + mc^2} \end{bmatrix} e^{i\left(\frac{p}{\hbar}x - \frac{E}{\hbar}t\right)};$$

$$\Psi = A\begin{bmatrix} \uparrow \dfrac{p_z c}{E^- - mc^2} \\ \downarrow \dfrac{p_x c + ip_y c}{E^- - mc^2} \\ \uparrow 1 \\ \downarrow 0 \end{bmatrix} e^{i\left(\frac{p}{\hbar}x - \frac{E}{\hbar}t\right)}; \quad \Psi = A\begin{bmatrix} \uparrow \dfrac{p_x c - ip_y c}{E^- - mc^2} \\ \downarrow \dfrac{-p_z c}{E^- - mc^2} \\ \uparrow 0 \\ \downarrow 1 \end{bmatrix} e^{i\left(\frac{p}{\hbar}x - \frac{E}{\hbar}t\right)}; \tag{10.53}$$

Arrows and signs have been temporarily included to remind us of which states correspond to the positive and negative energies that would result from each solution, and the spin of each state.

10.3 ANTIMATTER

Dirac was well aware that the presence of negative energy states was a significant problem in his equation. Why would an electron with positive energy not simply just fall into the negative energy states? In 1929, he surmised that these negative energy states do exist, but they were all full of electrons, undetectable by us, extending down to infinite negative energy. The Pauli exclusion principle would therefore prevent positive energy electrons from falling into it. This profusion of electrons, extending down to negative infinity energy, became known as the Dirac sea, with the positive energy electrons that we normally see floating on it. This model was widely criticised.

However, it did lead Dirac to a fruitful line of enquiry. He considered what would happen if an electron at the top of the negative energy sea were to gain positive energy, and thus make a transition from the sea into a positive energy state. The resulting hole left in the negative energy sea would behave like a positively charged hole – much like we think of the structure of a P type semiconductor.

Such a positively charged hole would act just like a particle – it having inertia and mass, the possibility of movement, etc., by virtue of the actions of the other electrons that surround it. In looking from afar, one would not be able to tell if the hole was a real particle or a hole. Thus, the promotion of a negative energy electron to the positive state would result in two particles – the electron (negatively charged but now with positive energy), and the hole (positively charged but with negative energy).

Dirac originally thought that these holes might somehow be positively charged protons, but of course the mass of a proton was very much larger than that of an electron, and so in 1931, this idea was abandoned and the hole was considered a new kind of particle with the same mass as an electron, but with positive charge. He called this new particle an anti-electron. Dirac said that we should not normally expect to see such particles in nature since they would disintegrate upon contact with a real electron and convert their mass, and that of the electron, to a photon. Only in a high vacuum would anti-electrons survive long enough to be detectable.

This startling prediction – made before any experimental supporting evidence, was one of the most significant discoveries of quantum physics.

In 1932, Carl Anderson observed the tracks of anti-electrons in cosmic rays using a cloud chamber – a vacuum system where the path of particles could be enhanced by the condensation of droplets made in the saturated vapour above dry ice. The particle was named the positron. The existence of this experimentally observed particle meant that it was no longer necessary to have a sea of negative electrons of negative energy for there to be a hole in; the positron existed as a particle in its own right.

Despite this success, there was still the problem of explaining just how sub-atomic particles can appear and disappear – such as when a photon creates an electron–positron pair (pair production), or when a positron and electron combine to form a photon. It seemed as though the basic assumption that the probability of finding an electron "somewhere" in a volume V could not be 100%. Quantum theory needed something more to explain the sometimes-fleeting nature of particles, and it was the theory of quantum electrodynamics, or QED, the first of the quantum field theories, that provided the answers.

10.4 NATURAL UNITS

Because some terms in the above derivations are often repetitive, it is customary to normalise quantities into what are called "natural units". In natural units, we replace c and $\hbar$ with 1 (See Table 10.1). This forces everything which has a dimension (length, mass, velocity, etc.) to be expressed in terms of one dimension – energy. Energy expressed in electron volts is the usual choice: specifically, MeV or GeV, where 1 GeV = 1.6×10^{-10} kg m^2 s^{-2}.

TABLE 10.1 Conversion from natural units to SI units

	NATURAL UNITS	*CONVERSION FROM SI UNITS*
Mass	GeV	$m\dfrac{c^2}{1.6\times10^{-19}}$
Length	GeV-1	$l\dfrac{1.6\times10^{-19}}{\hbar c}$
Time	GeV-1	$t\dfrac{1.6\times10^{-19}}{\hbar}$

For example, the mass, or energy of an electron, is thus expressed:

$$\begin{aligned}
E &= mc^2 \\
&= 9.1096\times10^{-31}\left(2.99793\times10^{8}\right)^2 \\
&= 8.19\times10^{-14}\ \text{J} \\
1\ \text{eV} &= 1.6022\times10^{-19}\ \text{J} \\
E &= 5.11\times10^{5}\ \text{eV} \\
&\approx 0.5\ \text{MeV}
\end{aligned} \tag{10.54}$$

In natural units, both distance and time are also measured in units of energy. Planck's constant has units of energy × time: $E \times T$. To obtain time on its own, we must divide by energy which we will show as dimension E:

$$\begin{aligned}
\hbar &= ET \\
T &= \frac{\hbar}{E} \\
&= \left(\frac{1}{E}\right)\hbar
\end{aligned} \tag{10.55}$$

The energy term is $1/E$, and so if we wish to express time t in terms of an energy unit t_E, then we have:

$$t_E = \frac{t}{\hbar} = \frac{1}{E} \tag{10.56}$$

That is, in natural units, time has the dimension $1/E$, or eV^{-1}.

For example, if the time $t = 1$ second has to be converted to natural units, we have:

$$\begin{aligned}
t_{\left(\text{eV}^{-1}\right)} &= \frac{(2\pi)1.6\times10^{-19}}{6.626\times10^{-34}} \\
&= 1.52\times10^{24}\ \text{GeV}^{-1}
\end{aligned} \tag{10.57}$$

Distances can also be converted to natural units. The product of $\hbar$ and c has the dimensions $E \times L$, and so:

$$\hbar c = \frac{ETL}{T} = EL$$
$$L = \left(\frac{1}{E}\right)\hbar c \tag{10.58}$$

To express length l in natural units ($1/E$), the conversion factor is $1/\hbar c$ for Joules^{-1}, or $1.6 \times 10^{-19}/\hbar c$ for GeV^{-1}.

Table 10.1 gives a summary of the conversion from natural units to SI units.

In natural units, we can convert back at the end of an analysis by inserting the required factors of c and $\hbar$ to make dimensions compatible on both sides of the final equation. It is found that this can only be done in a unique manner, and so nothing is lost.

It is useful to keep in mind that these fundamental constants c and $\hbar$ have dimensions of length, L, time, T and mass, M according to:

$$c = LT^{-1}$$
$$\hbar = L^2 MT^{-1} \tag{10.59}$$

In natural units, the relativistic energy equation is:

$$E^2 = p^2 + m^2 \tag{10.60}$$

and the Klein–Gordon equation becomes:

$$-\frac{\partial^2 \Psi}{\partial t^2} = -\nabla^2 \Psi + m^2 \Psi \tag{10.61}$$

This is often written:

$$-\frac{\partial^2 \Psi}{\partial t^2} + \nabla^2 \Psi - m^2 \Psi = 0 \tag{10.62}$$

Or, using the d'Alembertian:

$$\left(\square + m^2\right)\Psi(x) = 0 \tag{10.63}$$

10.5 SINGLE PARTICLE DIRAC EQUATION

As we have seen previously, the Dirac relativistic wave equation is written:

$$\gamma_0 i\hbar c \frac{\partial \Psi}{\partial ct} + \gamma_1 i\hbar c \frac{\partial \Psi}{\partial x} + \gamma_2 i\hbar c \frac{\partial \Psi}{\partial y} + \gamma_3 i\hbar c \frac{\partial \Psi}{\partial z} - mc^2 \Psi = 0 \tag{10.64}$$

In natural units this becomes:

$$\gamma_0 i\frac{\partial\Psi}{\partial t}+\gamma_1 i\frac{\partial\Psi}{\partial x}+\gamma_2 i\frac{\partial\Psi}{\partial y}+\gamma_3 i\frac{\partial\Psi}{\partial z}-m\Psi=0 \tag{10.65}$$

The differentials can be written with a numeric label μ which goes from 0 to 3:

$$\begin{aligned}\frac{\partial}{\partial t}&=\partial_t=\partial_0\\ \frac{\partial}{\partial x}&=\partial_x=\partial_1\\ \frac{\partial}{\partial y}&=\partial_y=\partial_2\\ \frac{\partial}{\partial z}&=\partial_z=\partial_3\end{aligned} \tag{10.66}$$

Thus, the Dirac equation is written:

$$\gamma_0 i\partial_0\Psi+\gamma_1 i\partial_1\Psi+\gamma_2 i\partial_2\Psi+\gamma_3 i\partial_3\Psi-m\Psi=0 \tag{10.67}$$

Or:

$$\gamma_0 i\partial_0\Psi+\gamma_1 i\partial_1\Psi+\gamma_2 i\partial_2\Psi+\gamma_3 i\partial_3\Psi=m\Psi \tag{10.68}$$

or even more compactly:*,†

$$i\gamma_\mu\partial_\mu\Psi=m\Psi \tag{10.69}$$

and this is what is engraved, in honour of Dirac, on the floor in front of Newton's tomb in Westminster Abbey.

* In our discussion so far, we have ignored the issue of what is called covariance and contravariance which is why you will see the subscript μ sometimes written as a superscript. This will be explained in a later section.

† μ goes from 0 to 3. For the gamma matrices, these are just numerical indexes. For the differentials, these are the time and space derivatives.

Probability Flow

11

11.1 INTRODUCTION

Say we draw out a definite volume V in space within which we might wish to calculate the probability of finding a particle such as an electron or a photon. At some point in time, we might find the particle inside the volume, and at other times, elsewhere. We imagine that, much like the situation of a volume enclosing a charge, that a probability wave (which in a sense carries the particle) might flow into the volume at one end and flow out of the volume at the other. An analogy might be the movement of a crime wave through a city, or a heat wave across a continent. These movements of probability constitute a probability current.

11.2 PROBABILITY CURRENT

The amplitude of the wave function is found from the solution to the time-independent Schrödinger equation. The arbitrary constant A in the general solution $\psi(x)$ could be found by a process of normalisation. That is, the total probability P of finding the electron between $-\infty$ and $+\infty$ is one.

$$1 = \int_{-\infty}^{\infty} P dx = \int_{-\infty}^{\infty} \Psi^* \Psi dx \tag{11.1}$$

P is the probability density function. For example, the probability that an electron is located within a small increment Δx around x at time t is:

$$P(x,t)\Delta x \tag{11.2}$$

For the discussion to follow, we will refer to P as the "probability".

The rate of flow of probability per unit time is called a "probability current" j.

$$j = \frac{dP}{dt} \tag{11.3}$$

The probability current *density* $\mathbf{J}$ is the probability crossing a unit cross-sectional area per unit time. $\mathbf{J}$ is a vector quantity since its value depends on the direction of current with respect to the orientation of the area under consideration.

$$\mathbf{J} = \frac{j}{\mathbf{A}} \qquad j = \mathbf{J} \cdot \mathbf{A} \tag{11.4}$$

The probability flow across the surface which encloses a volume V is expressed as a time rate of change of probability of finding the particle within the volume. That is, the time rate of increase or decrease of probability within the volume must equal the net flow of probability across the surface of the volume.

The net probability current j is found from the summation of the probability current density over all the elemental surface areas $d\mathbf{A}$:

$$j = \oint\!\!\oint_S \mathbf{J} \cdot d\mathbf{A} \tag{11.5}$$

Within the volume, we might have a net increase or decrease of probability. The time rate of change of probability within the volume can be expressed as:

$$\frac{di}{dt} = \iiint_V P\, dV \tag{11.6}$$

P is the probability density which we can calculate from the Schrödinger equation.

Since the net probability current must equal the time rate of change of probability, we have:

$$\oint\!\!\oint_S \mathbf{J} \cdot d\mathbf{A} = -\frac{\partial}{\partial t} \iiint_V P\, dV \tag{11.7}$$

The negative sign means that a decrease in the time rate of change of probability within the volume results in an outward positive current through the surface – assuming of course that we have a conservation of particles.

By the divergence theorem:

$$\oint\!\!\oint_S \mathbf{J} \cdot d\mathbf{A} = \iiint_V \vec{\nabla} \cdot \mathbf{J}\, dV \tag{11.8}$$

and so:

$$\iiint_V \vec{\nabla} \cdot \mathbf{J} dV = -\frac{\partial}{\partial t} \iiint_V P\, dV \tag{11.9}$$

The terms inside the integral must therefore be equal, and so:

$$\vec{\nabla} \cdot \mathbf{J} = -\frac{\partial P}{\partial t} \tag{11.10}$$

That is, the probability current density diverging from the volume is equal to the time rate of decrease of the probability density within it. This is a continuity equation. It represents a local conservation of charge.

We can calculate $\mathbf{J}$ from the Schrödinger equation.

$$\frac{\partial}{\partial t} \iiint_V P dV = \frac{\partial}{\partial t} \iiint_V \Psi^* \Psi\, dV = \iiint_V \left(\Psi \frac{\partial \Psi^*}{\partial t} + \Psi^* \frac{\partial \Psi}{\partial t} \right) dV \tag{11.11}$$

Now, in one dimension, and with the potential function set to zero, the Schrödinger equation becomes:

$$\begin{gathered} -\frac{\hbar^2}{2m} \frac{\partial^2 \Psi(x,t)}{\partial x^2} = i\hbar \frac{\partial \Psi(x,t)}{\partial t} \\ \frac{\partial \Psi(x,t)}{\partial t} = \frac{i\hbar}{2m} \frac{\partial^2 \Psi(x,t)}{\partial x^2} \end{gathered} \tag{11.12}$$

That is, the first time derivatives can be written as second space derivatives, and so, in one dimension, our volume V becomes a linear range x_1 to x_2:

$$\frac{\partial}{\partial t}\int_{x_1}^{x_2} P dx = \frac{\partial}{\partial t}\int_{x_1}^{x_2} \Psi^*\Psi dx = \frac{i\hbar}{2m}\int_{x_1}^{x_2} \Psi\frac{\partial^2\Psi^*}{\partial x^2} + \Psi^*\frac{\partial^2\Psi}{\partial x^2}dx \tag{11.13}$$

This can be integrated by parts (see Appendix 11) to give:

$$\frac{\partial}{\partial t}\int_{x_1}^{x_2} P dx = \frac{i\hbar}{2m}\left[\Psi^*\frac{\partial\Psi}{\partial x} - \Psi\frac{\partial\Psi^*}{\partial x}\right]_{x_1}^{x_2} \tag{11.14}$$

J is the net probability crossing a unit cross-sectional area per unit time, and this equals the time rate of change of the probability density that the particle is within the volume V. **J** is a function of both x and t.

That is:

$$\frac{\partial}{\partial t}\int_{x_1}^{x_2} P dx = -\left(\mathbf{J}(x_2,t) + \mathbf{J}(x_1,t)\right) = -\frac{i\hbar}{2m}\left[\Psi^*\frac{\partial\Psi}{\partial x} - \Psi\frac{\partial\Psi^*}{\partial x}\right]_{x_1}^{x_2}$$
$$\mathbf{J} = -\frac{i\hbar}{2m}\left[\Psi^*\frac{\partial\Psi}{\partial x} - \Psi\frac{\partial\Psi^*}{\partial x}\right] \tag{11.15}$$

The negative sign means that a decrease in the time rate of change of probability density within the volume results in an outward positive probability current density through the surface – assuming of course that we have a conservation of the number of particles. This conservation of particle number will become very important when we later move into the realm of quantum field theory.

The differential operators can be replaced by the vector differential operator, and so in three dimensions, the probability current density is written:

$$\mathbf{J}(\mathbf{r},t) = -\frac{i\hbar}{2m}\left(\Psi^*\nabla\Psi - \Psi\nabla\Psi^*\right) \tag{11.16}$$

where $\mathbf{r}$ is a position vector which defines the point in space where we wish to calculate the probability current density.

The probability current density can also be written in terms of the momentum operator:

$$\mathbf{J} = \frac{1}{2m}\left[\Psi^*\hat{p}\Psi - \Psi\hat{p}\Psi^*\right] \tag{11.17}$$

The continuity equation $\vec{\nabla}\cdot\mathbf{J} = -\frac{\partial P}{\partial t}$ is written:

$$\frac{\partial}{\partial x}J_x + \frac{\partial}{\partial y}J_y + \frac{\partial}{\partial y}J_z = -\frac{\partial P}{\partial t} \tag{11.18}$$

or:

$$\frac{\partial P}{\partial t} + \frac{\partial}{\partial x}J_x + \frac{\partial}{\partial y}J_y + \frac{\partial}{\partial y}J_z = 0 \tag{11.19}$$

This is the divergence in four dimensions. That is:

$$\nabla \cdot J_\mu = 0 \tag{11.20}$$

where:

$$\begin{aligned} J_0 &= P \\ J_{1,2,3} &= J_{x,y,z} \end{aligned} \tag{11.21}$$

The significance of this is that probability is conserved. That is, the Schrödinger equation is a one-particle equation, and that particle is neither created nor destroyed. $\Psi(x,t)$ is a one-particle wave function.

Remembering that P is the probability density and **J** is the probability current density, J_μ is commonly referred to as just the probability current, or the four-current.

11.3 THE ADJOINT DIRAC EQUATION

As we have seen previously, the Dirac relativistic wave equation is written:

$$i\gamma_0\hbar c\frac{\partial\Psi}{\partial ct} + i\gamma_1\hbar c\frac{\partial\Psi}{\partial x} + i\gamma_2\hbar c\frac{\partial\Psi}{\partial y} + i\gamma_3\hbar c\frac{\partial\Psi}{\partial z} - mc^2\Psi = 0 \tag{11.22}$$

and in natural units this becomes:

$$i\gamma_0\frac{\partial\Psi}{\partial t} + i\gamma_1\frac{\partial\Psi}{\partial x} + i\gamma_2\frac{\partial\Psi}{\partial y} + i\gamma_3\frac{\partial\Psi}{\partial z} - m\Psi = 0 \tag{11.23}$$

or:

$$i\gamma_0\partial_0\Psi + i\gamma_1\partial_1\Psi + i\gamma_2\partial_2\Psi + i\gamma_3\partial_3\Psi - m\Psi = 0 \tag{11.24}$$

or more compactly:

$$\begin{aligned} i\gamma_\mu\partial_\mu\Psi &= m\Psi \\ i\gamma_\mu\partial_\mu\Psi - m\Psi &= 0 \end{aligned} \tag{11.25}$$

where the time and spatial components are:

$$\partial_0 = \partial_t; \partial_1 = \partial_x; \partial_2 = \partial_y; \partial_3 = \partial_z \tag{11.26}$$

The Dirac equation is a partial differential matrix equation. The solution to the equation is a set of four component wave functions, each of which comprises a weighted sum of individual wave functions $\psi_{1,2,3,4}$. The wave function solution Ψ is not a four-vector. The components which make up the solution ψ do not represent x, y, z and t dimensions.

For now, we will begin with the Dirac equation written as:

$$i\gamma_1\frac{\partial\Psi}{\partial x} + i\gamma_2\frac{\partial\Psi}{\partial y} + i\gamma_3\frac{\partial\Psi}{\partial z} - m\Psi = -i\gamma_0\frac{\partial\Psi}{\partial t} \tag{11.27}$$

To go further, we now require the Hermitian conjugate of the Dirac equation. This is found by taking the complex conjugate transpose of Ψ:

$$\bar{\Psi} = \left(\Psi^*\right)^{\mathrm{T}} \tag{11.28}$$

We begin with the first term of the Dirac equation. Written out in full matrix form (without the factor *i*), this is:

$$\gamma_0 \frac{\partial \Psi}{\partial t} = \begin{bmatrix} 1 & 0 & 0 & 0 \\ 0 & 1 & 0 & 0 \\ 0 & 0 & -1 & 0 \\ 0 & 0 & 0 & -1 \end{bmatrix} \begin{bmatrix} \frac{\partial}{\partial t}\Psi_1 \\ \frac{\partial}{\partial t}\Psi_2 \\ \frac{\partial}{\partial t}\Psi_3 \\ \frac{\partial}{\partial t}\Psi_4 \end{bmatrix} \tag{11.29}$$

Remember that because this is the product of two matrices, the order in which the matrices are written is important.

Now, the transpose of the product of two matrices is the product of the transpose of each of these matrices in reverse order. That is:

$$(\mathbf{AB})^{\mathrm{T}} = \mathbf{B}^{\mathrm{T}}\mathbf{A}^{\mathrm{T}} \tag{11.30}$$

As well, the transpose of the summation of several matrices is the summation of the transpose of each:

$$(\mathbf{A}+\mathbf{B})^{\mathrm{T}} = \mathbf{A}^{\mathrm{T}} + \mathbf{B}^{\mathrm{T}} \tag{11.31}$$

And so:

$$\begin{aligned} \left(\gamma_0 \frac{\partial \Psi}{\partial t}\right)^{\mathrm{T}} &= \begin{bmatrix} \frac{\partial}{\partial t}\Psi_1 \\ \frac{\partial}{\partial t}\Psi_2 \\ \frac{\partial}{\partial t}\Psi_3 \\ \frac{\partial}{\partial t}\Psi_4 \end{bmatrix}^{\mathrm{T}} \begin{bmatrix} 1 & 0 & 0 & 0 \\ 0 & 1 & 0 & 0 \\ 0 & 0 & -1 & 0 \\ 0 & 0 & 0 & -1 \end{bmatrix}^{\mathrm{T}} \\ &= \begin{bmatrix} \frac{\partial}{\partial t}\Psi_1 & \frac{\partial}{\partial t}\Psi_2 & \frac{\partial}{\partial t}\Psi_3 & \frac{\partial}{\partial t}\Psi_4 \end{bmatrix} \begin{bmatrix} 1 & 0 & 0 & 0 \\ 0 & 1 & 0 & 0 \\ 0 & 0 & -1 & 0 \\ 0 & 0 & 0 & -1 \end{bmatrix} \\ &= \frac{\partial}{\partial t}\Psi^{\mathrm{T}}\gamma_0{}^{\mathrm{T}} \\ &= \frac{\partial}{\partial t}\Psi^{+}\gamma_0 \end{aligned} \tag{11.32}$$

Similarly, for the other terms:

$$\left(\gamma_1 \frac{\partial \Psi}{\partial x}\right)^{\mathrm{T}} = -\frac{\partial}{\partial x}\Psi^{\mathrm{T}}\gamma_1$$
$$\left(\gamma_2 \frac{\partial \Psi}{\partial y}\right)^{\mathrm{T}} = -\frac{\partial}{\partial y}\Psi^{\mathrm{T}}\gamma_2 \tag{11.33}$$
$$\left(\gamma_3 \frac{\partial \Psi}{\partial z}\right)^{\mathrm{T}} = -\frac{\partial}{\partial z}\Psi^{\mathrm{T}}\gamma_3$$

Now reversing the signs of the i factors, the Hermitian conjugate Dirac equation is thus:

$$-i\frac{\partial \Psi^+}{\partial x}\gamma_1{}^+ - i\frac{\partial \Psi^+}{\partial y}\gamma_2{}^+ - i\frac{\partial \Psi^+}{\partial z}\gamma_3{}^+ - m\Psi^+ = i\frac{\partial \Psi^+}{\partial t}\gamma_0{}^+ \tag{11.34}$$

Note also the order of writing the matrices in the conjugate equation has been reversed. The $^+$ superscript indicates the Hermitian conjugate of the matrices.

It is a property of the $\gamma_{1,2,3}$ matrices that:

$$\gamma^+ = \gamma_0\gamma_\mu\gamma_0 = -\gamma_\mu \text{ and } -\gamma_\mu\gamma_0 = \gamma_0\gamma_\mu \tag{11.35}$$

For example:

$$\gamma_0\gamma_1 = \begin{bmatrix} 1 & 0 & 0 & 0 \\ 0 & 1 & 0 & 0 \\ 0 & 0 & -1 & 0 \\ 0 & 0 & 0 & -1 \end{bmatrix} \begin{bmatrix} 0 & 0 & 0 & 1 \\ 0 & 0 & 1 & 0 \\ 0 & -1 & 0 & 0 \\ -1 & 0 & 0 & 0 \end{bmatrix} = \begin{bmatrix} 0 & 0 & 0 & 1 \\ 0 & 0 & 1 & 0 \\ 0 & -1 & 0 & 0 \\ -1 & 0 & 0 & 0 \end{bmatrix}$$

$$\gamma_0\gamma_1\gamma_0 = \begin{bmatrix} 0 & 0 & 0 & 1 \\ 0 & 0 & 1 & 0 \\ 0 & -1 & 0 & 0 \\ -1 & 0 & 0 & 0 \end{bmatrix} \begin{bmatrix} 1 & 0 & 0 & 0 \\ 0 & 1 & 0 & 0 \\ 0 & 0 & -1 & 0 \\ 0 & 0 & 0 & -1 \end{bmatrix} = \begin{bmatrix} 0 & 0 & 0 & -1 \\ 0 & 0 & -1 & 0 \\ 0 & 1 & 0 & 0 \\ 1 & 0 & 0 & 0 \end{bmatrix} \tag{11.36}$$
$$= -\gamma_1 = \gamma^+$$

Thus:

$$-i\frac{\partial \Psi^+}{\partial x}(-\gamma_1) - i\frac{\partial \Psi^+}{\partial y}(-\gamma_2) - i\frac{\partial \Psi^+}{\partial z}(-\gamma_3) - m\Psi^+ = i\frac{\partial \Psi^+}{\partial t}\gamma_0 \tag{11.37}$$

The adjoint of Ψ is defined as:

$$\bar{\Psi} = \Psi^+\gamma_0 \tag{11.38}$$

And so, multiplying through by γ_0:

$$-i\frac{\partial \Psi^+}{\partial x}(-\gamma_1\gamma_0) - i\frac{\partial \Psi^+}{\partial y}(-\gamma_2\gamma_0) - i\frac{\partial \Psi^+}{\partial z}(-\gamma_3\gamma_0) - m\Psi^+\gamma_0 = i\frac{\partial \Psi^+}{\partial t}\gamma_0\gamma_0 \tag{11.39}$$

$$-i\frac{\partial\Psi^+}{\partial x}(\gamma_0\gamma_1)-i\frac{\partial\Psi^+}{\partial y}(\gamma_0\gamma_2)-i\frac{\partial\Psi^+}{\partial z}(\gamma_0\gamma_3)-m\Psi^+\gamma_0=i\frac{\partial\Psi^+}{\partial t}\gamma_0\gamma_0$$

$$-i\frac{\partial\bar{\Psi}}{\partial x}(\gamma_1)-i\frac{\partial\bar{\Psi}}{\partial y}(\gamma_2)-i\frac{\partial\bar{\Psi}}{\partial z}(\gamma_3)-m\bar{\Psi}=i\frac{\partial\bar{\Psi}}{\partial t}\gamma_0$$

Multiplying through by −1 gives us the adjoint Dirac equation:

$$i\frac{\partial\bar{\Psi}}{\partial x}\gamma_1+i\frac{\partial\bar{\Psi}}{\partial y}\gamma_2+i\frac{\partial\bar{\Psi}}{\partial z}\gamma_3+m\bar{\Psi}=-i\frac{\partial\bar{\Psi}}{\partial t}\gamma_0 \tag{11.40}$$

In anticipation of what is to follow, we now multiply the Dirac equation by the adjoint Ψ on the left, and multiply the adjoint Dirac equation by Ψ on the right, and divide through by i:

$$\bar{\Psi}\left(\gamma_1\frac{\partial\Psi}{\partial x}+\gamma_2\frac{\partial\Psi}{\partial y}+\gamma_3\frac{\partial\Psi}{\partial z}-m\Psi\right)=-\bar{\Psi}\gamma_0\frac{\partial\Psi}{\partial t} \tag{11.41}$$

and:

$$\left(\frac{\partial\bar{\Psi}}{\partial x}\gamma_1+\frac{\partial\bar{\Psi}}{\partial y}\gamma_2+\frac{\partial\bar{\Psi}}{\partial z}\gamma_3+m\bar{\Psi}\right)\Psi=-\frac{\partial\bar{\Psi}}{\partial t}\gamma_0\Psi \tag{11.42}$$

Adding these together, we have:

$$\bar{\Psi}\left(\gamma_1\frac{\partial\Psi}{\partial x}+\gamma_2\frac{\partial\Psi}{\partial y}+\gamma_3\frac{\partial\Psi}{\partial z}-m\Psi\right)+\left(\frac{\partial\bar{\Psi}}{\partial x}\gamma_1+\frac{\partial\bar{\Psi}}{\partial y}\gamma_2+\frac{\partial\bar{\Psi}}{\partial z}\gamma_3+m\bar{\Psi}\right)\Psi=-\bar{\Psi}\gamma_0\frac{\partial\Psi}{\partial t}+-\frac{\partial\bar{\Psi}}{\partial t}\gamma_0\Psi \tag{11.43}$$

Using the product rule for derivatives on the right-hand side and eliminating the mass terms, we obtain:

$$\begin{gathered}\bar{\Psi}\gamma_1\frac{\partial\Psi}{\partial x}+\bar{\Psi}\gamma_2\frac{\partial\Psi}{\partial y}+\bar{\Psi}\gamma_3\frac{\partial\Psi}{\partial z}+\frac{\partial\bar{\Psi}}{\partial x}\gamma_1\Psi+\frac{\partial\bar{\Psi}}{\partial y}\gamma_2\Psi+\frac{\partial\bar{\Psi}}{\partial z}\gamma_3\Psi=-\frac{\partial}{\partial t}\left(\bar{\Psi}\gamma_0\Psi\right)\\ \left(\bar{\Psi}\gamma_1\frac{\partial\Psi}{\partial x}+\frac{\partial\bar{\Psi}}{\partial x}\gamma_1\Psi\right)+\left(\bar{\Psi}\gamma_2\frac{\partial\Psi}{\partial y}+\frac{\partial\bar{\Psi}}{\partial y}\gamma_2\Psi\right)+\left(\gamma_3\frac{\partial\Psi}{\partial z}+\frac{\partial\bar{\Psi}}{\partial z}\gamma_3\Psi\right)=-\frac{\partial}{\partial t}\left(\bar{\Psi}\gamma_0\Psi\right)\end{gathered} \tag{11.44}$$

Note, we can eliminate the mass terms because:

$$-\bar{\Psi}m\Psi+m\bar{\Psi}\Psi=-m\bar{\Psi}\Psi+m\bar{\Psi}\Psi=0 \tag{11.45}$$

Now, by the product rule:

$$\begin{aligned}\frac{\partial\left(\bar{\Psi}\gamma_1\Psi\right)}{\partial x}&=\bar{\Psi}\gamma_1\frac{\partial\Psi}{\partial x}+\Psi\frac{\partial\bar{\Psi}}{\partial x}\gamma_1\\&=\bar{\Psi}\gamma_1\frac{\partial\Psi}{\partial x}+\frac{\partial\bar{\Psi}}{\partial x}\gamma_1\Psi\end{aligned} \tag{11.46}$$

and so:

$$\frac{\partial(\bar{\Psi}\gamma_1\Psi)}{\partial x}+\frac{\partial(\bar{\Psi}\gamma_2\Psi)}{\partial y}+\frac{\partial(\bar{\Psi}\gamma_3\Psi)}{\partial z}=-\frac{\partial}{\partial t}(\bar{\Psi}\gamma_0\Psi) \tag{11.47}$$

That is:

$$\vec{\nabla}\cdot(\bar{\Psi}\gamma_\mu\Psi)=-\frac{\partial}{\partial t}(\bar{\Psi}\gamma_0\Psi) \tag{11.48}$$

Comparing with the equation of continuity:

$$\vec{\nabla}\cdot\mathbf{J}=-\frac{\partial P}{\partial t} \tag{11.49}$$

then the probability current can be expressed:

$$\begin{aligned} J_i &= \bar{\Psi}\gamma_i\Psi; \quad i=1,2,3 \\ P &= \bar{\Psi}\gamma_0\Psi \end{aligned} \tag{11.50}$$

In the above, we've taken a somewhat long-hand approach to the derivations, but we can also proceed in more compact notation. The Dirac equation is:

$$(i\gamma_\mu\partial_\mu - m)\Psi = 0 \tag{11.51}$$

and the adjoint Dirac equation:

$$\bar{\Psi}(i\overleftarrow{\partial}_\mu\gamma_\mu + m) = 0 \tag{11.52}$$

In the above, $\bar{\Psi}$ has been written on the left which implies that the differential within the brackets acts to the left. We will see that ordering the terms this way leads to a very elegant form of the final result.

Multiplying the Dirac equation by $\bar{\Psi}$ on the left and the adjoint Dirac equation by Ψ on the right, we obtain:

$$\bar{\Psi}(i\gamma_\mu\partial_\mu - m)\Psi = 0 \tag{11.53}$$

and:

$$\bar{\Psi}(i\overleftarrow{\partial}_\mu\gamma_\mu + m)\Psi = 0 \tag{11.54}$$

Allowing the differential operator to act to the left:

$$i\partial_\mu\bar{\Psi}\gamma_\mu\Psi + \bar{\Psi}m\Psi = 0 \tag{11.55}$$

Adding the two, eliminating the mass terms and dividing by i gives:

$$\begin{aligned} i\partial_\mu\bar{\Psi}\gamma_\mu\Psi + \bar{\Psi}m\Psi + \bar{\Psi}i\gamma_\mu\partial_\mu\Psi - \bar{\Psi}m\Psi = 0 \\ \partial_\mu\bar{\Psi}\gamma_\mu(\Psi) + \bar{\Psi}\gamma_\mu(\partial_\mu\Psi) = 0 \end{aligned} \tag{11.56}$$

And by the product rule for differentials, this becomes:

$$\partial_\mu \left(\bar{\Psi} \gamma_\mu \Psi \right) = 0 \tag{11.57}$$

This has the same form as the four-gradient of the current density:

$$\nabla \cdot J_\mu = 0 \tag{11.58}$$

Thus:*

$$J_\mu = \bar{\Psi} \gamma_\mu \Psi; \ \mu = 0,1,2,3 \tag{11.59}$$

The four-vector probability current J_μ combines the probability current density **J** (which comes from the spatial components) and the probability density P (which comes from the time-like components of the continuity equation).

* Remember that the γ matrices are not four-vectors.

Wave Functions and Spinors

12

12.1 PARTICLES

In complex exponential form, the electron wave function $\Psi(x,t)$ is written in much the same way as any other wave:

$$\Psi(x,t) = Ae^{i(kx-\omega t)} \tag{12.1}$$

Using the de Broglie relationship, we can write the wave function in terms of momentum and energy. In one dimension, this is written:

$$\Psi(x,t) = Ae^{i\left(\frac{p_x}{\hbar}x-\frac{E}{\hbar}t\right)} \tag{12.2}$$

Extended to three dimensions, the combined exponent of the wave function has the form of a dot product between momentum and position.

$$\Psi(\vec{x},t) = Ae^{i\left(\frac{p_x}{\hbar}\cdot\vec{x}-\frac{E}{\hbar}t\right)} \tag{12.3}$$

In four dimensions, and in natural units:

$$x_\mu = (t,\vec{x}) \tag{12.4}$$

and:

$$p_\mu = (E,\vec{p}) = (E,p_x,p_y,p_z) \tag{12.5}$$

and so, in keeping with the required minus signs for the dot product between two four-vectors, we have:

$$p_\mu \cdot x_\mu = Et - p_x x - p_y y - p_z z \tag{12.6}$$

But, in all our wave functions, we have kept the order $kx - \omega t$, and so we need an additional minus sign so that:

$$\Psi = Ae^{-ip_\mu \cdot x_\mu} \tag{12.7}$$

The amplitude A has particular significance. On its own, it is a normalisation factor, but it is more than that. As we've seen in Chapter 10, the relativistic wave function for an electron is a four-component function.

For the electron-at-rest case, the solution to the wave equation is a collection of four, four-component wave functions:

$$\Psi = A\begin{bmatrix}1\\0\\0\\0\end{bmatrix}e^{-iEt};\ \Psi = A\begin{bmatrix}0\\1\\0\\0\end{bmatrix}e^{-iEt};\ \Psi = A\begin{bmatrix}0\\0\\1\\0\end{bmatrix}e^{-iEt};\ \Psi = A\begin{bmatrix}0\\0\\0\\1\end{bmatrix}e^{-iEt} \tag{12.8}$$

The last two solutions are for the electron negative energy states, where:

$$E = -mc^2$$

That is, when E is determined from those solutions, the result is a negative number.

For the constant velocity case, the four wave function solutions to the four-component Dirac wave equations are:

$$\begin{aligned}
&\Psi = A_1\begin{bmatrix}1\\0\\\dfrac{p_z}{E+m}\\\dfrac{p_x+ip_y}{E+m}\end{bmatrix}e^{-i(p_\mu\cdot x_\mu)};\ \Psi = A_2\begin{bmatrix}0\\1\\\dfrac{p_x-ip_y}{E+m}\\\dfrac{-p_z}{E+m}\end{bmatrix}e^{-i(p_\mu\cdot x_\mu)};\\
&\Psi = A_3\begin{bmatrix}\dfrac{p_z}{E-m}\\\dfrac{p_x+ip_y}{E-m}\\1\\0\end{bmatrix}e^{-i(p_\mu\cdot x_\mu)};\ \Psi = A_4\begin{bmatrix}\dfrac{p_x-ip_y}{E-m}\\\dfrac{-p_z}{E-m}\\0\\1\end{bmatrix}e^{-i(p_\mu\cdot x_\mu)}
\end{aligned} \tag{12.9}$$

For the positive energy states, with the constants A_1 and A_2, there is a contribution from the negative energy state of opposite spin, and for the negative energy states, there is a contribution from the positive energy state of opposite spin.

The four Dirac wave equations each have a wave function solution. That is, there are four solutions, each of which has a four-component amplitude term multiplied by an exponential phase component.

The constants A are normalisation factors. They are equal to $\sqrt{E+m}$.*

So, for the wave function, in terms of the four-momentum, we write:

$$\Psi = u(p_\mu)e^{-ip_\mu\cdot x_\mu} \tag{12.10}$$

The amplitude term is written $u(p_\mu)$ because they are functions of both E and p. The four-component $u(p_\mu)$ is called a spinor, to highlight the fact that it is not four-vector.

What then do we calculate as the probability density?

$$\begin{aligned}
\Psi^*\Psi &= u(p_\mu)^* e^{+ip_\mu\cdot x_\mu}u(p_\mu)e^{-ip_\mu\cdot x_\mu}\\
&= u(p_\mu)^* u(p_\mu)
\end{aligned} \tag{12.11}$$

* This arises from a normalisation condition of having $2E$ particles per unit volume of space.

The Dirac solution, the relativistic wave equation, incorporates spin as part of the probability density.

In the relativistic quantum mechanics studied here, we have dealt with a free electron, not a bound electron. Even so, for a free electron, there is still a periodicity in space bound up in the four-momentum.

Four-vectors represent a set of four equations. It is only when there is a differential operator, or a dot product involved, that they combine to form a scalar. Thus, the four-dimensional form of the wave function is equivalent to:

$$\begin{aligned}\Psi &= u(p_\mu)e^{-ip_\mu \cdot x_\mu}\\ &= u(p_\mu)e^{i(p_x x + p_y y + p_z z - Et)}\end{aligned} \tag{12.12}$$

Or, written as components:

$$\begin{aligned}\Psi_t(t) &= e^{-Et}\\ \Psi_x(x) &= \psi(x) = u(p_x)e^{i(p_x x)}\\ \Psi_y(y) &= \psi(y) = u(p_y)e^{i(p_y y)}\\ \Psi_z(z) &= \psi(z) = u(p_z)e^{i(p_z z)}\end{aligned} \tag{12.13}$$

It is convenient to define an adjoint wave function as:

$$\bar{\Psi} = \bar{u}(p_\mu)e^{ip_\mu \cdot x_\mu} \tag{12.14}$$

where the sign on the exponent is now reversed. The amplitude term, $\bar{\mu}(p_\mu)$ is called the adjoint spinor. The adjoint of a matrix, like the spinors, is the complex conjugate transpose matrix.

Note that:

$$\begin{aligned}\bar{\Psi}\Psi &= \bar{u}(p_\mu)e^{ip_\mu \cdot x_\mu}u(p_\mu)e^{-ip_\mu \cdot x_\mu}\\ &= \bar{u}(p_\mu)u(p_\mu)\end{aligned} \tag{12.15}$$

and so, the probability current density can be economically written:

$$J_\mu = \bar{u}\gamma_\mu u \tag{12.16}$$

J_μ is commonly referred to as the four-vector current. Even though the spinors are not four-vectors, the probability current is a four-vector because of the way in which it transforms under a Lorentz transformation. This current will have particular significance when it comes to interpreting the interactions in quantum electrodynamics.

12.2 DIRAC SPINORS

Previously, the Dirac equation was developed in somewhat of a long-handed way:

$$\gamma_0 i\hbar c\frac{\partial\Psi}{\partial ct} + \gamma_1 i\hbar c\frac{\partial\Psi}{\partial x} + \gamma_2 i\hbar c\frac{\partial\Psi}{\partial y} + \gamma_3 i\hbar c\frac{\partial\Psi}{\partial z} - mc^2\Psi = 0 \tag{12.17}$$

Because the multiplying γ coefficients are 4×4 matrices, the Dirac equation is really a system of four simultaneous linear partial differential equations with plane wave solutions.

In natural units, the Dirac equation is:

$$i\gamma_\mu\partial_\mu\Psi - m\Psi = 0 \tag{12.18}$$

This is the Dirac equation for a single particle, such as an electron. In fact, it is applicable for any particle with a half-integer spin. Particles with a half-integer spin are called fermions and include particles, such as electrons, protons and neutrons.

Consider the wave function:

$$\Psi = u\left(p_\mu\right)e^{-ip_\mu \cdot x_\mu} \tag{12.19}$$

Now, taking the derivative with respect to x_μ:

$$\frac{\partial\Psi}{\partial x_\mu} = -ip_\mu\Psi \tag{12.20}$$

Substituting into the Dirac equation, and eliminating the common exponential term:

$$\begin{aligned} i\gamma_\mu\partial_\mu\Psi - m\Psi &= 0 \\ -i\gamma_\mu ip_\mu\Psi - m\Psi &= 0 \\ -i\gamma_\mu ip_\mu ue^{-ip_\mu x_\mu} - mue^{-ip_\mu x_\mu} &= 0 \\ \gamma_\mu p_\mu u - mu &= 0 \\ \left(\gamma_\mu p_\mu - m\right)u &= 0 \end{aligned} \tag{12.21}$$

This is the Dirac equation in terms of the four-momentum p_μ. The term u is the set of Dirac spinors with a normalisation factor A. Although the subscript μ for the gamma matrices signifies the four-space components t, x, y, z, the gamma matrices themselves do not comprise a four-vector.

To see how this works, consider again the electron-at-rest case. For the electron at rest, the space terms $\mu = 1, 2, 3$ drop out, and $p_\mu = E$, and so we have:

$$\Psi\left(t\right) = ue^{iEt} \tag{12.22}$$

And so, by the Dirac equation:

$$\begin{aligned} \left(\gamma_0 E - m\right)u &= 0 \\ \gamma_0 Eu &= mu \end{aligned} \tag{12.23}$$

Now, we must remember that the wave function Ψ will have four components (not to be confused with t, x, y, z): $\Psi_{1,2,3,4}$. We might call the associated spinors, $u_{1,2,3,4}$.

Multiplying through by γ_0, we obtain:

$$E\begin{bmatrix} 1 & 0 & 0 & 0 \\ 0 & 1 & 0 & 0 \\ 0 & 0 & -1 & 0 \\ 0 & 0 & 0 & -1 \end{bmatrix}\begin{bmatrix} u_1 \\ u_2 \\ u_3 \\ u_4 \end{bmatrix} = m\begin{bmatrix} u_1 \\ u_2 \\ u_3 \\ u_4 \end{bmatrix} \tag{12.24}$$

$$E\begin{bmatrix} u_1 \\ u_2 \\ -u_3 \\ -u_4 \end{bmatrix} = m\begin{bmatrix} u_1 \\ u_2 \\ u_3 \\ u_4 \end{bmatrix}$$

Or more explicitly:

$$\begin{aligned} Eu_1 &= mu_1 \\ Eu_2 &= mu_2 \\ Eu_3 &= -mu_3 \\ Eu_4 &= -mu_4 \end{aligned} \tag{12.25}$$

That is, there are two solutions with $E = m$ (in natural units) and two with $E = -m$, the latter being the negative energy solutions.

The four solutions are thus:

$$\Psi = \begin{bmatrix} 1 \\ 0 \\ 0 \\ 0 \end{bmatrix} ue^{-i\frac{E}{\hbar}t}; \ \Psi = \begin{bmatrix} 0 \\ 1 \\ 0 \\ 0 \end{bmatrix} ue^{-i\frac{E}{\hbar}t}; \ \Psi = \begin{bmatrix} 0 \\ 0 \\ 1 \\ 0 \end{bmatrix} ue^{-i\frac{E}{\hbar}t}; \ \Psi = \begin{bmatrix} 0 \\ 0 \\ 0 \\ 1 \end{bmatrix} ue^{-i\frac{E}{\hbar}t} \tag{12.26}$$

Do not confuse the four-component wave function spinors with the four wave function solutions. In the electron-at-rest case, each spinor does have only one of the wave function components contained within it, but as we saw for the constant-velocity case, a spinor may contain contributions from each of the component wave functions. The four wave functions Ψ have not been numbered to avoid this confusion.

12.3 ANTIPARTICLES

For the constant velocity case, the Dirac spinors are:

$$u_1 = A\begin{bmatrix} 1 \\ 0 \\ \dfrac{p_z}{E+m} \\ \dfrac{p_x + ip_y}{E+m} \end{bmatrix}; \ u_2 = A\begin{bmatrix} 0 \\ 1 \\ \dfrac{p_x - ip_y}{E+m} \\ \dfrac{-p_z}{E+m} \end{bmatrix}; \ u_3 = A\begin{bmatrix} \dfrac{p_z}{E-m} \\ \dfrac{p_x + ip_y}{E-m} \\ 1 \\ 0 \end{bmatrix}; \ u_4 = A\begin{bmatrix} \dfrac{p_x - ip_y}{E-m} \\ \dfrac{-p_z}{E-m} \\ 0 \\ 1 \end{bmatrix} \tag{12.27}$$

The negative energy spinors u_3 and u_4 correspond to the wave functions for antiparticles. That is, E in the solutions u_3 and u_4 is understood to be negative.

From the point of view of a particle, an antiparticle has negative energy and travels backwards in time. From the point of view of an antiparticle, its energy is positive and it moves forward in time. We might want to take this latter point of view to avoid remembering whether E is to be a positive or negative number. For the two negative energy solutions, the signs of p and E are reversed so that the wave functions are now:

$$\Psi' = v\left(p_\mu\right)e^{+ip_\mu \cdot x_\mu} \tag{12.28}$$

Where now:

$$v_1 = A\begin{bmatrix} \dfrac{p_x - ip_y}{E+m} \\ \dfrac{-p_z}{E+m} \\ 0 \\ 1 \end{bmatrix}; \quad v_2 = A\begin{bmatrix} \dfrac{p_z}{E+m} \\ \dfrac{p_x + ip_y}{E+m} \\ 1 \\ 0 \end{bmatrix} \tag{12.29}$$

Spinor u_4 is now v_1, and spinor u_3 is now v_2. This happens because if the previous analysis were repeated with $-E$, the spin changes as well as the sign. Note that comparing u_1 and v_1, and also u_2 and v_2, the sign of p is also reversed. The energy E is now understood to be positive.

In terms of the wave functions, the two original negative energy solutions are now related to the two new positive antiparticle solutions:

$$\begin{aligned} u_4\left(-p_\mu\right)e^{-ip_\mu \cdot x_\mu} &= v_1\left(p_\mu\right)e^{ip_\mu \cdot x_\mu} \\ u_3\left(-p_\mu\right)e^{-ip_\mu \cdot x_\mu} &= v_2\left(p_\mu\right)e^{ip_\mu \cdot x_\mu} \end{aligned} \tag{12.30}$$

$$-p_\mu = \left(-E,\ -\vec{p}\right) = \left(-E, -p_x, -p_y, -p_z\right) \tag{12.31}$$

For antiparticles, the Dirac equation is written:

$$\left(\gamma_\mu p_\mu + m\right)v = 0 \tag{12.32}$$

noting the + sign in front of the m.

The two particle positive energy solutions are:

$$u_1 = A_3\begin{bmatrix} 0 \\ 1 \\ \dfrac{p_x - ip_y}{E+m} \\ \dfrac{-p_z}{E+m} \end{bmatrix}; \quad u_2 = A_4\begin{bmatrix} 1 \\ 0 \\ \dfrac{p_z}{E+m} \\ \dfrac{p_x + ip_y}{E+m} \end{bmatrix} \tag{12.33}$$

The adjoint spinors are:

$$\bar{u} = \left(u^*\right)^{\mathrm{T}} \gamma_0 = u^+\gamma_0 \tag{12.34}$$

So, the adjoint spinors are now a row matrix of the form:

$$\bar{u} = \left[u_1^* \quad u_2^* \quad -u_3^* \quad -u_4^*\right] \tag{12.35}$$

For example:

$$\bar{u}_1 = A_3\begin{bmatrix} 0 & 1 & \dfrac{p_x + ip_y}{E+m} & \dfrac{-p_z}{E+m} \end{bmatrix} \tag{12.36}$$

and:

$$\bar{u}_2 = A_4\begin{bmatrix} 1 & 0 & \dfrac{p_z}{E+m} & \dfrac{p_x - ip_y}{E+m} \end{bmatrix} \tag{12.37}$$

As far as probabilities go, the Dirac spinors are the important quantities. When a particle such as an electron travels from place to place, it will be the spinor that determines the amplitude for the electron to take a particular path. The spinor contains within it the energy of the particle and its state of spin. The exponential term adds phase information. The phase information is important when the amplitudes, the probabilities, for different paths have to be added together.

A positive energy antiparticle going forward in time is interpreted as a negative energy particle travelling backwards in time.

Classical Field Theory 13

13.1 CLASSICAL FIELD THEORY

A field can be thought of as some quantity which is assigned to every point in space.

A scalar field (e.g. temperature) is represented by a single value, which might be a function of x, y, z. In matrix form, we write:

$$T = \left[T(x,y,z)\right] \tag{13.1}$$

A vector field (e.g. the electric field) is represented by three components, $\mathbf{E} = (E_x, E_y, E_z)$, where each of these components may themselves be a function of position x, y, z. That is:

$$\begin{aligned} E_x &= f(x,y,z) \\ E_y &= g(x,y,z) \\ E_z &= h(x,y,z) \end{aligned} \tag{13.2}$$

For a vector, the three components E_x, E_y and E_z form the vector field **E**. Each component is a vector quantity aligned with the relevant coordinate axis. Often, however, we are interested in the magnitude of the component vectors and the unit vectors **i**, **j**, **k** are only included when these components have to be interpreted as vectors.

In the mechanical system of Section 9.6, the particle, a ball, had a height (labelled x) above the ground which was a function of t. In that example, the action S was computed as the particle moved through the one degree of freedom and determined the path from the integration of the Lagrangian over time.

In the following sections, the coordinate x is replaced with the value of the field ϕ. Using the principle of least action, the dynamical properties, that is, the equations of motion of the field, can be determined.

Does a field have dynamical properties? Yes. When we see an animated barometric weather map showing the movement of isobars over a continent, we see movement of the values of the field in space and time. There are also strong field gradients where the isobars are bunched close together. In the most general sense, the dynamics of a field follow the principle of least action. And for this, we need the Lagrangian. Why? Because we need to find the equations of motion. We may not know them beforehand.

13.2 ACTION

The principle of least action seeks to obtain the minimum of an integral involving the sum of the differences between two functions as we proceed from one point in space to another. In mechanics, these functions are usually expressed in terms of kinetic and potential energy, and the integration is done with respect to time:

$$L = E_{KE} - E_{PE}$$

$$S = \int_{t_1}^{t_2} \left(E_{KE} - E_{PE} \right) dt \tag{13.3}$$

In Section 9.6 it was shown that if we vary the path of the particle from say x_0 to x, the value of the action S may change from S_0 to S'. If the variation had resulted in a linear (first-order) change in S, then this new path is not on the minimum path. Any difference between the new path and the true minimum path is second order. So, by analysing the first-order changes only, the condition for the minimum path is when the first order change is zero. The Lagrangian in that case was:

$$S = \int_{t_1}^{t_2} \left(\frac{1}{2} m \left(\frac{dx}{dt} \right)^2 - mgx \right) dt \tag{13.4}$$

That is, we have a function involving the first derivative of a quantity x with respect to time, and a function involving the quantity x directly. Further, in that example, the function involving the derivative was a quadratic function $(dx/dt)^2$. In general terms:

$$L = L\left(x(t), x'(t)\right) \tag{13.5}$$

Now, instead of the quantity x being the distance above the ground, x is now treated as if it were a field. The variable x is replaced by the value of the field $\phi(t, x, y, z)$. We would like to know the dynamics of this field, that is, its equations of motion. We need to compute the change in action if we vary the field a little bit and determine the path of least action. The procedure will be exactly the same as that of the mechanical system of Section 9.6.

The Lagrangian is written in terms of the value of the field and the first derivative of the field with respect *to each of the four coordinates*.* In four dimensions:

$$\mathcal{L} = \mathcal{L}(\varphi, \nabla\varphi) \tag{13.6}$$

where $\phi = \phi(t, x, y, z)$ and ∇ is the four-differential operator. We could write ct instead of t, but for the most part we will work in natural units. In space-time, time is now a distance, and so in this context, the Lagrangian is formed from a derivative with respect to four-distance rather than time.

The action along a four-dimensional path C within the field is thus:

$$\begin{aligned} S &= \int_C \mathcal{L}(\varphi, \nabla\varphi)\, dt, dx, dy, dz \\ &= \int_C \mathcal{L}(\varphi, \nabla\varphi)\, d^4x \end{aligned} \tag{13.7}$$

The 4 superscript on the differential signifies the four dimensions attached to what is now the general variable x. Whenever x appears in the integrand, we will just write x for x_μ since the d^4x will remind us of the four dimensions. If x appears outside the integrand, we will write x_μ.

* When the action integral is taken over four-dimensional space, the Lagrangian is more formally referred to as the Lagrangian density and given the symbol $\mathfrak{L}$. When the action integral is done with respect to time, the Lagrangian is given the symbol L and referred to as the Lagrangian. The term Lagrangian is, however, usually used for both L and $\mathfrak{L}$.

13.3 THE LAGRANGIAN

Having written the general form for the action S, it is of value to look at the Lagrangian itself. If we are talking about energies, then the Lagrangian is the difference in kinetic and potential energies. Let us define the kinetic term of the Lagrangian as being proportional to $\frac{1}{2}(\nabla\varphi)^2$ and the potential term $V(\phi)$. Constants of proportionality can be added later if needed.

By convention, a factor of ½ in included in the Lagrangian so that the constant of proportionality, for a mechanical system, is the mass m. In the case of a field, the mass has no real physical significance, it is just a constant. The ½ factor could be left out and put in later, but it is included here, even for a field, by convention, since the resulting Euler–Lagrange equations lead to familiar forms of Newtonian equations of motion.

Now, remembering that we need the square of the first derivative, the Lagrangian becomes:

$$\mathcal{L} = \frac{1}{2}(\nabla\varphi)^2 - V(\varphi) \tag{13.8}$$

In four dimensions, it should be kept in mind that the kinetic differential term is a time derivative followed by a three-gradient vector. That is, in natural units, the kinetic term is:

$$\begin{aligned}\frac{1}{2}(\nabla\varphi)^2 &= \frac{1}{2}\left(\frac{\partial\varphi}{\partial t}\right)^2 - \frac{1}{2}\left(\vec{\nabla}\varphi\right)^2 \\ &= \frac{1}{2}\left[\left(\frac{\partial\varphi}{\partial t}\right)^2 - \left(\frac{\partial\varphi}{\partial x}\right)^2 - \left(\frac{\partial\varphi}{\partial y}\right)^2 - \left(\frac{\partial\varphi}{\partial z}\right)^2\right]\end{aligned} \tag{13.9}$$

and so:

$$\mathcal{L} = \frac{1}{2}\left(\frac{\partial\varphi}{\partial t}\right)^2 - \frac{1}{2}\left(\vec{\nabla}\varphi\right)^2 - V(\varphi) \tag{13.10}$$

Let the value of the field vary a small amount γ from a selected value $\phi_0(t, x, y, z)$. If ϕ_0 is on the true minimum path in S, then only second order variations in S with γ will be present.

For the kinetic term:

$$\frac{1}{2}\left(\nabla(\varphi_0 + \gamma)\right)^2 = \frac{1}{2}\left((\nabla\varphi_0)^2 + 2\nabla\varphi_0\nabla\gamma + (\nabla\gamma)^2\right) \tag{13.11}$$

The second order term in γ can be eliminated in the above, and the kinetic term is thus:

$$\frac{1}{2}\left(\nabla(\varphi_0 + \gamma)\right)^2 = \frac{1}{2}\left((\nabla\varphi_0)^2 + 2\nabla\varphi_0\nabla\gamma\right) \tag{13.12}$$

In general, the potential function is not a constant, but a function of the value of the field, $V(\phi)$. Along the minimum path, this might be expressed as a polynomial:

$$V(\varphi_0) = V_0 + aV(\varphi_0) + bV(\varphi_0)^2 + cV(\varphi_0)^3 + \ldots \tag{13.13}$$

For a small deviation γ from the minimum path, the potential function takes the form of a Taylor series.

$$V(\varphi_0+\gamma)=V(\varphi_0)+\frac{\gamma}{1!}\nabla_\varphi V(\varphi_0)+\frac{\gamma^2}{2!}\nabla_\varphi{}^2V(\varphi_0)+\ldots+\frac{\gamma^n}{n!}\nabla_\varphi{}^nV(\varphi_0)+\ldots \tag{13.14}$$

where:

$$\nabla V_\varphi(\varphi_0)=\frac{\partial V_\varphi(\varphi_0)}{\partial\varphi_0};\ \nabla_\varphi{}^2(\varphi_0)=\frac{\partial V_\varphi{}^2(\varphi_0)}{\partial\varphi_0{}^2}+\ldots \tag{13.15}$$

Note that here, the differentials are with respect to ϕ and there is a subscript placed on the differential operator to remind us.

Because we are only interested in the first order terms in γ, then the potential term we want is:

$$V(\varphi_0+\gamma)=V(\varphi_0)+\gamma\nabla_\varphi V(\varphi_0) \tag{13.16}$$

Combining the kinetic and potential terms, the full expression for the action with the small perturbation γ is thus:

$$S'=\int\left[\frac{1}{2}\left((\nabla\varphi_0)^2+2\nabla\varphi_0\nabla\gamma-V(\varphi_0)\right)-\gamma\nabla_\varphi V(\varphi_0)\right]d^4x \tag{13.17}$$

With some rearrangement, this becomes:

$$\begin{aligned}S'&=\int_C\left[\frac{1}{2}\left((\nabla\varphi_0)^2-V(\varphi_0)\right)+\nabla\varphi_0\nabla\gamma-\gamma\nabla_\varphi V(\varphi_0)\right]d^4x\\&=\int_C\left[\frac{1}{2}\left((\nabla\varphi_0)^2-V(\varphi_0)\right)\right]+\int_C d^3x\left[\nabla\varphi_0\nabla\gamma-\gamma\nabla_\varphi V(\varphi_0)\right]d^4x\end{aligned} \tag{13.18}$$

The first term in the above is the action S_0 that would have been calculated as the true minimum. The difference in action δS between S' and S_0 is thus the second term:

$$\begin{aligned}\delta S&=\int_C\left[\nabla\varphi_0\nabla\gamma-\gamma\nabla_\varphi V(\varphi_0)\right]d^4x\\&=0\end{aligned} \tag{13.19}$$

Because this expression came about from a consideration of the *first order terms only*, we know that at ϕ_0 which gives the minimum value of S, then there will be no *first order* change in action, and so the difference δS will be zero.

We are looking for the value of ϕ_0, that which gives the true minimum in S. This is done by integrating by parts.

$$\int u\,dv=uv-\int v\,du \tag{13.20}$$

$$\begin{aligned}&u=f(x);\ du=f'(x)dx\\&v=g(x);\ dv=g'(x)dx\end{aligned} \tag{13.21}$$

Thus:

$$u = \nabla\varphi_0;\ v = \gamma;\ du = \nabla^2\varphi_0 d^4x;\ dv = \nabla\gamma dx$$

$$\int_C [\nabla\varphi_0 \nabla\gamma] d^4x = [\nabla\varphi_0\gamma]_{x1}^{x2} - \int_C \gamma\, \nabla^2\varphi_0 d^4x \tag{13.22}$$

And so:

$$\delta S = [\nabla\varphi_0\ \gamma]_{x1}^{x2} - \int_C \gamma\, \nabla^2\varphi_0 d^4x - \int_C \gamma\nabla_\varphi V(\varphi_0)\ d^4x \tag{13.23}$$

Now, the first term in the above contains the quantity γ and is to be evaluated at the two end points of the path. But, at the start and end of the path, there is no difference between the perturbed path and the minimum path so that at these points, x_1 and x_2, $\gamma = 0$. So, the difference in action becomes:

$$\begin{aligned}\delta S &= -\int_C \gamma\, \nabla^2\varphi_0 d^4x - \int_C \gamma\nabla_\varphi V(\varphi_0)\ d^4x \\ &= -\nabla^2\varphi_0 \int_C \gamma\ d^4x - \nabla_\varphi V(\varphi_0)\int_C \gamma\ d^4x \\ &= \left(-\nabla^2\varphi_0 - \nabla_\varphi V(\varphi_0)\right)\int_C \gamma\ d^4x\end{aligned} \tag{13.24}$$

And so:

$$\nabla^2\varphi + \nabla_\varphi V(\varphi) = 0 \tag{13.25}$$

This is the equation of motion for our scalar field $\phi(t, x, y, z)$ for the condition of least action ($\phi = \phi_0$).

We might set the constant of proportionality for the kinetic term (which we have so far left out) as a quantity m.

Also, V was earlier defined as: $V(\varphi) = -mg\varphi$, and so:

$$\begin{aligned}\nabla_\varphi V(\varphi) &= -mg\nabla_\varphi\varphi \\ &= -mg\frac{\partial\varphi}{\partial\varphi} \\ &= -mg\end{aligned} \tag{13.26}$$

Then:

$$m\nabla^2\varphi - mg = 0 \tag{13.27}$$

which is very similar to that obtained earlier for the mechanical system in Section 9.6. In that example, the derivative was with respect to time, but here, we have the differential operator with the derivatives taken with respect to t, x, y and z.

In space-time, the Lagrangian is not the difference in kinetic and potential energy terms as such. This is why it is said the Lagrangian comprised a "kinetic term" and a "potential term". In four dimensions, the kinetic term contains space derivatives that we might expect to be more associated with a potential energy. Kinetic energy is usually associated with time derivatives.

13.4 THE EULER–LAGRANGE EQUATION

The equation of motion for the field can also be written in terms of the Lagrangian. In general terms, the Lagrangian is:

$$\begin{aligned}\mathcal{L} &= (\nabla\varphi)^2 - V(\varphi)\\ &= \frac{1}{2}\left(\frac{\partial\varphi}{\partial x_\mu}\right)^2 - V(\varphi)\end{aligned} \tag{13.28}$$

Taking the derivatives:

$$\frac{\partial\mathcal{L}}{\partial\varphi} = -\frac{\partial V(\varphi)}{\partial\varphi} \tag{13.29}$$

and also:

$$\frac{\partial\mathcal{L}}{\partial(\partial\varphi/\partial x_\mu)} = \frac{\partial\varphi}{\partial x_\mu} \tag{13.30}$$

Thus, going back to the equation for δS:

$$\delta S = \int_C \left[\frac{\partial\varphi_0}{\partial x}\frac{\partial\gamma}{\partial x} - \gamma\frac{\partial V(\varphi_0)}{\partial\varphi}\right]d^4x \tag{13.31}$$

This becomes:

$$\delta S = \int_C \left[\frac{\partial\mathcal{L}}{\partial(\partial\varphi_0/\partial x)}\frac{\partial\gamma}{\partial x} + \gamma\frac{\partial\mathcal{L}}{\partial\varphi_0}\right]d^4x \tag{13.32}$$

Integrating by parts:

$$\begin{aligned}&u = \frac{\partial\mathcal{L}}{\partial(\partial\varphi_0/\partial x)};\ v = \frac{\partial\gamma}{\partial x};\ du = \frac{\partial\mathcal{L}}{\partial(\partial\varphi_0/\partial x)}d^3x;\ dv = \frac{\partial\gamma}{\partial x}dx\\ &\delta S = \left[\frac{\partial\mathcal{L}}{\partial(\partial\varphi_0/\partial x_\mu)}\frac{\partial\gamma}{\partial x_\mu}\right]_{x_{\mu 1}}^{x_{\mu 2}} - \int_C \frac{\partial\gamma}{\partial x}\frac{\partial\mathcal{L}}{\partial(\partial\varphi_0/\partial x)}d^4x + \int_C \gamma\frac{\partial\mathcal{L}}{\partial\varphi_0}d^4x\end{aligned} \tag{13.33}$$

Remember we put a μ subscript when the term x appears outside the integral sign to remind us this is a four-dimensional term. Applying the boundary conditions:

$$\delta S = -\int_C \frac{\partial \gamma}{\partial x} \frac{\partial \mathcal{L}}{\partial(\partial \varphi_0/\partial x)} d^4x + \int_C \gamma \frac{\partial \mathcal{L}}{\partial \varphi_0} d^4x \tag{13.34}$$

$$= 0$$

And so:

$$\begin{aligned} 0 &= -\int_C \gamma \frac{\partial}{\partial x} \frac{\partial \mathcal{L}}{\partial(\partial \varphi_0/\partial x)} d^4x + \int_C \gamma \frac{\partial \mathcal{L}}{\partial \varphi_0} d^4x \\ &= \left(-\frac{\partial}{\partial x_\mu} \frac{\partial \mathcal{L}}{\partial(\partial \varphi_0/\partial x_\mu)} + \frac{\partial L}{\partial \varphi_0} \right) \int_C \gamma \, d^4x \\ &= \frac{\partial \mathcal{L}}{\partial \varphi_0} - \frac{\partial}{\partial x_\mu} \frac{\partial \mathcal{L}}{\partial(\partial \varphi_0/\partial x_\mu)} \end{aligned} \tag{13.35}$$

Thus:

$$0 = \frac{\partial \mathcal{L}}{\partial \varphi} - \frac{\partial}{\partial x_\mu} \frac{\partial \mathcal{L}}{\partial(\nabla \varphi)} \tag{13.36}$$

This is often written:

$$\frac{\partial}{\partial x_\mu} \frac{\partial \mathcal{L}}{\partial(\nabla \varphi)} - \frac{\partial \mathcal{L}}{\partial \varphi} = 0 \tag{13.37}$$

This is called the Euler–Lagrange equation.

The Euler–Lagrange equation is the equation of motion for the field for the condition of least action ($\phi = \phi_0$). From it, we can develop equations of motion for any Lagrangian on the basis of the principle of least action.

The Euler–Lagrange equation is of far-reaching importance. Let us examine some examples.

For a second order Lagrangian, the Lagrangian is:

$$\mathcal{L} = \frac{1}{2}(\nabla \varphi)^2 - V(\varphi) \tag{13.38}$$

The required differentials are:

$$\frac{\partial \mathcal{L}}{\partial \varphi} = -\frac{\partial V(\varphi)}{\partial \varphi} \tag{13.39}$$

and:

$$\frac{\partial}{\partial x_\mu} \frac{\partial \mathcal{L}}{\partial(\nabla \varphi)} = \nabla^2 \varphi \tag{13.40}$$

Substituting into the Euler–Lagrange equation:

$$-\nabla^2 \varphi - \frac{\partial V(\varphi)}{\partial \varphi} = 0 \tag{13.41}$$

and separating the time and space dimensions gives an equation of motion:

$$-\frac{\partial^2 \varphi}{\partial t^2} + \vec{\nabla}^2 \varphi - \frac{\partial V(\varphi)}{\partial \varphi} = 0$$
$$\frac{\partial^2 \varphi}{\partial t^2} - \frac{\partial^2 \varphi}{\partial x^2} - \frac{\partial^2 \varphi}{\partial y^2} - \frac{\partial^2 \varphi}{\partial z^2} + \frac{\partial V(\varphi)}{\partial \varphi} = 0 \tag{13.42}$$

Note the time derivative here. The quantity $\partial^2 \varphi / \partial t^2$ is the acceleration of the field.

Going back to the example of a falling ball, we replace ϕ by the coordinate x, and x was given as a function of t. The subscript μ becomes 0. The dimension x_μ becomes t and $\nabla \varphi$ is ∇x where:

$$\nabla x = \frac{\partial x}{\partial t} \tag{13.43}$$

That is, for our example of Section 9.6, it is as if the field (which we now call x) is a function of time only – there being no "space" terms (although in this case, x is a height above the ground, but x could mean anything, e.g. temperature).

The Lagrangian was:

$$L = \frac{1}{2} m \left(\frac{dx}{dt} \right)^2 - mgx \tag{13.44}$$

And so, the equation of motion is found from the Euler–Lagrange equation:

$$\frac{\partial L}{\partial x} = -\frac{\partial V(x)}{\partial x} = -mg$$
$$\frac{\partial L}{\partial (dx/dt)} = m \frac{\partial x}{\partial t} \tag{13.45}$$

Note that the right-hand side of the above is the momentum. This is called the conjugate momentum although this derivative does not always correspond to the product of a mass times a velocity. It depends on the nature of the potential. The conjugate momentum is given the symbol $\pi(x)$, and we will have cause to use this notation in a later section.

Therefore:

$$\frac{\partial}{\partial x_\mu} \frac{\partial L}{\partial (\nabla x)} = m \frac{\partial}{\partial t} \frac{\partial x}{\partial t} = m \frac{\partial^2 x}{\partial t^2} \tag{13.46}$$

Thus:

$$m \frac{\partial^2 x}{\partial t^2} - mg = 0 \tag{13.47}$$

which is the same as before.

If we have a zero-potential function, and going back to conventional units by inserting c^2, and if ϕ is the displacement of a particle in simple harmonic motion, then:

$$-\frac{1}{c^2} \frac{\partial^2 \varphi}{\partial t^2} + \vec{\nabla}^2 \varphi = 0 \tag{13.48}$$

which is the equation of a travelling wave, a wave equation where the velocity of the wave in the medium is the term c. This may well be the speed of sound, or the speed of water waves, etc. The wave becomes an oscillation in the field, and the Euler–Lagrange equation, which contains the Lagrangian, describes the motion of the field – that is, the motion of the waves within the field.

Einstein's relativistic energy equation shows that the potential term will be proportional to m^2. As well, at the minimum action, the potential varies in the second order terms in $\gamma = (\phi - \phi_0)$ only, and so:

$$V(\varphi) = V(\varphi_0) + \frac{1}{2}(\varphi - \varphi_0)^2 \nabla_\varphi^2 V(\varphi_0) \tag{13.49}$$

If it is assumed that the minimum potential is zero (i.e. $h = 0$ or ground level for the case of our falling ball) then:

$$V(\varphi) = \frac{1}{2}(\varphi - \varphi_0)^2 \left(\nabla_\varphi^2 V(\varphi_0)\right) \tag{13.50}$$

We can't take this analogy any further but note the ϕ^2 dependence.

If we return for a moment to the case of a particle of mass m, we might set the potential function to be:

$$\begin{aligned} V(\varphi) &= \frac{1}{2}m^2\varphi^2 \\ \frac{\partial V(\varphi)}{\partial \varphi} &= m^2\varphi \end{aligned} \tag{13.51}$$

And so, by the Euler–Lagrange equation:

$$-\frac{\partial^2 \varphi}{\partial t^2} + \vec{\nabla}^2\varphi - m^2\varphi = 0 \tag{13.52}$$

which is the Klein–Gordon equation. In the Klein–Gordon equation, the square of the value of the field ϕ is interpreted to be a probability density function Ψ. Notice that the Lagrangian density itself is in the form of the Klein–Gordon equation:

$$\mathcal{L} = \frac{1}{2}(\nabla\varphi)^2 - V(\varphi) \tag{13.53}$$

Now setting the potential function to be:

$$\begin{aligned} V(\varphi) &= \frac{1}{2}m\omega^2\varphi^2 \\ \frac{\partial V(\varphi)}{\partial \varphi} &= m\omega^2\varphi \end{aligned} \tag{13.54}$$

And so:

$$-\frac{\partial^2 \varphi}{\partial t^2} + \vec{\nabla}^2\varphi - m\omega^2\varphi = 0 \tag{13.55}$$

which is the equation of motion for a harmonic oscillator with ϕ taking the place of the displacement x.

We already know that for an electron of rest mass m, the solution of the Klein–Gordon equation is a series of complex wave functions, but here in this section, we started off with a field, ϕ, and there was no mention of a wave function. It is as if the wave function for a particle has become a field function. In the time dimensions, velocities and accelerations of the particle have become velocities and accelerations of the field.

13.5 LAGRANGIAN FOR A FREE PARTICLE

In Section 9.6, we analysed the motion of a particle of mass m in the gravitational field, but this was actually a combination of two things. What we had was a field (the gravitational field) and a particle. For a particle of mass m we might ask, what is the Lagrangian for a particle when it moves from place to place in the absence of a field? That is, what is the Lagrangian when a particle just moves from place to place in space-time? That is, what is the Lagrangian for an interval in space-time?

In four-dimensional space-time, a distance is given by:

$$s^2 = c^2t^2 - \left(x^2 + y^2 + z^2\right) \tag{13.56}$$

Or:

$$s^2 = c^2t^2 - \vec{x}^2 \tag{13.57}$$

where the arrow indicates the three-vector x, y and z. This terminology allows us to proceed in two dimensions with $\vec{x}$ on the horizontal axis and ct on the vertical axis. Written in four dimensions as a four-position:

$$\begin{aligned} s^2 &= c^2t^2 - \vec{x}^2 \\ &= x_\mu{}^2 \end{aligned} \tag{13.58}$$

A particle at some point on the path from A to B can be characterised by its four-position and its four-velocity.

The path from A to B is divided up into a series of increments. Action is a local phenomenon. The particle does not know in advance where point B is. The only way the particle knows where to proceed is to proceed a small step which minimises the action compared to other steps in the nearby vicinity.

As shown in Figure 13.1, the particle moves from A to B. Now, previously we've always said that the Lagrangian is formed from the difference of the kinetic energy and the potential energy – but energies are not invariant quantities. The kinetic energy of a particle may be quite different for a stationary observer compared to that measured by an observer moving with the particle. We would like our principle of least action to be stated in an invariant form, and so we cannot proceed with kinetic and potential energies explicitly for the Lagrangian. In the non-relativistic limit, the Lagrangian is the difference between the kinetic and potential energies. In the relativistic case, we will see that the Lagrangian for a free particle is not a difference in *kinetic* and *potential* energies but a difference in energies – it's just that the kinetic term contains space integrals as well as a time integral, and so we would not usually call this kinetic energy.

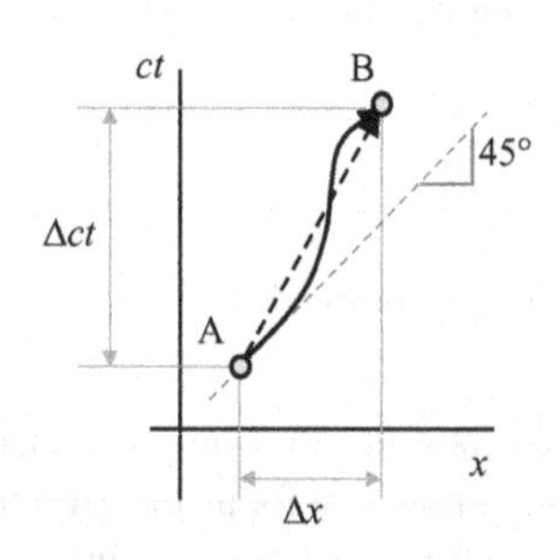

FIGURE 13.1 Possible path taken by a particle between two points A and B. Dotted line at 45° shows limit imposed by the velocity of light.

We might therefore say that the proper time would be a good candidate for the interval over which the action will be computed. Observers in different reference frames will agree on their calculation of the proper time for each increment along the path of the particle. The proper time is the time recorded by an observer travelling with the particle so from that point of view, the particle is at rest, and the measurement of time, in that coordinate system, is along the *ct* axis.

Since:

$$t' = t\sqrt{1 - v^2/c^2} \tag{13.59}$$

Then notice also that:

$$dt' = dt\sqrt{1 - v^2/c^2} \tag{13.60}$$

The action becomes:

$$S = \int_{t_1}^{t_2} dt' = \int_{t_1}^{t_2} \sqrt{1 - v^2/c^2}\, dt \tag{13.61}$$

The velocity in this expression is the three-vector:

$$\begin{gathered} v^2 = v_x^2 + v_y^2 + v_z^2 \\ v_x = \frac{dx}{dt};\ v_y = \frac{dy}{dt};\ v_z = \frac{dz}{dt} \\ \vec{v} = \frac{d\vec{x}}{dt} \end{gathered} \tag{13.62}$$

All the quantities in the action integral are now with respect to the local coordinate system, that is, for some particular observer. By convention, a factor of $-m$ is included in the Lagrangian so that the final equations of motion are recognisable as Newton's equations of motion in the non-relativistic limit. The multiplication of the Lagrangian by a constant does not alter the value of the Lagrangian.

The Lagrangian, in natural units, is thus:

$$L = -m\sqrt{1 - \vec{v}^2} = -m\sqrt{1 - (d\vec{x}/dt)^2} \tag{13.63}$$

This Lagrangian is Lorentz invariant. That is, we can translate or rotate the reference frame and obtain the same result. Restoring units by adding *c*, we write:

$$L = -mc^2\sqrt{1 - \vec{v}^2/c^2} \tag{13.64}$$

This then is the Lagrangian for a free particle.

The time component of the four-momentum is found from the time component of the four-velocity:

$$u_0 = \frac{c}{\sqrt{1 - v^2/c^2}} \tag{13.65}$$

And so:

$$p_0 = mu_0 = \frac{mc}{\sqrt{1 - v^2/c^2}} \tag{13.66}$$

We've seen before that there is an association between time and energy. In the Schrödinger equation, the energy operator is a time differential. Thus, it may be no surprise to learn that the time component of the four-momentum is equivalent to an energy in the low velocity regime.

From the binomial theorem (Appendix 8):

$$\begin{aligned} mu_0 &= \frac{mc}{\sqrt{1-\vec{v}^2/c^2}} \\ &= mc\left(1+\frac{1}{2}\frac{\vec{v}^2}{c^2}\right) \\ &= \frac{1}{c}\left(mc^2+\frac{1}{2}m\vec{v}^2\right) \\ mu_0c &= mc^2+\frac{1}{2}mv^2 \end{aligned} \tag{13.67}$$

which, remembering that here, m is the rest mass, and we are in the low velocity regime, we recognise as the total energy of the particle.

$$E=\frac{1}{2}m_0v^2+m_0c^2 \quad v<<c \tag{13.68}$$

Total energy is conserved in relativistic motion. For the motion of a particle, it is easier to describe this with reference to the momentum rather than the energy. When the particle moves along the increment, the four-momentum is conserved.

In the non-relativistic limit, then the Lagrangian can be written in terms of the differences between kinetic and potential energies and so:

$$\begin{aligned} L &= -mc^2\sqrt{1-\vec{v}^2/c^2} \\ &= -mc^2\left(1-\frac{1}{2}\frac{v^2}{c^2}\right) \quad v<<c \\ &= \frac{1}{2}mv^2-mc^2 \end{aligned} \tag{13.69}$$

m is the rest mass, and we might regard mc^2 as the particle's rest mass energy.

Now, when forming the equations of motion from the Euler–Lagrange equation, we often need to take the derivative of the Lagrangian with respect to the velocity:

$$\begin{aligned} L &= -m\sqrt{1-\vec{v}^2/c^2} \\ \frac{\partial L}{\partial \vec{v}} &= \frac{1}{2}\frac{-m}{\sqrt{1-\vec{v}^2/c^2}}\left(-2\frac{\vec{v}}{c^2}\right) \\ &= \frac{1}{c^2}\frac{m\vec{v}}{\sqrt{1-\vec{v}^2/c^2}} \\ &= \frac{1}{c^2}m\vec{u} \end{aligned} \tag{13.70}$$

That is, this derivative gives the three-momentum. In the Euler–Lagrange equation, this is now differentiated with respect to time, and we obtain the time rate of change of momentum which is a force. But, in this case, the derivative of the above with respect to time, for a constant velocity, is zero. That is, the particle is moving in a straight line and no force is acting upon it. The shortest path is therefore the straight line between the two points A and B.

In four dimensions, we will later see that the zero term is equivalent to the time rate of change of energy, which is power, just as the zero term in the four-momentum is energy.

13.6 LAGRANGIAN FOR A FREE PARTICLE IN A SCALAR FIELD

Now we ask, what happens when the particle moves from place to place within a field? The field acts upon the particle, and the presence of the particle affects the local value of the field. By the principle of least action, the Lagrangian for each, the field and the particle, must combine so that the field and the particle move in a way which minimises the action.

For a free particle, the Lagrangian was found to be:

$$L = -mc^2\sqrt{1-\vec{v}^2/c^2} \tag{13.71}$$

If we have a particle in a field, then we need to include the field in the Lagrangian in order to compute the action. We might, for a simple approximation, say that the Lagrangian for the field just adds to that of the particle as if it were an extra mass. Thus, the combined Lagrangian is:

$$\begin{aligned} L &= -m\sqrt{1-\vec{v}^2/c^2} + \varphi(\vec{x},t)\sqrt{1-\vec{v}^2/c^2} \\ &= -\left(m+\varphi(\vec{x},t)\right)\sqrt{1-\vec{v}^2/c^2} \end{aligned} \tag{13.72}$$

But we need some way of quantifying the strength of the coupling between the field and the particle. This is done by including a coupling constant, g. It describes the strength of the interaction between the particle and the field. For example, in the case of a charged particle in an electric field, g might be the charge on the particle.

For a mass m in a gravitational field, g might be the gravitational acceleration constant. The combined Lagrangian becomes:

$$L = -\left(m+g\varphi(\vec{x},t)\right)\sqrt{1-\vec{v}^2/c^2} \tag{13.73}$$

This Lagrangian may describe how the presence of a field affects the motion of the particle – by adding extra mass. The equations of motion that follow are the same as those which would come from the Euler–Lagrange equation but now the mass $-m$ is replaced by the term $-\left(m+g\phi(\vec{x},t)\right)$.

13.7 LAGRANGIAN FOR THE DIRAC FIELD

Just as we constructed the field equations from those of a particle, we can do much the same thing with the electron. That is, we wish to find the relativistic equations that represent the probability field, or the Dirac field, for an electron.

We know that the Lagrangian has to be a scalar for it to be Lorentz invariant. In the context of the Dirac equations, we mean that the Lagrangian will be expressed as the magnitude of a complex number. As with many formulations of the Lagrangian, we try something until the correct equations of motion are derived from the Euler–Lagrange equation.

A suitable expression might be:

$$\mathcal{L} = \bar{\Psi}\left(i\gamma_\mu \partial_\mu - m\right)\Psi \tag{13.74}$$

Applying the Euler–Lagrange equation with respect to $\bar{\Psi}$:

$$\frac{\partial}{\partial x_\mu}\frac{\partial \mathcal{L}}{\partial\left(\partial\bar{\Psi}/\partial x_\mu\right)} - \frac{\partial \mathcal{L}}{\partial \bar{\Psi}} = 0 \tag{13.75}$$

We have:

$$\begin{aligned} \frac{\partial}{\partial x_\mu}\frac{\partial \mathcal{L}}{\partial\left(\partial\bar{\Psi}/\partial x_\mu\right)} &= 0 \\ \frac{\partial \mathcal{L}}{\partial \bar{\Psi}} &= \left(i\gamma_\mu \partial_\mu - m\right)\Psi \end{aligned} \tag{13.76}$$

And so:

$$\left(i\gamma_\mu \partial_\mu - m\right)\Psi = 0 \tag{13.77}$$

which is the required Dirac equation.

Applying the Euler–Lagrange equation with respect to Ψ:

$$\begin{aligned} \frac{\partial \mathcal{L}}{\partial\left(\partial\Psi/\partial x_\mu\right)} &= \bar{\Psi}\left(i\gamma_\mu\right) \\ \frac{\partial}{\partial x_\mu}\bar{\Psi}\left(i\gamma_\mu\right) &= \bar{\Psi}\overleftarrow{\partial}_\mu i\gamma_\mu \\ \frac{\partial \mathcal{L}}{\partial \Psi} &= -\bar{\Psi}m \end{aligned} \tag{13.78}$$

And so:

$$\bar{\Psi}\left(i\gamma_\mu \partial_\mu + m\right) = 0 \tag{13.79}$$

which is the adjoint Dirac equation.

The proposed Lagrangian is therefore consistent with the Dirac equation, and so we accept the Lagrangian:

$$\mathcal{L} = \bar{\Psi}\left(i\gamma_\mu \partial_\mu - m\right)\Psi \tag{13.80}$$

The associated wave functions are:

$$\Psi\left(x_u\right) = u\left(p\right)e^{-ip_\mu \cdot x_\mu} \tag{13.81}$$

where μ goes from 0 to 4. The wave functions and the Dirac equation are not four vectors in the sense that here, μ does not represent the four coordinates of space-time.

Lorentz Invariance 14

14.1 INTRODUCTION

One of the issues so far ignored is that of the invariance of all these equations. In both special and general relativity, physical quantities like distance, time, mass, momentum, velocity and energy had to be presented in such a way so that physical laws remained unchanged when viewed from another frame of reference, that is, equations should be expressed so that they give the same result under a translation or rotation of reference frames. This requirement led to Einstein to his theory of special relativity.

In special relativity, we encountered a change of reference frames when one set of axes S′ was moving with respect to another S.

$$\varphi(x_\mu) = \varphi'(x'_\mu) \tag{14.1}$$

For a scalar, the value of the scalar is invariant under a change of axes. For example, we would expect that the temperature at a point to be the same whether or not we are moving in relation to that point. Or that the number of slices in a loaf of bread is the same from the point of view of a pedestrian looking at the loaf through the shop window compared to what we see if we happen to be driving past in a car. These quantities are invariant. As well, since any function of a scalar is a scalar, then that function is invariant. That is, the number of slices in two loaves of bread is invariant. As well, if we have two scalars and multiply them together, we obtain another scalar. Say there are 12 loaves of bread to a box. Then the total number of slices in the box would be 12 times the number of slices in a loaf. The total number of slices in a box is a scalar and is invariant no matter who is looking at them.

For a vector, we cannot expect a direct one-to-one transformation since vectors, by their very nature, depend upon their direction with respect to the coordinate axes. For example, the velocity of a particle in one frame of reference will be different in another frame of reference.

Earlier, we studied the Lorentz transformations for time and displacement where the displacement occurred at a constant velocity only along the x axis:

$$\begin{aligned} t' &= \frac{t - xv/c^2}{\sqrt{1 - v^2/c^2}} \\ x' &= \frac{x - vt}{\sqrt{1 - v^2/c^2}} \\ y' &= y \\ z' &= z \end{aligned} \tag{14.2}$$

These equations show how time and displacement transform from one reference frame to another. These transformation equations are used to describe how physical quantities that depend on time and displacement also transform.

In general, we want to be able to express equations in an invariant way in all four axes, not just in t and x. To do this, it is largely a matter of notation.

14.2 TRANSFORMATIONS

In Galileo's time, if we wished to transform say, the x coordinate, from one reference frame to another, then this was accomplished by:

$$x' = x - Vt \tag{14.3}$$

As shown in Figure 8.11, if the S′ reference frame is moving with velocity V with respect to the S reference frame, this Galilean transformation equation transforms a value x in an S frame of reference to a coordinate x' in an S′ reference frame and vice versa. At some instant, a particle might have a coordinate $x = 12.5$ in the S frame of reference, and a coordinate $x' = 2.5$ in the S′ frame of reference. The transformation equation transforms the value 12.5 to 2.5, and the transformation depends on the velocity of the S′ reference frame and the time at which the coordinates are measured. Coordinates are not invariant. They change from one reference frame to another.

A consequence of the Galilean transformation of coordinates is that the distance measured between two points in one frame of reference is the same as that measured between the same two points in another frame of reference. Although coordinates are not the same in the two frames of reference, the distance between two points is invariant; it didn't vary with a change of reference frames. Coordinates transform, but distances do not.

In relativistic terms, things are not so straightforward. The transformation of coordinates is given by:

$$x' = \frac{x - vt}{\sqrt{1 - v^2/c^2}} \tag{14.4}$$

A consequence of this relativistic transformation is that the distance between two points measured in one frame of reference is different from that measured between the same two points in another frame of reference. The distance between two points in space in relativistic terms is not invariant. We might ask ourselves, is there is a distance in space-time which is invariant? A distance that all observers can agree upon as to its value?

Such a distance is an *interval* in space-time s. It is invariant. That is:

$$s^2 = c^2t^2 - x^2 - y^2 - z^2 = c^2t'^2 - x'^2 - y'^2 - z'^2 \tag{14.5}$$

In the frame of reference where the particle is at rest, an interval in that reference frame is vertical, along the ct axis. Each observer, using their own values of x, y, x and t for a point in space-time, will calculate the same value for the interval.

In relativistic terms, time is not invariant; it also transforms. The transformation between one frame of reference and another is given by:

$$t' = \frac{t - xv/c^2}{\sqrt{1 - v^2/c^2}} \tag{14.6}$$

There is, however, a measurement of time which is invariant, and that is the time in the coordinate system in which the particle is at rest. This is called the proper time and is the quantity t'. In space-time, this time interval is completely time-like and is vertical along the ct axis.

When the particle is at the zero coordinate O′ in the S′ reference frame, then the x coordinate (in the S reference frame) is then vt and so:

$$t' = \frac{t - tv^2/c^2}{\sqrt{1-v^2/c^2}} = t\sqrt{1-v^2/c^2} \tag{14.7}$$

The above transformation transforms a time interval time in the S′ reference frame to a time in the S frame of reference and vice versa.

Another invariant quantity is the rest mass of a particle. This is the mass as it appears in the frame of reference at which the particle is at rest – that is, the S′ reference frame. This same mass is calculated from the relativistic mass as it appears in any other frame of reference.

Yet another invariant quantity is any scalar. For example, we expect that the temperature of a point in space-time would not depend on the frame of reference of the observer.

The transformation equations above are the Lorentz transformations for space and time for two coordinate systems where one is moving with velocity v in the x direction relative to the other.

From these, we can develop other transformation equations. For momentum, we have:

$$p' = \frac{p\left(1 - v/v'\right)}{\sqrt{\left(1-v^2/c^2\right)}} \tag{14.8}$$

where v' is the velocity of the particle in the S′ reference frame.

And the four-velocity u_μ as:

$$u_o = \frac{c}{\sqrt{1-v^2/c^2}}$$
$$u_i = \frac{v_i}{\sqrt{1-v^2/c^2}} \tag{14.9}$$

where:

$$v^2 = v_x^2 + v_y^2 + v_z^2 = \vec{v}^2 \tag{14.10}$$

It was shown above that an interval in space-time is invariant. The interval (squared) is calculated from the square of the four-vector components. The same is true for the square or multiplication of any four-vector quantity with another. That is:

$$A_\mu B_\mu = A_0 B_0 - A_1 B_1 - A_2 B_2 - A_3 B_3 \tag{14.11}$$

The result is a scalar. A scalar is invariant; it does not transform with a change of coordinate frames. We will find that Lorentz invariant quantities in our equations are formed from four-vector scalar products.

14.3 CONTRAVARIANT AND COVARIANT NOTATION

In four dimensions, physical quantities have so far been written in terms of a time-like component and a space-like component. For example, the four-position is written:

$$x_\mu = \left(ct, \vec{x}\right) \tag{14.12}$$

The four-velocity u_μ is:

$$u_\mu = (c, \vec{v}) \tag{14.13}$$

and the four-momentum is written:

$$p_\mu = (E, \vec{p}) \tag{14.14}$$

In all of these, the index μ has been written as a subscript. Four-dimensional vector quantities are written as ordered by their dimensions, usually t, x, y, z. Following a convention introduced by Einstein, we will now label the four dimensions with a *superscript*. As before, the components are $\mu = 0, 1, 2, 3$ which stand for t, x, y and z. For example, we write the four-vector position as:

$$\begin{aligned} x^\mu &= (t, \vec{x}) \\ &= (x^0, x^1, x^2, x^3) \\ &= (t, x, y, z) \end{aligned} \tag{14.15}$$

When the four-dimensional vector is written with an upper index, it is called a contravariant form of the vector. A quantity written with a lower index is the covariant form of the four-vector. There is a geometric interpretation to this, and it has to do with the way vectors transform between reference frames, but we will not pursue this issue here. For now, we just want to focus on the notation.

There is a matrix called the "metric", or the metric tensor. It is a 4 × 4 matrix. It is written:*

$$g_{\mu\upsilon} = \begin{bmatrix} 1 & 0 & 0 & 0 \\ 0 & -1 & 0 & 0 \\ 0 & 0 & -1 & 0 \\ 0 & 0 & 0 & -1 \end{bmatrix} \tag{14.16}$$

It carries the signs of the Minkowskian geometry in four-dimensional space-time. It is called the space-time signature.

For scalars, it doesn't matter if the index is written as a superscript or a subscript, a scalar will have the same value. There are no minus signs involved. For a three-vector, likewise, it doesn't matter if the index is a superscript or a subscript, there are no minus signs. The only significance is four-vectors.

Let us consider a four-vector A^μ with time and spatial components:

$$A^\mu = (t, \vec{A}) \tag{14.17}$$

In matrix form, we write this as a single column matrix:

$$A^\mu = \begin{bmatrix} A^0 \\ A^1 \\ A^2 \\ A^3 \end{bmatrix} = \begin{bmatrix} A^t \\ A^x \\ A^y \\ A^z \end{bmatrix} \tag{14.18}$$

Now we form the product:

* An alternative form of the metric tensor has the signs reversed and there are good arguments for either definition. For the sake of consistency with our previous definition of the space-time interval, we will stay with the form of $g_{\mu\nu}$ above.

$$g_{\mu\upsilon}A^{\nu} = \begin{bmatrix} 1 & 0 & 0 & 0 \\ 0 & -1 & 0 & 0 \\ 0 & 0 & -1 & 0 \\ 0 & 0 & 0 & -1 \end{bmatrix} \begin{bmatrix} A^0 \\ A^1 \\ A^2 \\ A^3 \end{bmatrix} = \begin{bmatrix} A_0 \\ -A_1 \\ -A_2 \\ -A_3 \end{bmatrix} = A_{\mu} \tag{14.19}$$

ν is just a dummy index over which the sums are taken for each value (row) of μ. That is, we hold μ (the row in g) constant and then sum over products of the ν's (the columns in g, rows in A) to get the μ (row) term in the result vector. Then increment μ and sum over the ν's again, etc. In summation notation, we can write this product as:

$$g_{uv}A^{\nu} = \sum_{\mu=0}^{3}\sum_{\nu=0}^{3} g_{\mu\upsilon}A^{\nu} = A_{\mu} \tag{14.20}$$

When the metric tensor acts on a contravariant four-vector (upper index) the result is a covariant vector (lower index). The covariant form, the result vector, contains minus signs on the space components. That is:

$$g_{uv}A^{\nu} = A_{\mu} \tag{14.21}$$

where:

$$\begin{aligned} A^{\nu} &= \left[A^t, A^x, A^y, A^z\right] \\ A_{\mu} &= \left[A_t, -A_x, -A_y, -A_z\right] \end{aligned} \tag{14.22}$$

If we form the product:

$$g_{\mu\upsilon}A_{\nu} = \begin{bmatrix} 1 & 0 & 0 & 0 \\ 0 & -1 & 0 & 0 \\ 0 & 0 & -1 & 0 \\ 0 & 0 & 0 & -1 \end{bmatrix} \begin{bmatrix} A^0 \\ -A^1 \\ -A^2 \\ -A^3 \end{bmatrix} = \begin{bmatrix} A_0 \\ A_1 \\ A_2 \\ A_3 \end{bmatrix} = A^{\mu} \tag{14.23}$$

The metric tensor $g_{\mu\nu}$ has the effect of raising or lowering the index. Raising or lowering the index introduces a change of sign on the space components.

Now, we take the dot product of the two four-vectors, one contravariant and the other covariant:

$$\begin{aligned} A^{\mu}A_{\mu} &= \begin{bmatrix} A^0 & A^1 & A^2 & A^3 \end{bmatrix} \begin{bmatrix} A_0 \\ -A_1 \\ -A_2 \\ -A_3 \end{bmatrix} \\ &= \left[\left(A^0\right)^2 - \left(A^1\right)^2 - \left(A^2\right)^2 - \left(A^3\right)^2\right] \end{aligned} \tag{14.24}$$

The result is a single number, a scalar. This is not the same as the previously defined dot product of two four-vectors. The minus signs are now associated with the contravariant term.

For example, an interval s in space-time is written in terms of a four-vector x_μ:

$$s^2 = \sum_{\mu,\nu} x^\mu g_{\mu\upsilon} x^\nu \tag{14.25}$$

ν is the dummy index over which the sum is performed.

The $g_{\mu\nu}$ matrix lowers the index on x^ν, and so this is equivalent to:

$$s^2 = \sum_{\mu} x^\mu x_\mu \tag{14.26}$$

The x_μ term is now covariant (it contains minus signs on the space components), and so the sum is the dot product:

$$s^2 = c^2t^2 - x^2 - y^2 - z^2 \tag{14.27}$$

This type of sum occurs often in particle physics, and Einstein introduced the convention that the sum is implied whenever there is an upper index followed by a lower index, or a lower index followed by an upper index; therefore, we leave the summation symbol out and just write $x^\mu x_\mu$ or $x_\mu x^\mu$. That is:

$$x^\mu x_\mu = \sum_{\mu} x^\mu x_\mu = x_0^2 - x_1^2 - x_2^2 - x_3^2 \tag{14.28}$$

The sum is called a contraction. In the above example, the vector x is contracted with itself.

If we have two four-vectors A and B, then it is found that:

$$A^\mu B_\mu = A_\mu B^\mu = A_0B^0 - A_1B^1 - A_2B^2 - A_3B^3 \tag{14.29}$$

As we've seen previously, differentials can be written in a shorthand notation:

$$\frac{\partial}{\partial t} = \partial_t = \partial_0; \ \frac{\partial}{\partial x} = \partial_x = \partial_1; \ \frac{\partial}{\partial y} = \partial_y = \partial_2; \ \frac{\partial}{\partial z} = \partial_z = \partial_3 \tag{14.30}$$

When we talk about vector differentiation, then we should be careful about whether the vector is being used in covariant or contravariant form. The notation for the derivative of a contravariant vector is:

$$\frac{\partial}{\partial x^\mu} = \partial_\mu \tag{14.31}$$

When a scalar is differentiated in four dimensions, the result is a four-vector, that is, the four-gradient. For example, if $\phi(x^\mu)$ is a scalar field, then, in contravariant notation:

$$\nabla^\mu \varphi = \frac{\partial \varphi}{\partial x^\mu} = \left[\frac{\partial \varphi}{\partial t}, \frac{\partial \varphi}{\partial x}, \frac{\partial \varphi}{\partial y}, \frac{\partial \varphi}{\partial z}\right] = A^\mu \tag{14.32}$$

This is written in relativistic form as:

$$\nabla^\mu \varphi = \frac{\partial \varphi}{\partial x^\mu} = \partial_\mu \varphi \tag{14.33}$$

This is "contrary" to our previous definition of the four-differential operator. Here, there are no minus signs on the space differentials.

On the other hand, in covariant notation, the four-differential operator is:

$$\nabla_\mu \varphi = \frac{\partial \varphi}{\partial x_\mu} = \left[\frac{\partial \varphi}{\partial t}, -\frac{\partial \varphi}{\partial x}, -\frac{\partial \varphi}{\partial y}, -\frac{\partial \varphi}{\partial z}\right] = \partial^\mu \varphi = A_\mu \tag{14.34}$$

The covariant four-differential operator is written:

$$\nabla_\mu = \partial^\mu = \frac{\partial}{\partial t} - \frac{\partial}{\partial x} - \frac{\partial}{\partial y} - \frac{\partial}{\partial z} \tag{14.35}$$

The d'Alembertian operator is:

$$\partial^\mu \partial_\mu = \frac{\partial^2}{\partial t^2} - \frac{\partial^2}{\partial x^2} - \frac{\partial^2}{\partial y^2} - \frac{\partial^2}{\partial z^2} \tag{14.36}$$

Now consider the product:

$$\frac{\partial \varphi}{\partial x_\mu} \frac{\partial \varphi}{\partial x^\mu} = \partial^\mu \varphi \, \partial_\mu \varphi \tag{14.37}$$

where ϕ is a scalar. This is a dot product between two four-vectors (i.e. the covariant gradient of one times the contravariant gradient of the other). The result is a scalar:

$$\left[\frac{\partial \varphi}{\partial t}, -\frac{\partial \varphi}{\partial x}, -\frac{\partial \varphi}{\partial y}, -\frac{\partial \varphi}{\partial z}\right] \cdot \left[\frac{\partial \varphi}{\partial t}, \frac{\partial \varphi}{\partial x}, \frac{\partial \varphi}{\partial y}, \frac{\partial \varphi}{\partial z}\right] = \left[\left(\frac{\partial \varphi}{\partial t}\right)^2 - \left(\frac{\partial \varphi}{\partial x}\right)^2 - \left(\frac{\partial \varphi}{\partial y}\right)^2 - \left(\frac{\partial \varphi}{\partial z}\right)^2\right] \tag{14.38}$$

This is of the form of the kinetic term of the Lagrangian for the scalar field, where:

$$\begin{aligned} (\nabla \varphi)^2 &= \frac{1}{2}\left(\frac{\partial \varphi}{\partial t}\right)^2 - \frac{1}{2}\left(\vec{\nabla} \varphi\right)^2 \\ &= \frac{1}{2}\left[\left(\frac{\partial \varphi}{\partial t}\right)^2 - \left(\frac{\partial \varphi}{\partial x}\right)^2 - \left(\frac{\partial \varphi}{\partial y}\right)^2 - \left(\frac{\partial \varphi}{\partial z}\right)^2\right] \end{aligned} \tag{14.39}$$

In other words, the Lagrangian is a scalar and so is Lorentz invariant.

Thus, in relativistic notation, the Lagrangian is thus written:

$$\mathcal{L} = \frac{1}{2} \partial^\mu \varphi \, \partial_\mu \varphi \tag{14.40}$$

For differentials of four-vectors, we have:

$$\begin{aligned} \nabla_\mu \cdot A_\mu &= \partial^\mu A_\mu = \frac{\partial A_\mu}{\partial x_\mu} \\ &= \frac{\partial}{\partial t} A_t - \frac{\partial}{\partial x}(-A_x) - \frac{\partial}{\partial y}(-A_y) - \frac{\partial}{\partial z}(-A_z) \\ &= \frac{\partial A_t}{\partial t} + \frac{\partial A_x}{\partial x} + \frac{\partial A_y}{\partial y} + \frac{\partial A_z}{\partial z} \end{aligned} \tag{14.41}$$

This is the four-divergence. On the other hand:

$$\begin{aligned}\nabla_\mu \cdot A^\mu &= \partial^\mu A^\mu = \frac{\partial A^\mu}{\partial x_\mu} \\ &= \frac{\partial A_t}{\partial t} - \frac{\partial A_x}{\partial x} - \frac{\partial A_y}{\partial y} - \frac{\partial A_z}{\partial z}\end{aligned} \tag{14.42}$$

As well:

$$\begin{aligned}\nabla^\mu \cdot A_\mu &= \partial_\mu A_\mu = \frac{\partial A_\mu}{\partial x^\mu} \\ &= \frac{\partial A_t}{\partial t} - \frac{\partial A_x}{\partial x} - \frac{\partial A_y}{\partial y} - \frac{\partial A_z}{\partial z}\end{aligned} \tag{14.43}$$

and:

$$\begin{aligned}\nabla^\mu \cdot A^\mu &= \partial_\mu A^\mu = \frac{\partial A_\mu}{\partial x^\mu} \\ &= \frac{\partial A_t}{\partial t} + \frac{\partial A_x}{\partial x} + \frac{\partial A_y}{\partial y} + \frac{\partial A_z}{\partial z}\end{aligned} \tag{14.44}$$

There is a very important observation to be made in the above. Whenever there is a repeated index, then this represents a sum over that index. That is, $\partial_\mu A_\mu$ is a sum of the derivatives (the divergence) whereas $\partial_\mu A_\nu$ is not a sum, but a set of derivatives for each combination of μ and ν.

Scalars are not written with an index; they are just a number. However, the value of a scalar may depend upon a four-vector, e.g. $\phi(t, x, y, z) = \phi(x^\mu)$.

A vector is written with one index. Scalars and three-vectors can be written with either an upper or a lower index; there is no significance to the placement of the index. If a four-vector is written with an upper index, then this is a vector in contravariant form. If a four-vector is written with a lower index, it is a covariant vector.

Now, consider two contravariant four-vectors A^μ and B^μ. There are 16 ways in which to multiply the components of A^μ and B^μ together. These combinations, this set of products, are called a tensor. It is naturally written in matrix form:

$$T^{\mu\nu} = \begin{bmatrix} A^0B^0 & A^0B^1 & A^0B^2 & A^0B^3 \\ A^1B^0 & A^1B^1 & A^1B^2 & A^1B^3 \\ A^2B^0 & A^2B^1 & A^2B^2 & A^2B^3 \\ A^3B^0 & A^3B^1 & A^3B^2 & A^3B^3 \end{bmatrix} \tag{14.45}$$

The tensor is not the "product" of two vectors. It is a *combination* of coefficients in an ordered set. We shall see later that the combination of the electric and magnetic fields can be written in tensor form. A tensor is written with two indices. In general terms, we might write: $T^{\mu\nu}$ or $T_{\mu\nu}$ or even $T^\mu{}_\nu$.

To express a contravariant tensor in covariant form, we need two applications of the metric tensor.

$$g_{\mu\nu} T^{\mu\nu} g_{\mu\nu} = T_{\mu\nu} \tag{14.46}$$

In general, the components of a tensor are not necessarily symmetric. If the tensor is symmetric, then this is a special case and the indices can be interchanged:

$$T^{\mu\nu} = T^{\nu\mu} \tag{14.47}$$

If when the indices are reversed, the components change sign, then this is an anti-symmetric tensor.

Note that:

$$g_{\mu\nu}g_{\mu\nu} = \begin{bmatrix} 1 & 0 & 0 & 0 \\ 0 & -1 & 0 & 0 \\ 0 & 0 & -1 & 0 \\ 0 & 0 & 0 & -1 \end{bmatrix}\begin{bmatrix} 1 & 0 & 0 & 0 \\ 0 & -1 & 0 & 0 \\ 0 & 0 & -1 & 0 \\ 0 & 0 & 0 & -1 \end{bmatrix}$$

$$g^{\mu\nu} = g_{\mu\nu}g_{\mu\nu}g_{\mu\nu} = \begin{bmatrix} 1 & 0 & 0 & 0 \\ 0 & 1 & 0 & 0 \\ 0 & 0 & 1 & 0 \\ 0 & 0 & 0 & 1 \end{bmatrix}\begin{bmatrix} 1 & 0 & 0 & 0 \\ 0 & -1 & 0 & 0 \\ 0 & 0 & -1 & 0 \\ 0 & 0 & 0 & -1 \end{bmatrix} \tag{14.48}$$

That is:

$$g_{\mu\nu} = g^{\mu\nu} \tag{14.49}$$

14.4 TRANSFORMATION MATRIX

The covariant notation becomes important when we express transformations in matrix terms. The Lorentz transformations can be expressed in terms of a transformation matrix. In natural units, for a Lorentz transformation along the x axis we write:

$$\begin{bmatrix} t' \\ x' \\ y' \\ z' \end{bmatrix} = \begin{bmatrix} \frac{1}{\sqrt{1-v^2}} & \frac{-v}{\sqrt{1-v^2}} & 0 & 0 \\ \frac{-v}{\sqrt{1-v^2}} & \frac{1}{\sqrt{1-v^2}} & 0 & 0 \\ 0 & 0 & 1 & 0 \\ 0 & 0 & 0 & 1 \end{bmatrix}\begin{bmatrix} t \\ x \\ y \\ z \end{bmatrix} \tag{14.50}$$

For example:

$$\begin{aligned} t' &= \frac{t}{\sqrt{1-v^2}} - \frac{vx}{\sqrt{1-v^2}} + 0 + 0 \\ &= \frac{t - vx}{\sqrt{1-v^2}} \end{aligned} \tag{14.51}$$

and:

$$\begin{aligned} x' &= \frac{-vt}{\sqrt{1-v^2}} + \frac{x}{\sqrt{1-v^2}} + 0 + 0 \\ &= \frac{x - vt}{\sqrt{1-v^2}} \end{aligned} \tag{14.52}$$

In matrix terms:

$$A'^{\mu} = L^{\mu}_{\nu}A^{\nu} \tag{14.53}$$

where L is the Lorentz transformation matrix (also given the symbol Λ) and we remember that matrices are written with two indices. Here, A^{μ} and A^{ν} are contravariant vectors. The transformation above shows how a contravariant vector A_{μ} in reference frame S would appear as A'_{μ} in another reference frame S′ in terms of the quantities t and x in the S reference frame where S′ has a constant velocity along the x axis with respect to S.

Note that the L matrix is written with an upper index μ to match the transformed matrix A'^{μ}. The index ν is used as a dummy index for the sum. When we see two indices, one lower and one upper, it is a sum.

For a covariant transformation, the transformation matrix is given the symbol M, so that:

$$A'_{\mu} = M^{\nu}_{\mu}A_{\nu} \tag{14.54}$$

The M matrix is given a lower index μ to match the result matrix A'_{μ} and then we have ν as the dummy index over which the sum is performed.

The relationship between L and M is:

$$M^{\nu}_{\mu} = g_{\mu\nu}L^{\mu}_{\nu}g_{\mu\nu} \tag{14.55}$$

Thus:

$$M = \begin{bmatrix} \frac{1}{\sqrt{1-v^2}} & \frac{v}{\sqrt{1-v^2}} & 0 & 0 \\ \frac{v}{\sqrt{1-v^2}} & \frac{1}{\sqrt{1-v^2}} & 0 & 0 \\ 0 & 0 & 1 & 0 \\ 0 & 0 & 0 & 1 \end{bmatrix} \tag{14.56}$$

And so, since:

$$A'_{\mu} = M^{\nu}_{\mu}A_{\nu} \tag{14.57}$$

then:

$$\begin{bmatrix} t' \\ x' \\ y' \\ z' \end{bmatrix} = \begin{bmatrix} \frac{1}{\sqrt{1-v^2}} & \frac{v}{\sqrt{1-v^2}} & 0 & 0 \\ \frac{v}{\sqrt{1-v^2}} & \frac{1}{\sqrt{1-v^2}} & 0 & 0 \\ 0 & 0 & 1 & 0 \\ 0 & 0 & 0 & 1 \end{bmatrix} \begin{bmatrix} t \\ -x \\ -y \\ -z \end{bmatrix}$$

$$t' = \frac{t}{\sqrt{1-v^2}} - \frac{vx}{\sqrt{1-v^2}}$$

$$= \frac{t - vx}{\sqrt{1 - v^2}}$$

$$x' = \frac{vt}{\sqrt{1-v^2}} + \frac{-x}{\sqrt{1-v^2}} \tag{14.58}$$

$$-x = \frac{x - vt}{\sqrt{1-v^2}}$$

$$y' = y$$

$$z' = z$$

We can see from the above that a four-vector, whether it be in contravariant or covariant form, transforms from one coordinate system to another via the L and M matrices.

We might ask, how does a tensor transform? If we have a tensor T in the S reference frame, how does this appear as T' in the S′ reference frame? Previously we constructed a tensor from two four-vectors A^{μ} and B^{μ}. We know how vectors transform according to the L and M matrices.

$$A'^{\mu} = L^{\mu}_{\nu} A^{\nu}$$
$$B'^{\mu} = L^{\mu}_{\nu} B^{\nu} \tag{14.59}$$

And so, showing the multiplication as:

$$A'^{\mu} B'^{\mu} = L^{\mu}_{\nu} A^{\nu} L^{\mu}_{\nu} B^{\nu} \tag{14.60}$$

The transformation equation becomes:

$$T'^{\mu\nu} = L^{\mu}_{\nu} L^{\mu}_{\nu} T^{\mu\nu} \tag{14.61}$$

For the covariant case, we might write:

$$T_{\mu\nu} = \begin{bmatrix} A_0B_0 & A_0B_1 & A_0B_2 & A_0B_3 \\ A_1B_0 & A_1B_1 & A_1B_2 & A_1B_3 \\ A_2B_0 & A_2B_1 & A_2B_2 & A_2B_3 \\ A_3B_0 & A_3B_1 & A_3B_2 & A_3B_3 \end{bmatrix} \tag{14.62}$$

And so:

$$T'_{\mu\nu} = M^{\nu}_{\mu} M^{\nu}_{\mu} T_{\mu\nu} \tag{14.63}$$

For the most part in this book, we have always written four-vector quantities as covariant quantities with a lower index and ignoring the physical significance. However, there are occasions when we will need to be more precise.

The ability to express physical quantities in a Lorentz invariant way enables physical laws to be generally applicable. In relativity, the part played by four-vector quantities is an essential part of this process. In general, Lorentz invariant quantities are usually formed from the scalar, or dot product, of four-vectors. That is, the square of "something".

The Electromagnetic Field

15

15.1 INTRODUCTION

The electric field **E** and the magnetic field **B** at a point are, in general, functions of t, x, y, z.

$$\mathbf{E} = \mathbf{E}(t, x, y, z))$$
$$\mathbf{B} = \mathbf{B}(t, x, y, z)) \tag{15.1}$$

That is, at every point in space (x, y, z) there may be an electric and/or a magnetic field. Both the **E** field and the **B** field are vector fields. In the case of the electric field, the vector points in the direction in which a positive test charge would move if placed at that point. In the case of the magnetic field, the direction of the vector is perpendicular to the direction of motion of the test charge and the direction from which the force acts upon the test charge by the right-hand rule. The magnitude of each field is usually written as E and B. Both magnitude and direction of the **E** and **B** fields may be a function of time as well as position.

When fields vary with time, the change in magnitude or direction can be determined by taking the derivative of the function that describes the field. That is, the gradient. We might take the derivative with respect to time, or with respect to a direction – a spatial rate of change of a function. The value of a field may change at a different rate depending on which direction the derivative is taken. The spatial derivative can be found from the differential operator ∇.

In this chapter, we deal with the electromagnetic field which is either stationary or moving with a constant velocity. The electric field arises from the presence of a charged particle. The magnetic field arises from the uniform motion of a charged particle. It is only when we have an accelerated charged particle that we obtain an electromagnetic *wave*.

We begin with Maxwell's equations in differential form in ordinary units.

Gauss' law (electric charge):

$$\vec{\nabla} \cdot \mathbf{E} = \frac{\rho}{\varepsilon_0} \tag{15.2}$$

Gauss' law (magnetism):

$$\vec{\nabla} \cdot \mathbf{B} = 0 \tag{15.3}$$

Faraday's law:

$$\vec{\nabla} \times \mathbf{E} = -\frac{\partial}{\partial t}\mathbf{B} \tag{15.4}$$

Ampere's law:

$$\vec{\nabla} \times \mathbf{B} = \mu_0 \mathbf{J} + \mu_0 \varepsilon_0 \frac{\partial}{\partial t}\mathbf{E} \tag{15.5}$$

Note that the first two equations are concerned with static fields, while the last two equations contain **B** and **E** fields whose amplitudes change with time.

In the absence of any charges or currents, Maxwell's equations take on an almost symmetric form:

$$\begin{aligned}
\vec{\nabla}\cdot\mathbf{E} &= 0 \\
\vec{\nabla}\times\mathbf{E} &= -\frac{\partial}{\partial t}\mathbf{B} \\
\vec{\nabla}\cdot\mathbf{B} &= 0 \\
\vec{\nabla}\times\mathbf{B} &= \mu_0\varepsilon_0\frac{\partial}{\partial t}\mathbf{E}
\end{aligned} \tag{15.6}$$

In the presence of a charge, or a moving charge (i.e. an electric current), we go back to the original equations with the additional terms on the right-hand sides.

Can we have an electric field without a charge? Yes. In an electromagnetic wave. If the wave is travelling along the z axis, then there is an electric field **E** oscillating sinusoidally say in the y direction and this is accompanied by a magnetic field oscillating sinusoidally in the x direction. Somewhere out in space, there is a volume through which are oscillating electric and magnetic fields travelling together in step at the speed of light. The oscillating charge that created the field in the first place was left many thousands of kilometres behind seconds ago.

Our aim is to find a way in which the **E** and **B** fields transform from one reference frame to another. We are then able to find the Lorentz invariant Lagrangian for the dynamics of the electromagnetic field and from this, determine the interaction between the field and a charged particle such as an electron.

15.2 THE SCALAR POTENTIAL

Consider a positive charge q_2 which is moved from point A to point B in a static electric field created by q_1 located at the origin as shown in Figure 15.1(a). We wish to compute the work done on (positive) or by (negative) the charge during this movement.

From Coulomb's law, we have:

$$\begin{aligned}
\mathbf{F} &= \frac{q_1 q_2 \mathbf{u}}{4\pi\varepsilon_0 r^2} \\
&= q_2\mathbf{E}
\end{aligned} \tag{15.7}$$

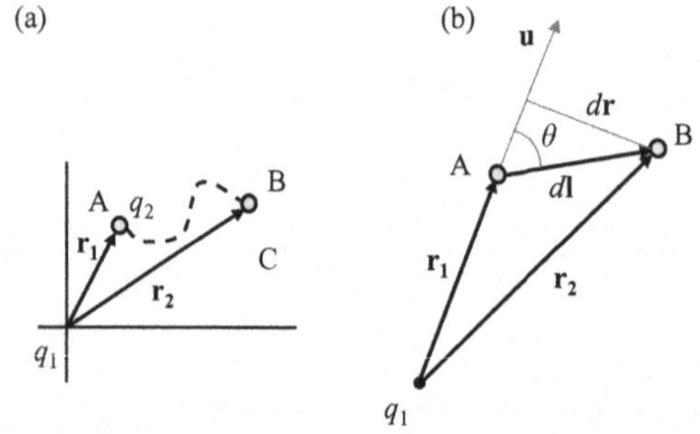

(a) A q_2 B C $\mathbf{r}_1$ $\mathbf{r}_2$ q_1

(b) $\mathbf{u}$ dr θ A B $d\mathbf{l}$ $\mathbf{r}_1$ $\mathbf{r}_2$ q_1

FIGURE 15.1 (a) A path between points A and B taken by a charge in an electric field. (b) Position vectors for A and B.

where:

$$\mathbf{E} = \frac{q_1}{4\pi\varepsilon_0 r^2}\mathbf{u} \tag{15.8}$$

By definition, the electrostatic potential ϕ is the work *per unit charge* in moving the charge through a distance $d\mathbf{l}$. The electrical potential is usually given in the units of volts.

The change in electric potential is computed from the line integral from A to B along the curve C. That is, we take the sum of the dot product of **E** and $d\mathbf{l}$ along the curve. Let $\Delta\phi = \phi_2 - \phi_1$ be the change in potential from A to B. Then:

$$\Delta\varphi = -\mathbf{E}\cdot d\mathbf{l} \tag{15.9}$$

As shown in Figure 15.1(b), the vectors $\mathbf{r}_1$ and $\mathbf{r}_2$ are the position vectors of the two end points A and B from the origin. The change in potential is thus:

$$\begin{aligned}
\varphi_2 - \varphi_1 &= -\int_A^B \mathbf{E}.d\mathbf{l} \\
&= -\int_A^B \frac{q_1}{4\pi\varepsilon_0 r^2}\mathbf{u}.d\mathbf{l} \\
&= -\frac{q_1}{4\pi\varepsilon_0}\int_A^B \frac{1}{r^2}\cos\theta\, dl \\
&= -\frac{q_1}{4\pi\varepsilon_0}\int_{r_1}^{r_2} \frac{1}{r^2}\, dr \\
&= \frac{q_1}{4\pi\varepsilon_0}\left[\frac{1}{r_2} - \frac{1}{r_1}\right]
\end{aligned} \tag{15.10}$$

This result shows that $\Delta\phi$ is independent of path, but only depends on r_1 and r_2 (i.e. the magnitude of the position vectors at the end points A and B).

The electric potential is a scalar. Now, let the vector field **E** be the gradient of a scalar field, or scalar potential $\phi(x, y, z)$ such that:

$$\mathbf{E} = -\vec{\nabla}\varphi \tag{15.11}$$

then:

$$-\Delta\varphi = \int \mathbf{E}\cdot d\mathbf{l} = \int_{P_1}^{P_2} \vec{\nabla}\varphi \cdot d\mathbf{r}$$

$$= \int_{P_1}^{P_2}\left(\frac{\partial\varphi}{\partial x}\mathbf{i} + \frac{\partial\varphi}{\partial y}\mathbf{j} + \frac{\partial\varphi}{\partial z}\mathbf{k}\right)\cdot\left(dx\mathbf{i} + dy\mathbf{j} + dz\mathbf{k}\right)$$

$$= \int_{P_1}^{P_2} \frac{\partial \varphi}{\partial x} dx + \frac{\partial \varphi}{\partial y} dy + \frac{\partial \varphi}{\partial z} dz \tag{15.12}$$

$$= \int_{P_1}^{P_2} d\varphi$$

$$= \varphi(x_2, y_2, z_2) - \varphi(x_1, y_1, z_1)$$

If the end points are at the same location in space (i.e. the line integral is taken around a closed curve), then:

$$\oint \mathbf{E}.d\mathbf{l} = 0 \tag{15.13}$$

These results indicate that the vector field **E** is a conservative field. In the absence of any charges (i.e. just focussing on the field **E** in empty space), by Stokes' law:

$$\vec{\nabla} \times \mathbf{E} = 0 \tag{15.14}$$

What this means is that the difference in the value of a scalar field between two points is equal to the line integral of the gradient of that scalar field along any path between the two points.

To determine the value of the electric field **E**, we essentially take the derivative of the electric potential $\mathbf{E} = -\vec{\nabla}\varphi$ which is of course a vector.

The scalar potential between two points is written:

$$\varphi_2 - \varphi_1 = \frac{q_1}{4\pi\varepsilon_0}\left[\frac{1}{r_2} - \frac{1}{r_1}\right] \tag{15.15}$$

We might wonder if it is possible to state the potential at a point $\phi(x, y, z)$. If we take the reference point at infinity, then:

$$\varphi(x, y, z) = \frac{q}{4\pi\varepsilon_0 r} \tag{15.16}$$

That is, the scalar potential is a scalar field. It is a function of position only for a static field.

15.3 THE VECTOR POTENTIAL

For a magnetic field, Gauss' law states that:*

$$\vec{\nabla} \cdot \mathbf{B} = 0 \tag{15.17}$$

As well, from Ampere's law, in the absence of any changing electric field:

$$\vec{\nabla} \times \mathbf{B} = \mu_0 \mathbf{J} \tag{15.18}$$

* The arrow indicates the use of the three-dimensional operator.

This is slightly different from the case of an electric field. Maxwell's equations are not quite symmetric in their expressions of **E** and **B** fields and this arises due to the physical existence of isolated electric charges whereas there are no isolated magnetic "charges".

Just as **E** is the gradient of a scalar potential, we would like to express **B** as the gradient of some kind of potential. This potential cannot be a *scalar* potential because if it was, the divergence would not be zero and the curl would be zero. In this case, the curl of a magnetic field is not zero.

Since the divergence of **B** *is* zero, then this means that **B** must be equal to the curl of another field. This other field is called a vector potential field and is given the symbol **A**. That is:

Since:

$$\vec{\nabla}\cdot\mathbf{B}=0 \tag{15.19}$$

then:

$$\vec{\nabla}\cdot\left(\vec{\nabla}\times\mathbf{A}\right)=0 \tag{15.20}$$

and so:

$$\mathbf{B}=\vec{\nabla}\times\mathbf{A} \tag{15.21}$$

A is the vector potential. It is a three-vector.

15.4 MAXWELL'S EQUATIONS IN POTENTIAL FORM

15.4.1 Maxwell's Equations and the Vector Potential

As shown in the previous section, the electric field **E** can be represented by the gradient of a *scalar* potential ϕ.

$$\mathbf{E}=-\vec{\nabla}\varphi \tag{15.22}$$

From Gauss' law, we have the divergence of **E** being equal to the charge density:

$$\vec{\nabla}\cdot\mathbf{E}=\rho \tag{15.23}$$

In the case of a steady magnetic field, we have **B** expressed as the curl of a vector potential **A**.

$$\mathbf{B}=\vec{\nabla}\times\mathbf{A} \tag{15.24}$$

And from Gauss' law for magnetism:

$$\vec{\nabla}\cdot\mathbf{B}=0 \tag{15.25}$$

If we make the substitutions into Faraday's law, we obtain:

$$\vec{\nabla}\times\mathbf{E} = -\frac{\partial}{\partial t}\mathbf{B}$$

$$= -\frac{\partial}{\partial t}\vec{\nabla}\times\mathbf{A}$$

$$0 = \vec{\nabla}\times\mathbf{E} + \frac{\partial}{\partial t}\vec{\nabla}\times\mathbf{A} \tag{15.26}$$

$$= \vec{\nabla}\times\mathbf{E} + \vec{\nabla}\times\frac{\partial\mathbf{A}}{\partial t}$$

$$= \vec{\nabla}\times\left(\mathbf{E} + \frac{\partial\mathbf{A}}{\partial t}\right)$$

Now, in the above, the bracketed term is a vector and one for which the curl is equal to zero. That is, this vector must be the gradient of a scalar. The scalar is the scalar potential ϕ.

$$-\vec{\nabla}\varphi = \mathbf{E} + \frac{\partial\mathbf{A}}{\partial t} \tag{15.27}$$

or:

$$\mathbf{E} = -\vec{\nabla}\varphi - \frac{\partial\mathbf{A}}{\partial t} \tag{15.28}$$

This is Faraday's law.

This is of course different to the earlier definition of the scalar potential by the addition of the time differential of the vector potential. The expression $\mathbf{E} = -\vec{\nabla}\varphi$ only applies when the charge is stationary.

Gauss' law for electric charge may also be written in terms of the potentials ϕ and $\mathbf{A}$ as:

$$\vec{\nabla}\cdot\mathbf{E} = \frac{\rho}{\varepsilon_0}$$

$$\vec{\nabla}\cdot\left(-\vec{\nabla}\varphi - \frac{\partial\mathbf{A}}{\partial t}\right) = \frac{\rho}{\varepsilon_0} \tag{15.29}$$

And so, Gauss' law for electric charge is written:

$$-\vec{\nabla}^2\varphi - \frac{\partial}{\partial t}\vec{\nabla}\cdot\mathbf{A} = \frac{\rho}{\varepsilon_0} \tag{15.30}$$

At this point, we make a choice. It is an informed choice. We say that:

$$\vec{\nabla}\cdot\mathbf{A} = -\frac{1}{c^2}\frac{\partial\varphi}{\partial t} \tag{15.31}$$

The process is called choosing a gauge. In this case, it is called the Lorenz gauge.* The significance is that the final equations of motion for the electromagnetic field will be Lorentz invariant.

So, with the Lorenz gauge, we see that:

$$\frac{1}{c^2}\frac{\partial^2\varphi}{\partial t^2} - \vec{\nabla}^2\varphi = \frac{\rho}{\varepsilon_0} \tag{15.32}$$

* Named after Ludvig Lorenz, not Hendrik Lorentz.

where ρ is the electric charge density.

Turning now to Ampere's law:

$$\vec{\nabla}\times\mathbf{B} = \mu_0\mathbf{J} + \mu_0\varepsilon_0\frac{\partial}{\partial t}\mathbf{E} \tag{15.33}$$

Since:

$$\frac{1}{c^2} = \mu_0\varepsilon_0 \tag{15.34}$$

then:

$$\begin{aligned}
\vec{\nabla}\times\mathbf{B} &= \frac{1}{c^2\varepsilon_0}\mathbf{J} + \frac{1}{c^2}\frac{\partial}{\partial t}\mathbf{E} \\
c^2\left(\vec{\nabla}\times\mathbf{B}\right) &= \frac{\mathbf{J}}{\varepsilon_0} + \frac{\partial}{\partial t}\mathbf{E} \\
c^2\vec{\nabla}\times\left(\vec{\nabla}\times\mathbf{A}\right) &= \frac{\mathbf{J}}{\varepsilon_0} + \frac{\partial}{\partial t}\left(-\vec{\nabla}\varphi - \frac{\partial\mathbf{A}}{\partial t}\right) \\
c^2\vec{\nabla}\times\left(\vec{\nabla}\times\mathbf{A}\right) &= \frac{\mathbf{J}}{\varepsilon_0} - \frac{\partial}{\partial t}\vec{\nabla}\varphi - \frac{\partial^2\mathbf{A}}{\partial t^2}
\end{aligned} \tag{15.35}$$

Now, it is a property of the differential operator that:

$$\vec{\nabla}\times\vec{\nabla}\times\mathbf{A} = \vec{\nabla}\left(\vec{\nabla}\cdot\mathbf{A}\right) - \vec{\nabla}^2\mathbf{A} \tag{15.36}$$

And so, using the Lorenz gauge, Maxwell's equations are thus elegantly written in terms of the scalar and vector potentials. In natural units:

$$\frac{\partial^2\varphi}{\partial t^2} - \vec{\nabla}^2\varphi = \rho \tag{15.37}$$

and for the vector potential:

$$\frac{\partial^2\mathbf{A}}{\partial t^2} - \vec{\nabla}^2\mathbf{A} = \mathbf{J} \tag{15.38}$$

J is the current density.

15.4.2 The Four-Potential

Maxwell's equations can now be expressed in even a more condensed form. For this, we will work in natural units where:

$$\varepsilon_0\mu_0 = \frac{1}{c^2} = 1 \tag{15.39}$$

Remembering that the d'Alembertian is:

$$\nabla^2 = \frac{\partial^2}{\partial t^2} - \vec{\nabla}^2 \tag{15.40}$$

It was shown earlier that Maxwell's equations can be expressed in terms of the scalar and vector potentials:

$$\frac{\partial^2\varphi}{\partial t^2} - \vec{\nabla}^2\varphi = \rho \tag{15.41}$$

$$\frac{\partial^2\mathbf{A}}{\partial t^2} - \vec{\nabla}^2\mathbf{A} = \mathbf{J} \tag{15.42}$$

where ρ is the charge density and **J** is the current density.

The two equations are of a similar form and can be combined together to form a new single four-vector A_μ.

$$A_\mu = (\varphi, \mathbf{A}) \tag{15.43}$$

Combining these two, Maxwell's equations are thus encompassed by the single equation:

$$\nabla^2 A_\mu = J_\mu \tag{15.44}$$

Or:

$$\partial^\mu \partial_\mu A_\mu = J_\mu \tag{15.45}$$

A_μ is called the four-potential $(\varphi, \vec{A})$, and J_μ is the four-current $(\rho, \vec{J})$.

That is, in terms of Ampere's law:

$$\begin{aligned} J_0 &= \rho \\ \mathbf{J} &= \vec{\nabla}\times\mathbf{B} - \frac{\partial\mathbf{E}}{\partial t} \end{aligned} \tag{15.46}$$

The four-potential A_μ is itself a four-vector. Maxwell's equations in four-potential form are invariant.

The associated continuity equation is:

$$\frac{\partial}{\partial x}J_x + \frac{\partial}{\partial y}J_y + \frac{\partial}{\partial y}J_z = -\frac{\partial\rho}{\partial t} \tag{15.47}$$

or:

$$\partial_\mu J_\mu = 0 \tag{15.48}$$

J_0 is the charge density ρ, and $\mathbf{J}_1$, $\mathbf{J}_2$, $\mathbf{J}_3$ are the components $\mathbf{J}_x$, $\mathbf{J}_y$, $\mathbf{J}_z$ of the current density **J**.

Maxwell's equations written in this form show that rather than deal with the separate **E** and **B** fields, it becomes mathematically more convenient and elegant to work with the four-potential A_μ. If we need to go back to the fields **E** and **B**, we can find these from:

$$\mathbf{E} = -\vec{\nabla}\varphi - \frac{\partial\mathbf{A}}{\partial t}; \quad \mathbf{B} = \vec{\nabla}\times\mathbf{A} \tag{15.49}$$

The Lorenz gauge condition becomes:

$$\begin{aligned} \frac{\partial\varphi}{\partial t} - \vec{\nabla}\cdot\mathbf{A} &= 0 \\ \partial_\mu A^\mu &= 0 \end{aligned} \tag{15.50}$$

15.5 TRANSFORMATIONS OF THE FOUR-POTENTIAL

The four-potential is written:

$$A_\mu = \left(\varphi, \vec{A}\right) \tag{15.51}$$

and is a four-vector. The Lorentz transformations for the four-potential are therefore stated:

$$\begin{aligned} \varphi' &= \frac{\varphi - A_x v}{\sqrt{1-v^2}} \\ A'_x &= \frac{A_x - v\varphi}{\sqrt{1-v^2}} \\ A'_y &= A_y \\ A'_z &= A_z \end{aligned} \tag{15.52}$$

For a moving charge in a static field, with the S′ frame of reference moving with the charge (i.e. so that $V = v$), then the scalar potential in S′ is:

$$\varphi' = \frac{q}{4\pi\varepsilon_0 r'} \tag{15.53}$$

Since we are in the moving frame "looking back", the primed and unprimed quantities swap around, and so we have:

$$\varphi = \frac{\varphi'}{\sqrt{1-v^2}} \tag{15.54}$$

And therefore:

$$\varphi = \frac{q}{4\pi\varepsilon_0 r'} \frac{1}{\sqrt{1-v^2}} \tag{15.55}$$

What we would like to do is express ϕ in terms of its coordinate r rather than r'.

$$r'^2 = x'^2 + y'^2 + z'^2 \tag{15.56}$$

And the transformations are the Lorentz transformations:

$$\begin{aligned} x' &= \frac{x - vt}{\sqrt{1-v^2}} \\ y' &= y \\ z' &= z \end{aligned} \tag{15.57}$$

Thus:

$$r'^2 = x'^2 + y'^2 + z'^2$$
$$= \frac{(x - vt)^2}{1 - v^2} + y^2 + z^2 \tag{15.58}$$

Therefore:

$$\varphi = \frac{q}{4\pi\varepsilon_0} \frac{1}{\sqrt{\frac{(x - vt)^2}{1 - v^2} + y^2 + z^2}} \frac{1}{\sqrt{1 - v^2}} \tag{15.59}$$

Let us now look at the vector potential **A**.

$$A'_x = \frac{A_x - v\varphi}{\sqrt{1 - v^2}}$$
$$A'_y = A_y \tag{15.60}$$
$$A'_z = A_z$$

In the S′ reference frame, $A' = 0$. Thus:

$$0 = \frac{A_x - v\varphi}{\sqrt{1 - v^2}} \tag{15.61}$$

and so:

$$\mathbf{A}_x = \mathbf{v}\varphi \tag{15.62}$$

where **A** is the vector potential and ϕ is the scalar potential.

All these manipulations of Maxwell's equations show that they are already Lorentz invariant. While **E** and **B** might depend upon the frame of reference, expressed as a four-potential, Maxwell's equations are the same in any reference frame.

15.6 LAGRANGIAN FOR THE ELECTROMAGNETIC FIELD

15.6.1 The Lagrangian for a Field

In Chapter 13, the action for the dynamics of a scalar field was expressed:

$$S = \int_C \mathcal{L}(\varphi, \nabla\varphi) d^4x \tag{15.63}$$

where the Lagrangian, or more precisely, the Lagrangian density, was given as a second order expression:

$$\mathcal{L} = \frac{1}{2}(\nabla\varphi)^2 - V(\varphi) \tag{15.64}$$

In Chapter 14, it was found that the Lagrangian for a scalar field ϕ is a scalar, and as with all scalars, it is Lorentz invariant. The four-potential A_μ is a vector field, and so we expect that the Lagrangian for it will be in the form of a scalar.

The resulting equations of motion arise from the Euler–Lagrange equation which is expressed:

$$\frac{\partial}{\partial x_\mu}\frac{\partial \mathcal{L}}{\partial(\nabla\varphi)} - \frac{\partial \mathcal{L}}{\partial \varphi} = 0 \tag{15.65}$$

There is nothing to say that the Lagrangian must be a quadratic function of the derivative term, and the Euler–Lagrange equation holds just as well for a first order, linear Lagrangian. A linear form may indeed be the best initial choice when devising the Lagrangian for an unknown system.

15.6.2 The Lagrangian for the Electromagnetic Field

Often in physics, it is inefficient to determine mathematical relationships based on theoretical concepts alone. That is, experience with similar constructs and the results of experiment play a role. Nowhere is this more evident than in the formulation of Lagrangians. Sometimes the best approach is to make a guess of the simplest form and test the resulting equations of motion with those that might already be known from experimental results. Refine as necessary. In this book, we have the benefit of knowing that the Lagrangian for the electromagnetic field was established long ago, but it is still of value to see how it comes about.

In the present case, the electromagnetic field is represented by the four-potential A_μ. This is a combination of the scalar potential ϕ and the vector potential $\boldsymbol{A}$.

The action is computed on the basis of the sum of the Lagrangian over infinitesimal steps between two end points in space-time which are fixed. This is shown in Figure 15.2. Here, we have the path of the four-potential, which is a vector field, over a short distance dx^μ.

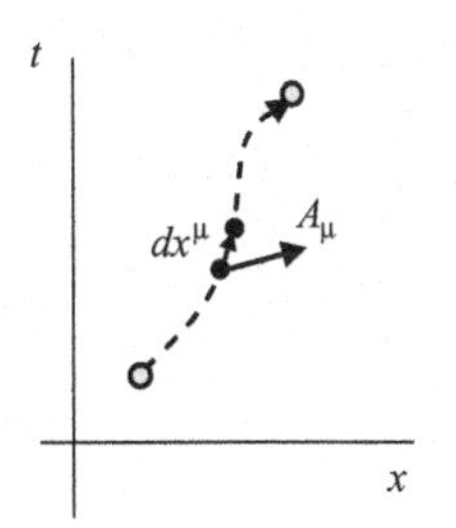

FIGURE 15.2 Incremental step along a path in space-time from one point to another showing position vector and electromagnetic field four-potential.

The principle of least action is local. The minimum action is computed as a series of infinitesimal steps along the path. There is no "action at a distance" in field theory.

A_μ is a vector field and can point in any direction. For the action to be invariant under Lorentz transformation, it is necessary that the Lagrangian be formulated as a scalar. In the present case, we have two vectors, A_μ and the four-position x^μ. The dot product between them, when summed over the path, would be a good candidate for the Lagrangian, and so, with the benefit of some foresight, the proposal is that the action be expressed:

$$S = \int A_\mu \cdot dx_\mu \tag{15.66}$$

This is a line integral along the path, that is, the component of A_μ that lies along the direction of dx_μ.*

But, to make a connection with established field equations such as Maxwell's equations, it is desirable to express the action with an integral over time, and so:

$$S = \int A_\mu \cdot \frac{dx_\mu}{dt} dt \tag{15.67}$$

The Lagrangian is expressed:

$$L = A_\mu \frac{dx_\mu}{dt} \tag{15.68}$$

The equations of motion come from the Euler–Lagrange equation. What term might we use in the Euler–Lagrange equation? It has to be that which involves the derivative in the Lagrangian, and so the Euler–Lagrange equation becomes:

$$\frac{\partial}{\partial t}\frac{\partial L}{\partial(\partial x_\mu/\partial t)} - \frac{\partial L}{\partial x_\mu} = 0 \tag{15.69}$$

Taking each term in turn:

$$\begin{aligned}\frac{\partial L}{\partial(\partial x_\mu/\partial t)} &= A_\mu \\ \frac{\partial}{\partial t}\frac{\partial L}{\partial(\partial x_\mu/\partial t)} &= \frac{\partial A_\mu}{\partial t}\end{aligned} \tag{15.70}$$

Now, A_μ is a function of x_μ, and so for the second term in the Euler–Lagrange equation, we have a function of a function:

$$\frac{\partial L}{\partial x_\mu} = \frac{\partial A_\nu}{\partial x_\mu}\frac{\partial x_\nu}{\partial t} \tag{15.71}$$

And so:

$$\frac{\partial A_\mu}{\partial t} - \frac{\partial A_\nu}{\partial x_\mu}\frac{\partial x_\nu}{\partial t} = 0 \tag{15.72}$$

Therefore:

$$\begin{aligned}0 &= \frac{\partial A_\nu}{\partial x_\mu}\frac{\partial x_\nu}{\partial t} - \frac{\partial A_\mu}{\partial t} \\ &= \frac{\partial A_\nu}{\partial x_\mu}\frac{\partial x_\nu}{\partial t} - \frac{\partial A_\mu}{\partial x_\nu}\frac{\partial x_\nu}{\partial t} \\ &= \left(\frac{\partial A_\nu}{\partial x_\mu} - \frac{\partial A_\mu}{\partial x_\nu}\right)\frac{\partial x_\nu}{\partial t} \\ &= \left(\partial_\mu A_\nu - \partial_\nu A_\mu\right)\partial_t x_\nu\end{aligned} \tag{15.73}$$

* dx_μ is of course the same as writing d^4x. We should probably also write this as x^μ but we can stay with a lower index if we remember to put a minus sign on the space derivatives.

Just as in the case of the mechanical system of Section 9.6, there is a convention attached to this Lagrangian. Instead of adding a constant of proportionality m, this time, the multiplicative constant is $-q$, which will be the charge on the particle which might find itself within the electromagnetic field. With this, the equation of motion becomes:

$$0 = -q\partial_t x_\nu \left(\partial_\mu A_\nu - \partial_\nu A_\mu\right) \tag{15.74}$$

Now, the term $\partial_t x_\nu$ is a velocity, and we might foresee that the bracketed term has a connection with the magnetic field **B**, leading to the well-known expression qvB. As well, when $\nu = 0$, the velocity term becomes $\partial t/\partial t = 1$ and, if the bracketed term has a connection with the electric field **E**, the equation of motion takes the form qE.

The equation of motion given above is not Lorentz invariant at this stage because we have differentials with respect to t, and not the proper time t'. We will address this issue in the next section.

The bracketed term above represented 16 possible combinations of the derivatives of the four-potential, and these are organised into a tensor $F_{\mu\nu}$. That is:

$$F_{\mu\nu} = \partial_\mu A_\nu - \partial_\nu A_\mu \tag{15.75}$$

$F_{\mu\nu}$ is called the electromagnetic field tensor.

15.7 THE ELECTROMAGNETIC FIELD TENSOR

15.7.1 The Electromagnetic Field Tensor

Consider the definition of the cross product in terms of the three-dimensional differential operator with the vector potential. That is, the curl of **A** is given by:

$$\vec{\nabla}\times\mathbf{A} = \begin{vmatrix} \mathbf{i} & \mathbf{j} & \mathbf{k} \\ \dfrac{\partial}{\partial x} & \dfrac{\partial}{\partial y} & \dfrac{\partial}{\partial z} \\ A_x & A_y & A_z \end{vmatrix} = \left(\frac{\partial A_z}{\partial y} - \frac{\partial A_y}{\partial z}\right)\mathbf{i} + \left(\frac{\partial A_x}{\partial z} - \frac{\partial A_z}{\partial x}\right)\mathbf{j} + \left(\frac{\partial A_y}{\partial x} - \frac{\partial A_x}{\partial y}\right)\mathbf{k} \tag{15.76}$$

This cross product is the vector field **B**. That is:

$$\mathbf{B} = \vec{\nabla}\times\mathbf{A} \tag{15.77}$$

and so the *magnitudes of the components* of the B vector are:*

$$\begin{aligned} B_x &= \frac{\partial A_z}{\partial y} - \frac{\partial A_y}{\partial z} \\ B_y &= \frac{\partial A_x}{\partial z} - \frac{\partial A_z}{\partial x} \\ B_z &= \frac{\partial A_y}{\partial x} - \frac{\partial A_x}{\partial y} \end{aligned} \tag{15.78}$$

* Note carefully what is being said here. Ordinarily, the derivative of a vector is another vector. The components **Ax**, **Ax**, **Az** are vectors. When the cross product is formed, the derivatives of the *magnitudes* of Ax, Ax, Az are used to obtain the magnitudes of the resulting vector. That is, we are now dealing with scalar quantities, those scalars representing the magnitudes of the unit vectors for the result of the vector cross product. This has considerable importance when it comes to applying the four-differential operator.

For the electric field, we have from Maxwell's equations:

$$\mathbf{E} = -\vec{\nabla}\varphi - \frac{\partial \mathbf{A}}{\partial t} \tag{15.79}$$

This is a vector equation. Breaking this up into x, y and z components:

$$\begin{aligned} -\vec{\nabla}\varphi_x &= -\frac{\partial \varphi}{\partial x} \\ -\vec{\nabla}\varphi_y &= -\frac{\partial \varphi}{\partial y} \\ -\vec{\nabla}\varphi_z &= -\frac{\partial \varphi}{\partial z} \end{aligned} \tag{15.80}$$

and so:*

$$\begin{aligned} E_x &= -\frac{\partial \varphi}{\partial x} - \frac{\partial A_x}{\partial t} \qquad -E_x = \frac{\partial \varphi}{\partial x} + \frac{\partial A_x}{\partial t} \\ E_y &= -\frac{\partial \varphi}{\partial y} - \frac{\partial A_y}{\partial t} \qquad -E_y = \frac{\partial \varphi}{\partial y} + \frac{\partial A_y}{\partial t} \\ E_z &= -\frac{\partial \varphi}{\partial z} - \frac{\partial A_z}{\partial t} \qquad -E_z = \frac{\partial \varphi}{\partial z} + \frac{\partial A_z}{\partial t} \end{aligned} \tag{15.81}$$

B_x, B_y, B_z and E_x, E_y, E_z are the six component magnitudes that describe the electromagnetic field. These are the elements of the electromagnetic field tensor $F_{\mu\nu}$. In covariant form, this is written:†

$$F_{\mu\nu} = \begin{bmatrix} 0 & -E_x & -E_y & -E_z \\ E_x & 0 & -B_z & B_y \\ E_y & B_z & 0 & -B_x \\ E_z & -B_y & B_x & 0 \end{bmatrix} \tag{15.82}$$

This is an antisymmetric tensor such that:

$$F_{\mu\nu} = -F_{\nu\mu} \tag{15.83}$$

and:

$$F_{\mu\mu} = 0 \tag{15.84}$$

The elements of $F_{\mu\nu}$ are such that:

$$F_{\mu\nu} = \partial_\mu A_\nu - \partial_\nu A_\mu \tag{15.85}$$

* Again, we must be careful about what is being said. When the differential operator acts on the scalar potential, the resulting gradient is a vector. When we subtract one vector from another, we subtract the magnitudes of each component, and so the quantities Ex, Ey, etc., in these equations are magnitudes, which are treated as scalars.

† There are several formulations of this F tensor, and the resulting equations differ, sometimes quite subtly. Here, we have stayed with Feynman's notation.

This tensor is a four-dimensional description of the electromagnetic field. The matrix elements for the E field contain differentials of the four-potential with respect to both space and time components, while those of the B field contain differentials with respect to space components only.

That is, we recognise as elements of the $F_{\mu\nu}$ tensor with μ, ν = 3,2; 1,3 and 2,1 for B_x, B_y and B_z respectively.

For example, take the term B_x in the cross product:

$$B_x = \frac{\partial A_z}{\partial y} - \frac{\partial A_y}{\partial z} \tag{15.86}$$

This is element $F_{3,2}$ in $F_{\mu\nu}$. Therefore, $\mu = 3$ stands for z and $\mu = 2$ stands for y, and so:*

$$\begin{aligned} F_{3,2} &= \partial_z A_y - \partial_y A_z \\ &= -\frac{\partial A_y}{\partial z} + \frac{\partial A_z}{\partial y} \\ &= \frac{\partial A_z}{\partial y} - \frac{\partial A_y}{\partial z} \\ &= B_x \end{aligned} \tag{15.87}$$

Now, for the $\mu = 0$ component of the electromagnetic tensor:

$$F_{0,i} = \partial_0 A_i - \partial_i A_0 \tag{15.88}$$

The i subscripts go from 1 to 3. So, in the electromagnetic tensor:

$$F_{0i} = -E_x, -E_y, -E_z = -\mathbf{E} \tag{15.89}$$

Remembering that:

$$\partial_0 = \frac{\partial}{\partial t}; \; A_0 = \varphi \tag{15.90}$$

then for $\mu = 0$ and $i = 1, 2, 3$:

$$\begin{aligned} F_{0,i} &= \partial_0 A_i - \partial_i A_0 \\ &= \frac{\partial}{\partial t} A_i + \frac{\partial \varphi}{\partial x_i} \\ &= -\mathbf{E} \end{aligned} \tag{15.91}$$

or:

$$\mathbf{E} = -\vec{\nabla}\varphi - \frac{\partial}{\partial t}\vec{A} \tag{15.92}$$

in accordance with Maxwell's equations in four-potential form.

It is also a property of the electromagnetic tensor that:

$$\partial_\mu F_{\nu\sigma} + \partial_\nu F_{\sigma\mu} + \partial_\sigma F_{\mu\nu} = 0 \tag{15.93}$$

* The covariant differential has a change of sign on the space derivatives.

This is called the Bianchi identity.

This results in 64 different equations, some of them duplicated, some the negative of the other and some where the terms cancel. We are left with just four unique equations, the first being:

$$0 = \partial_x B_x + \partial_y B_y + \partial_z B_z \tag{15.94}$$

This is the divergence of **B**:

$$0 = \vec{\nabla} \cdot \mathbf{B} \tag{15.95}$$

The remaining equations:

$$\begin{aligned} 0 &= \partial_t B_x - \partial_y E_z + \partial_z E_y \\ 0 &= \partial_t B_y + \partial_x E_z - \partial_z E_x \\ 0 &= \partial_t B_z - \partial_x E_y + \partial_y E_x \end{aligned} \tag{15.96}$$

These equate (with a change of sign on the space differentials) to:

$$\begin{aligned} \frac{\partial}{\partial t} B_x &= \frac{\partial E_z}{\partial y} - \frac{\partial E_y}{\partial z} \\ \frac{\partial}{\partial t} B_y &= \frac{\partial E_x}{\partial z} - \frac{\partial E_z}{\partial x} \\ \frac{\partial}{\partial t} B_z &= \frac{\partial E_y}{\partial x} - \frac{\partial E_x}{\partial y} \end{aligned} \tag{15.97}$$

which is:

$$\vec{\nabla} \times \mathbf{E} = -\frac{\partial}{\partial t} \mathbf{B} \tag{15.98}$$

In accordance with Maxwell's equations.

In contravariant form, the electromagnetic tensor $F^{\mu\nu}$ is:

$$\begin{aligned} g^{\mu\nu} F_{\mu\nu} g_{\mu\nu} &= \begin{bmatrix} 1 & 0 & 0 & 0 \\ 0 & -1 & 0 & 0 \\ 0 & 0 & -1 & 0 \\ 0 & 0 & 0 & -1 \end{bmatrix} \begin{bmatrix} 0 & -E_x & -E_y & -E_z \\ E_x & 0 & -B_z & B_y \\ E_y & B_z & 0 & -B_x \\ E_z & -B_y & B_x & 0 \end{bmatrix} \begin{bmatrix} 1 & 0 & 0 & 0 \\ 0 & -1 & 0 & 0 \\ 0 & 0 & -1 & 0 \\ 0 & 0 & 0 & -1 \end{bmatrix} \\ &= \begin{bmatrix} 0 & -E_x & -E_y & -E_z \\ -E_x & 0 & B_z & -B_y \\ -E_y & -B_z & 0 & B_x \\ -E_z & B_y & -B_x & 0 \end{bmatrix} \begin{bmatrix} 1 & 0 & 0 & 0 \\ 0 & -1 & 0 & 0 \\ 0 & 0 & -1 & 0 \\ 0 & 0 & 0 & -1 \end{bmatrix} \\ F^{\mu\nu} &= \begin{bmatrix} 0 & E_x & E_y & E_z \\ -E_x & 0 & -B_z & B_y \\ -E_y & B_z & 0 & -B_x \\ -E_z & -B_y & B_x & 0 \end{bmatrix} \end{aligned} \tag{15.99}$$

Note that for the tensor $F_{\mu\nu}$, raising an index on the electric field components has the net effect of introducing a change of sign on that component. This is because the **E** field components have derivatives in both space and time, and raising an index introduced a change of sign on a space index only. For the **B** components, there are only space derivatives, and when both are raised, there is no net sign change.

Let us now compute: $\partial_\mu F^{\mu\nu}$.

Taken over all combinations of μ and ν, 16 equations condense down to four different equations which equate to the four-current J_μ:

$$\begin{aligned}
&\frac{\partial E_x}{\partial x}+\frac{\partial E_y}{\partial y}+\frac{\partial E_z}{\partial z}=\vec{\nabla}\mathbf{E}=-\rho=-J_0\\
&\frac{\partial E_x}{\partial t}-\frac{\partial B_z}{\partial y}+\frac{\partial B_y}{\partial z}=\frac{\partial E_x}{\partial t}-\left(\frac{\partial B_z}{\partial y}-\frac{\partial B_y}{\partial z}\right)=\frac{\partial E_x}{\partial t}-\left(\vec{\nabla}_x\times B_x\right)=-J_1\\
&\frac{\partial E_y}{\partial t}+\frac{\partial B_z}{\partial x}-\frac{\partial B_x}{\partial z}=\frac{\partial E_y}{\partial t}-\left(\vec{\nabla}_y\times B_y\right)=-J_2\\
&\frac{\partial E_z}{\partial t}-\frac{\partial B_y}{\partial x}+\frac{\partial B_x}{\partial y}=\frac{\partial E_z}{\partial t}-\left(\vec{\nabla}_z\times B_z\right)=-J_3
\end{aligned} \tag{15.100}$$

since:

$$\begin{aligned}
&J_0=\rho\\
&\mathbf{J}=\vec{\nabla}\times\mathbf{B}-\frac{\partial\mathbf{E}}{\partial t}
\end{aligned} \tag{15.101}$$

Thus, we obtain Maxwell's equations in terms of the electromagnetic tensor:

$$-\partial_\mu F^{\mu\nu}=J_\mu \tag{15.102}$$

In the absence of any charges or currents, this becomes:

$$-\partial_\mu F^{\mu\nu}=0 \tag{15.103}$$

15.7.2 The Lagrangian for the Electromagnetic Field Tensor

Now, let us form the product:

$$F_{\mu\nu}F^{\mu\nu} \tag{15.104}$$

This is not a 4 × 4 matrix multiplication. To evaluate this product, we hold one index constant and sum over the product of that index the other and add all the terms together. It is the tensor equivalent to a vector dot product and, in this case, is called a tensor double inner product. The result is a scalar.

Holding μ constant and summing over ν, we have 16 combinations which reduce to:

$$\begin{aligned}
&F_{0\nu}F^{0\nu}=-E_x^2-E_y^2-E_z^2\\
&F_{1\nu}F^{1\nu}=-E_x^2+B_y^2+B_z^2\\
&F_{2\nu}F^{2\nu}=-E_y^2+B_x^2+B_z^2\\
&F_{3\nu}F^{3\nu}=-E_z^2+B_y^2+B_x^2
\end{aligned} \tag{15.105}$$

Adding all these components together, we obtain:

$$F_{\mu\nu}F^{\mu\nu} = -2E^2 + 2B^2 \tag{15.106}$$

It is a convention that a factor of −1/4 is applied, giving:

$$-\frac{1}{4}F_{\mu\nu}F^{\mu\nu} = \frac{E^2 - B^2}{2} \tag{15.107}$$

We will take this as our definition of the Lagrangian and then use the equations of motion to confirm the choice. That is, it is proposed that:

$$\mathcal{L} = -\frac{1}{4}F_{\mu\nu}F^{\mu\nu} \tag{15.108}$$

where:

$$F_{\mu\nu} = \partial_\mu A_\nu - \partial_\nu A_\mu \tag{15.109}$$

Since the electromagnetic field tensor is now expressed as differentials of the four-potential, the Euler–Lagrange equation can be written in terms of the four-potential:

$$\frac{\partial}{\partial x^\mu}\frac{\partial \mathcal{L}}{\partial\left(\partial A_\mu/\partial x^\mu\right)} = \frac{\partial \mathcal{L}}{\partial A_\mu} \tag{15.110}$$

There is one of the above equations for each of the four values of μ.

Taking an example, $\mu = 3(z)$ and $\nu = 2(y)$, this is element $F_{3,2}$ in $F_{\mu\nu}$ which is B_x. In $F^{\mu\nu}$, this is also B_x. The components of the electromagnetic tensor for this combination of μ and ν are, with the change of sign for space differentials:

$$\begin{aligned} F_{3,2} &= \partial_z A_y - \partial_y A_z \\ &= -\frac{\partial A_y}{\partial z} + \frac{\partial A_z}{\partial y} \\ &= \frac{\partial A_z}{\partial y} - \frac{\partial A_y}{\partial z} \end{aligned} \tag{15.111}$$

Also, when interchanging rows and columns, then, for the space–space components of F,

$$F_{3,2} = F^{2,3} \tag{15.112}$$

and so the contribution to the Lagrangian is:

$$\mathcal{L} = -\frac{1}{4}\left(\frac{\partial A_z}{\partial y} - \frac{\partial A_y}{\partial z}\right)\left(\frac{\partial A_z}{\partial y} - \frac{\partial A_y}{\partial z}\right) \tag{15.113}$$

By the product rule:

$$\frac{\partial \mathcal{L}}{\partial\left(\partial A_y/\partial z\right)} = -\frac{1}{4}\left[\left(\frac{\partial A_z}{\partial y} - \frac{\partial A_y}{\partial z}\right)(-1) + \left(\frac{\partial A_z}{\partial y} - \frac{\partial A_y}{\partial z}\right)(-1)\right] \tag{15.114}$$

$$= -\frac{1}{2}\left(\frac{\partial A_z}{\partial y} - \frac{\partial A_y}{\partial z}\right)$$

$$= -\frac{1}{2}F_{3,2}$$

$$= -\frac{1}{2}F^{3,2}$$

For an example involving the electric field, we might pick $\mu = 0(t)$ and $\nu = 2(y)$, element $F_{0,2}$, which is $-E_y$:

$$\begin{aligned} F_{0,2} &= \partial_t A_y - \partial_y A_t \\ &= \frac{\partial A_y}{\partial t} + \frac{\partial \varphi}{\partial y} \\ &= -E_y \end{aligned} \tag{15.115}$$

In this case, $F^{0,2}$ would be $+E_y$ and so for the time-space components of F, then:

$$F_{0,2} = -F^{0,2} \tag{15.116}$$

The contribution to the Lagrangian is:

$$\begin{aligned} \mathcal{L} &= -\frac{1}{4}F_{0,2}F^{0,2} \\ &= -\frac{1}{4}\left(\frac{\partial A_y}{\partial t} + \frac{\partial \varphi}{\partial y}\right)\left(-\frac{\partial \varphi}{\partial y} - \frac{\partial A_y}{\partial t}\right) \end{aligned} \tag{15.117}$$

By the product rule:

$$\begin{aligned} \frac{\partial \mathcal{L}}{\partial\left(\partial A_y/\partial t\right)} &= -\frac{1}{4}\left[\left(\frac{\partial A_y}{\partial t} + \frac{\partial \varphi}{\partial y}\right)(-1) + \left(-\frac{\partial \varphi}{\partial y} - \frac{\partial A_y}{\partial t}\right)(+1)\right] \\ &= -\frac{1}{4}\left[\left(-\frac{\partial A_y}{\partial t} - \frac{\partial \varphi}{\partial y}\right) + \left(-\frac{\partial \varphi}{\partial y} - \frac{\partial A_y}{\partial t}\right)\right] \\ &= -\frac{1}{4}\left[2\left(-\frac{\partial A_y}{\partial t} - \frac{\partial \varphi}{\partial y}\right)\right] \\ &= -\frac{1}{2}F^{0,2} \end{aligned} \tag{15.118}$$

But, in the total Lagrangian, that is, in the total of the 16 combinations of μ and ν, each differential appears again when μ and ν are swapped around. Thus, for the total Lagrangian:

$$\frac{\partial \mathcal{L}}{\partial\left(\partial A_\mu/\partial x^\mu\right)} = -F^{\mu\nu} \tag{15.119}$$

With the differential added:

$$\frac{\partial}{\partial x^{\mu}} \frac{\partial \mathcal{L}}{\partial\left(\partial A_{\mu}/\partial x^{\mu}\right)} = -\frac{\partial F^{\mu\nu}}{\partial x^{\mu}} = -\partial_{\mu} F^{\mu\nu} \tag{15.120}$$

The next step is to evaluate: $\partial\mathcal{L}/\partial A_{\mu}$ but since A_{μ} does not appear in the proposed expression for L, then this differential is zero.

Thus, the Euler–Lagrange equation of motion gives:

$$-\partial_{\mu} F^{\mu\nu} = 0 \tag{15.121}$$

This is Maxwell's equations in terms of the electromagnetic tensor in the absence of any charges or currents. That is, for the electromagnetic field, the equations of motion are Maxwell's equations.

The Lagrangian for the electromagnetic field, in the absence of any charges and currents, is therefore stated as:

$$\mathcal{L} = -\frac{1}{4} F_{\mu\nu} F^{\mu\nu} \tag{15.122}$$

With charges and currents, Maxwell's equations are:

$$-\partial_{\mu} F^{\mu\nu} = J_{\mu} \tag{15.123}$$

and so the differential that was equal to zero in the Lagrangian must be now:

$$\frac{\partial \mathcal{L}}{\partial A_{\mu}} = J_{\mu} \tag{15.124}$$

But, by Maxwell's equations expressed in terms of the four-potential:

$$J_{\mu} = \partial^{\mu}\partial_{\mu} A_{\mu} = \partial^{\mu}\left(\partial_{\mu} A_{\mu}\right) \tag{15.125}$$

Thus, the part of the Lagrangian that accounts for the presence of charges and currents is:

$$\frac{\partial \mathcal{L}}{\partial A_{\mu}} = \partial^{\mu}\left(\partial_{\mu} A_{\mu}\right)$$

$$\mathcal{L} = \int \left(\partial_{\mu} A_{\mu}\right) \partial^{\mu} \partial A_{\mu} \tag{15.126}$$

$$= \frac{1}{2}\left(\partial_{\mu} A_{\mu}\right)^{2}$$

The Lagrangian for the electromagnetic field with charges and currents is thus expressed:

$$\mathcal{L} = -\frac{1}{4} F_{\mu\nu} F^{\mu\nu} - \frac{1}{2}\left(\partial_{\mu} A_{\mu}\right)^{2} \tag{15.127}$$

where the equations of motion (Maxwell's equations) are:

$$-\partial_{\mu} F^{\mu\nu} = J_{\mu} \tag{15.128}$$

in the Lorenz gauge:

$$\partial_\mu A^\mu = 0 \tag{15.129}$$

The minus sign appears for the charges and currents term by convention.

15.8 CHARGED PARTICLE IN AN ELECTROMAGNETIC FIELD

When a charged particle interacts with the electromagnetic field, the strength of the coupling, or the interaction, depends upon the charge of the particle. When a particle interacts with the field, the action is that of the field and the particle, that is, the Lagrangians add together so that the total action may be computed.

For a free particle, the Lagrangian is:

$$L = -m\sqrt{1-\vec{v}^2/c^2} \tag{15.130}$$

Or, in natural units:

$$\begin{aligned} L &= -m\sqrt{1-\vec{v}^2} \\ &= -m\sqrt{1-\left(\partial x_i/\partial t\right)^2} \end{aligned} \tag{15.131}$$

where i goes from 1 to 3. Notice that this Lagrangian contains a differential of three space coordinates with respect to time. It also contains a constant of proportionality $-m$.

In Section 15.6.2, the Lagrangian for the electromagnetic field was stated as:

$$L = A_\mu \frac{dx_\mu}{dt} \tag{15.132}$$

By convention, a factor of $-q$ is included so that the final equations of motion are recognisable as Maxwell's equations. The multiplication of the Lagrangian by a constant does not alter the value of the Lagrangian. Separating into time and space components, the Lagrangian for the field can be thus written:

$$L = -qA_0 \frac{\partial x_0}{\partial t} - qA_i \frac{\partial x_i}{\partial t} \tag{15.133}$$

Since $\partial x_0/\partial t = 1$. A_0 is the scalar potential ϕ, and A_i is the vector potential.

The Lagrangian for the field is for the *field* dynamics – that is, for a moving field. For a stationary field, and a particle moving through it, from the particle's point of view, the field is moving in the opposite direction. Hence, the velocity here is that of the particle with respect to the field and the field with respect to the particle. Thus, we put a change of sign on the quantity $\partial x_i/\partial t$.

The total Lagrangian for a charged particle moving in an electromagnetic field is thus the addition of the individual Lagrangians, and so:

$$L = -m\sqrt{1-\left(\partial x_i/\partial t\right)^2} - qA_0 + qA_i \frac{\partial x_i}{\partial t} \tag{15.134}$$

We now need the Euler–Lagrange equation:

$$\frac{\partial}{\partial t}\frac{\partial L}{\partial(\partial x/\partial t)}=\frac{\partial L}{\partial x} \tag{15.135}$$

And so, we require:

$$\begin{aligned}\frac{\partial L}{\partial(\partial x_i/\partial t)}&=-m\frac{1}{2}\frac{-2(\partial x_i/\partial t)}{\sqrt{1-(\partial x_i/\partial t)^2}}+qA_i\\&=\frac{m(\partial x_i/\partial t)}{\sqrt{1-(\partial x_i/\partial t)^2}}+qA_i\\\frac{\partial}{\partial t}\frac{\partial L}{\partial((\partial x_i/\partial t))}&=\frac{\partial}{\partial t}\left[\frac{m(\partial x_i/\partial t)}{\sqrt{1-(\partial x_i/\partial t)^2}}\right]+q\frac{\partial}{\partial t}A_i\end{aligned} \tag{15.136}$$

Now, the four-potential A_μ is a four-vector and consists of the scalar potential ϕ and the vector potential **A**. The contribution to the action for the scalar potential depends only on the end points, and not on the path, whereas the value of the vector potential is a function of position and time. If we differentiate the vector potential **A** with respect to *t*, as we need to in the second term of the right-hand side above, then there is an implicit differentiation with respect to *x* since, as the particle moves through the field, its position *x* depends on *t*. That is, we have a function of a function. Thus:

$$\frac{\partial A_i}{\partial t}=\frac{\partial A_i}{\partial t}+\frac{\partial A_i}{\partial x_j}\frac{\partial x_j}{\partial t} \tag{15.137}$$

Thus:

$$\frac{\partial}{\partial t}\frac{\partial L}{\partial((\partial x_i/\partial t))}=\frac{\partial}{\partial t}\left[\frac{m(\partial x_i/\partial t)}{\sqrt{1-(\partial x_i/\partial t)^2}}\right]+q\frac{\partial A_i}{\partial t}+q\frac{\partial A_i}{\partial x_j}\frac{\partial x_j}{\partial t} \tag{15.138}$$

For the Euler–Lagrange equation, we now differentiate *L* with respect to x_i:*

$$\frac{\partial L}{\partial x_i}=-q\frac{\partial A_0}{\partial x_i}-q\frac{\partial x_j}{\partial t}\frac{\partial A_j}{\partial x_i} \tag{15.139}$$

Although it might seem like $\partial x_i/\partial t(\partial A_i/\partial x_i)=\partial A_i/\partial t$ we have reason to keep these differentials as separate terms in anticipation of what is to come. Thus:

$$\begin{aligned}\frac{\partial}{\partial t}\left[\frac{m(\partial x_i/\partial t)}{\sqrt{1-(\partial x_i/\partial t)^2}}\right]+q\frac{\partial A_i}{\partial t}+q\frac{\partial A_i}{\partial x_j}\frac{\partial x_j}{\partial t}&=-q\frac{\partial A_0}{\partial x_i}-q\frac{\partial x_j}{\partial t}\frac{\partial A_j}{\partial x_i}\\\frac{\partial}{\partial t}\left[\frac{m(\partial x_i/\partial t)}{\sqrt{1-(\partial x_i/\partial t)^2}}\right]&=q\left(-\frac{\partial A_i}{\partial t}-\frac{\partial A_0}{\partial x_i}\right)+q\frac{\partial x_j}{\partial t}\left(+\frac{\partial A_i}{\partial x_j}-\frac{\partial A_j}{\partial x_i}\right)\end{aligned} \tag{15.140}$$

* Note, A_0 is the scalar potential ϕ, and its contribution to the action is independent of the path so we don't need a term for the implicit differentiation with respect to *x*.

In the first bracketed term on the right-hand side of the above, A_0 is the scalar potential ϕ, and so that term is:

$$\mathbf{E} = -\vec{\nabla}\varphi - \frac{\partial \mathbf{A}_i}{\partial t} \tag{15.141}$$

For i going from 1 to 3, these are the components of the electric field as we saw earlier in Section 15.7. These are the elements of the $F_{\mu\nu}$ tensor with $\nu = 0$ and $\mu = 1, 2, 3$ for E_x, E_y and E_z respectively.

Multiplied by the charge, this term therefore represents the Coulomb force:

$$\mathbf{F} = q\mathbf{E} \tag{15.142}$$

The last bracketed term on the right-hand side has the charge times the velocity times the curl of the vector potential.

$$\mathbf{B} = \vec{\nabla} \times \mathbf{A} \tag{15.143}$$

where:

$$\vec{\nabla} \times \mathbf{A} = \begin{vmatrix} \mathbf{i} & \mathbf{j} & \mathbf{k} \\ \frac{\partial}{\partial x} & \frac{\partial}{\partial y} & \frac{\partial}{\partial z} \\ A_x & A_y & A_z \end{vmatrix} = \left(\frac{\partial A_z}{\partial y} - \frac{\partial A_y}{\partial z}\right)\mathbf{i} + \left(\frac{\partial A_x}{\partial z} - \frac{\partial A_z}{\partial x}\right)\mathbf{j} + \left(\frac{\partial A_y}{\partial x} - \frac{\partial A_x}{\partial y}\right)\mathbf{k} = B_x\mathbf{i} + B_y\mathbf{j} + B_z\mathbf{k} \tag{15.144}$$

That is, as we saw in Section 15.7, we recognise as elements of the $F_{\mu\nu}$ tensor with $\mu, \nu = 3{,}2$; $1{,}3$ and $2{,}1$ for B_x, B_y and B_z respectively. Thus, this term represents the magnetic force.

$$\mathbf{F} = q\mathbf{v} \times \mathbf{B} \tag{15.145}$$

The total equation implies a force on the left-hand side, and so the equation of motion appears to have the form:

$$\mathbf{F} = q(\mathbf{E} + \mathbf{v} \times \mathbf{B}) \tag{15.146}$$

We can express this equation in a Lorentz invariant way by multiplying through by $\partial t/\partial t'$, which is equivalent to $\partial x_0/\partial t'$.* Remembering to account for a change in sign in space derivatives when moving from three to four dimensions, and noting that $(\partial x_i/\partial t)^2$ is the three-velocity:

$$\vec{v}^2 = \left(\frac{\partial x}{\partial t}\right)^2 + \left(\frac{\partial y}{\partial t}\right)^2 + \left(\frac{\partial z}{\partial t}\right)^2 \tag{15.147}$$

we have in three dimensions:

$$\frac{\partial}{\partial t}\left[\frac{m(\partial x_i/\partial t)}{\sqrt{1-(\partial x_i/\partial t)^2}}\right] = q\left(-\frac{\partial A_i}{\partial t} - \frac{\partial A_0}{\partial x_i}\right) + q\frac{\partial x_j}{\partial t}\left(+\frac{\partial A_i}{\partial x_j} - \frac{\partial A_j}{\partial x_i}\right) \tag{15.148}$$

* Note that that $\partial t = \partial x_0$ and $\partial t/\partial t' = 1$.

going to:

$$\frac{\partial}{\partial t'}\left[\frac{m\vec{v}}{\sqrt{1-\vec{v}^2}}\right] = q\frac{\partial x_\nu}{\partial t'}\left(\frac{\partial A_\nu}{\partial x_\mu} - \frac{\partial A_\mu}{\partial x_\nu}\right) \tag{15.149}$$

in four dimensions. The first term on the right-hand side of the three-dimensional equation gets folded into the right-hand side of the four-dimensional equation when $\nu = 0$.

The right-hand side of the above is of the form of an invariant four-vector. The left-hand side is presently of the form of a three-vector. This implies the existence of a $\mu = 0$ term for the left-hand side which has physical significance matching $\mu = 0$ on the right-hand side.

Note that the left-hand side of the equation is of the form of the change of the *three* space components of the four-momentum with respect to the proper time. The zero component of the four-momentum is energy, and so we expect that the missing $\mu = 0$ component of the left-hand side is the derivative of the energy with respect to the proper time. The four-momentum is:

$$p_\mu = \left[E, p_x, p_y, p_z\right] \tag{15.150}$$

And the time rate of change of p_μ is therefore:

$$\frac{d}{dt'}p_\mu = \left(\frac{\partial E}{\partial t}, \frac{\partial \bar{p}}{\partial t}\right) \tag{15.151}$$

Here, E is the energy of the particle.

Now, we know that the zero component of the four-velocity is:

$$u_0 = \frac{1}{\sqrt{1-\vec{v}^2}} \tag{15.152}$$

and so the left-hand side of the Lorentz equation for $\mu = 0$ becomes:

$$\frac{\partial}{\partial t'}\left[\frac{m}{\sqrt{1-\vec{v}^2}}\right] = \frac{\partial}{\partial t'}mu_0 \tag{15.153}$$

which is the time rate of change of the *zero component* of the four-momentum, or, time rate of change of energy. That is, power.

To check this is consistent with the right-hand side of the Lorentz equation for $\mu = 0$, we can see that A_0 is the scalar potential ϕ, and x_0 is the time t. So, looking on the right-hand side only, for $\mu = 0$, and making an adjustment of sign for the space derivative to express it in terms of $\vec{\nabla}$:*

$$\begin{aligned} q\frac{\partial x_\nu}{\partial t'}\left(\frac{\partial A_\nu}{\partial x_\mu} - \frac{\partial A_\mu}{\partial x_\nu}\right) &= q\frac{\partial x_\nu}{\partial t'}\left(\frac{\partial A_\nu}{\partial x_0} - \frac{\partial A_0}{\partial x_\nu}\right) \\ &= q\frac{\partial x_\nu}{\partial t'}\left(+\vec{\nabla}\varphi + \frac{\partial A_j}{\partial t}\right) \\ &= -q\frac{\partial x_\nu}{\partial t'}\mathbf{E} \end{aligned} \tag{15.154}$$

* Note there's been a change of sign here for **E** because we have $\partial A_\nu/\partial t$ and $\partial\varphi/\partial x_\nu$ and not $\partial A_\mu/\partial t$ and $\partial\varphi/\partial x_\mu$. The sign change comes about because $F_{\mu\nu} = -F_{\nu\mu}$.

We can see that $q\mathbf{E}$ has the units of force, the Coulomb force, and force times velocity is power. For $\mu = 0$, both sides of the equation of motion have units of power.

In four dimensions, the equation of motion is stated:

$$\begin{aligned} m\frac{\partial u_\mu}{\partial t'} &= q\frac{\partial x_\nu}{\partial t'}\left(\frac{\partial A_\nu}{\partial x_\mu} - \frac{\partial A_\mu}{\partial x_\nu}\right) \\ &= q\frac{\partial x_\nu}{\partial t'}\left(\nabla_\mu A_\nu - \nabla_\nu A_\mu\right) \end{aligned} \tag{15.155}$$

On the left-hand side, the differential, with respect to the proper time, is the four-acceleration. The whole term on the left-hand side is the four-force f_μ. The velocity term $\partial x_\nu/\partial t'$ on the right-hand side is the four-velocity u_ν.

μ takes on values 0 to 3, and for each value of μ, ν goes from 0 to 3. We have therefore a set of four equations, one for each value of μ. Expressed as a sum over ν, we have:

$$\begin{aligned} m\frac{\partial u_\mu}{\partial t'} &= q\sum_{\nu=0}^{3} u_\nu\left(\nabla_\mu A_\nu - \nabla_\nu A_\mu\right) \\ f_\mu &= q\sum_{\nu=0}^{3} u_\nu F_{\mu\nu} \end{aligned} \tag{15.156}$$

This is the Lorentz force law, which we usually state as:

$$\mathbf{F} = q\left(\mathbf{E} + \mathbf{v}\times\mathbf{B}\right) \tag{15.157}$$

Whenever there is a repeated subscript, like ν in the above, the summation is usually implied, and so we write:

$$f_\mu = q u_\nu F_{\mu\nu} \tag{15.158}$$

15.9 TRANSFORMATIONS OF THE ELECTROMAGNETIC FIELD

Space and time quantities transform with a velocity in the x direction in accordance with the Lorentz transformations. We saw previously that the transformation matrix for motion in the x direction is:

$$M_{\mu\nu} = \begin{bmatrix} \frac{1}{\sqrt{1-v^2}} & \frac{v}{\sqrt{1-v^2}} & 0 & 0 \\ \frac{v}{\sqrt{1-v^2}} & \frac{1}{\sqrt{1-v^2}} & 0 & 0 \\ 0 & 0 & 1 & 0 \\ 0 & 0 & 0 & 1 \end{bmatrix} \tag{15.159}$$

The electromagnetic tensor transforms as:

$$F'_{\mu\nu} = M_{\mu\sigma} M_{\nu\tau} F_{\sigma\tau} \tag{15.160}$$

Note that we need two applications of the transformation matrix to account for the two indices of the F tensor.

Consider a charged particle moving along the x axis with velocity v in coordinate frame S as shown in Figure 15.3. There is a magnetic field B_z acting along the z axis.

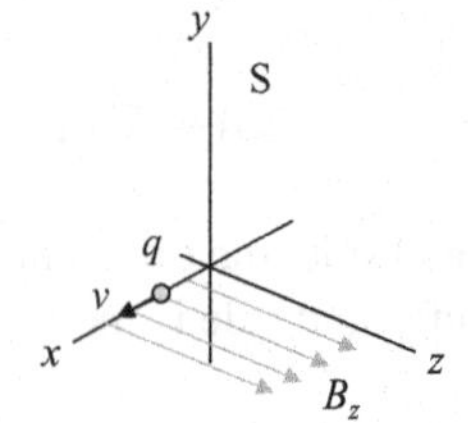

FIGURE 15.3 Charged particle moving in the x direction with a magnetic field pointing in the y direction in the S coordinate frame.

What can we say about the forces on the particle from the point of view of the S′ axis which travels along with the particle?

From the electromagnetic tensor, a magnetic field B_z is element $F_{2,1}$. That is, $\sigma = 2$, $\tau = 1$. So, we have:

$$F'_{\mu\nu} = M_{\mu,2}M_{\nu,1}F_{2,1} \tag{15.161}$$

Our next question is, what is the electromagnetic field in S′? We need to choose a direction. Let's say we choose the z axis. Let's also say that we want to know if there is a $+E$ field in that direction, that is, $+E_z$, in the S′ reference frame. This is element $F'_{3,0}$.

$$F'_{3,0} = M_{3,2}M_{0,1}F_{2,1} \tag{15.162}$$

Now, according to the transformation matrix:

$$M_{3,2} = 0$$
$$M_{0,1} = \frac{v}{\sqrt{1-v^2}} \tag{15.163}$$

Thus:

$$F'_{3,0} = 0\frac{v}{\sqrt{1-v^2}}F_{2,1} \tag{15.164}$$

And so we can say that there is no E_z field in the S′ coordinate frame. This should not surprise us since we know from our right-hand rule that we would expect there to be something happening only along the y axis. Let us now compute the $+E$ field along the y axis in the S′ frame. This is element $F'_{2,0}$.

$$F'_{2,0} = M_{2,2}M_{0,1}F_{2,1} \tag{15.165}$$

Now, according to the transformation matrix:

$$M_{2,2} = 1$$
$$M_{0,1} = \frac{1}{\sqrt{1-v^2}} \tag{15.166}$$

Thus:

$$F'_{2,0} = \frac{v}{\sqrt{1-v^2}} F_{2,1} \tag{15.167}$$

therefore:

$$E'_y = \frac{v}{\sqrt{1-v^2}} B_z \tag{15.168}$$

Now, if the particle were an electron, with a negative charge, we would expect that the electron (by the left-hand rule) would experience a downward force $F=qvB$. This is how it appears in the S reference frame. As we see above, in the S′ reference frame, there is no magnetic field, but instead an electric field E_y pointing upwards. The electron will thus experience the Coulomb force downwards according to the transformation from B_z in S to E_y in S′. The magnitude of the electric force depends upon v, which in this example, is along the x axis. The Lorentz transformation matrix above has been expressed in terms of velocities in the x direction.

Note the extraordinary circumstance that in the S frame of reference, there is a magnetic field B_z and no electric field, but in the S′ reference frame, for the exact same motion and force on the electron, there is an electric field E_y and no magnetic field. Hence the oft-said comment that "one person's magnetic field is another person's electric field".

15.10 THE ELECTROMAGNETIC WAVE

As shown in Chapter 3, when the **E** and **B** fields change with time in a periodic manner, there is created an electromagnetic wave. Now, if our axes are oriented such that the **E** field oscillates in the y direction and the **B** field oscillates in the x direction, the direction of travel of the wave is in the z direction. This was illustrated in Figure 3.9. Thus, $E_z=0$, $E_x=0$, $B_z=0$ and $B_y=0$.

The electric and magnetic fields are a function of z and t such that:

$$\begin{aligned} E(z,t) &= \varepsilon \sin(kz - \omega t) \\ B(z,t) &= \beta \sin(kz - \omega t) \end{aligned} \tag{15.169}$$

In exponential form:

$$\begin{aligned} E(z,t) &= \varepsilon e^{i(kz-\omega t)} \\ B(z,t) &= \beta e^{i(kz-\omega t)} \end{aligned} \tag{15.170}$$

The factors ε and β are the polarisation four-vectors and contain information about the amplitude of the wave relative to the coordinate axes.

When an electromagnetic wave propagates through space, say in the z axis direction, the oscillating electric field may be aligned with the x or the y axis and the associated magnetic field oscillating along the y or the x axis. This is usually how electromagnetic waves are illustrated in textbooks. There is no oscillation of the field along the direction of travel. If there were some component of E along the z axis, then in that direction, the Maxwell's equation, $\vec{\nabla} \cdot \mathbf{E} = 0$, for a field (in the absence of any charges or currents) would be violated since any differential of E_z would not equal zero. Similarly, for the **B** field, there is no **B** field in the direction of the z axis since then, $\vec{\nabla} \cdot \mathbf{B} = 0$ would be violated.

But there is nothing to say that the oscillating **E** and **B** fields should line up with a particular coordinate axis x or y. The polarisation vector gives the weighting between the x and y directions for the oscillating fields. If the coordinate axes were rotated so that say the **E** field was lined up with the y axis, then the polarisation vector for the x direction would be zero and that for the y direction would be a maximum, the magnitude of which depends on the amplitude of the field.

We can see that for a photon, the polarisation vector takes the place of the Dirac spinor. For a fermion (e.g. an electron), each of the four wave function solutions Ψ comprises a four-component spinor. The spinors contain information about the positive and negative energy states and the up and down half-integer spin states. A photon has integer spin. There are two "spin" states which can take the values −1 and +1. In physical terms, this represents the transverse polarisation of the electromagnetic wave.

It is a consequence of Maxwell's equations that the speed of the electromagnetic wave is, in free space, fixed at c.

$$\text{Now, } c = \frac{\omega}{2\pi}\lambda \text{ and } k = \frac{2\pi}{\lambda}, \tag{15.171}$$

and so:

$$(kx - \omega t) = \omega(z - ct) \tag{15.172}$$

In natural units, we can write:

$$\begin{aligned} E(z,t) &= \varepsilon \sin \omega(z-t) \\ B(z,t) &= -\beta \sin \omega(z-t) \end{aligned} \tag{15.173}$$

Maxwell's equations show that:

$$\frac{\partial}{\partial t}\mathbf{E} = \vec{\nabla}\times\mathbf{B} = \begin{vmatrix} \mathbf{i} & \mathbf{j} & \mathbf{k} \\ \frac{\partial}{\partial x} & \frac{\partial}{\partial y} & \frac{\partial}{\partial z} \\ B_x & B_y & B_z \end{vmatrix} = \left(\frac{\partial B_z}{\partial y} - \frac{\partial B_y}{\partial z}\right)\mathbf{i} + \left(\frac{\partial B_x}{\partial z} - \frac{\partial B_z}{\partial x}\right)\mathbf{j} + \left(\frac{\partial B_y}{\partial x} - \frac{\partial B_x}{\partial y}\right)\mathbf{k} \tag{15.174}$$

For our example wave travelling in the z direction, with **E** oriented in the y direction, this becomes:

$$\frac{\partial}{\partial t}E_y = \nabla \times B_y = \frac{\partial}{\partial z}B_x - \frac{\partial}{\partial x}B_z \tag{15.175}$$

But $B_z = 0$, and so:

$$\frac{\partial}{\partial t}E_y = \frac{\partial}{\partial z}B_x \tag{15.176}$$

Thus:

$$\begin{aligned} \frac{\partial}{\partial t}E_y &= -\varepsilon \cos \omega(z-t) \\ \frac{\partial}{\partial z}B_y &= \beta \cos \omega(z-t) \end{aligned} \tag{15.177}$$

Indicating that $\varepsilon = \beta$ as we might expect. That is, in natural units, the amplitude of the E and B fields are the same.

Although we have discussed the electromagnetic field at some length, we have so far had nothing much to say about the photon. Maxwell described the electromagnetic field in terms of field equations, which we have seen arise from the Euler–Lagrange equations of motion which, in turn, derive from the action principle.

Some years after Maxwell, the photon, the quantum of the electromagnetic field, was proposed by Planck and taken further by Einstein to explain experimental phenomena that were found to be inconsistent with the classical view.

Quantum mechanics of the 1920s explained the role of the electron in an atom in a probabilistic sense, and the addition of relativity resulted in the discovery of antimatter.

For the electromagnetic field, classical field theory was extended into a Lorentz invariant form via the notion of the four-potential.

We are now ready to quantise the classical field and find where a particle like a photon might come from.

The Quantum Field 16

16.1 INTRODUCTION

Although Dirac's relativistic equation predicted the existence of antiparticles, his equation was still a single-particle based approach where the particle remained intact. The number of particles is conserved. However, we know that mass and energy can be converted from one to the other. Particles can be created and annihilated. We need a theory which can account for this.

In the quantum particle theory, we have seen that physical quantities such as energy and momentum are replaced by operators and those operators act on the wave function. The wave functions are the solutions of the Schrödinger or Dirac wave equations. The boundary conditions imposed by the wave equation determines the allowable energies, the eigenvalues, of the wave function. The amplitude (squared) of the wave function gives the probability density for a particle to be in one of the allowable energy states. That is, for example, given an electron, we can determine the probability of it being located at a particular place or the probability of it having a specific energy or momentum.

For photons, we have an electromagnetic field. The wave equation is the set of Maxwell's equations. The wave function is the four-potential. However, the states are not quantised. It is a classical field. The photon is not a part of it. We know this is not a sufficient description of the electromagnetic field because of the observed quantum effects associated with the existence of photons.

The concept of a quantised field was implicit in Dirac's negative-energy sea. According to Dirac's idea, the vacuum of space-time is populated by an infinite number of negative energy, negatively charged electrons. When a positive energy electron is created, it leaves behind a positively charged hole which behaves as if it had positive energy, an anti-electron. Dirac thought originally that these anti-electrons were protons, but it was soon discovered that anti-electrons actually existed.

When a photon scatters off an electron, according to Dirac, the following steps occur:

1. We start with the photon, the initial positive energy electron, and a corresponding negative energy electron in the sea.
2. The negative energy electron absorbs the photon and acquires positive energy thus leaving behind a hole in the sea.
3. The original positive energy electron falls into the hole and emits a new photon.
4. We are left with the new positive energy electron and a new photon.

The sea is the field from which positive energy electrons are created. This is the beginning of the concept of a quantum field.

Just as in quantum particle theory, where physical quantities became quantum operators, in quantum field theory, we find that wave functions become operators and what are called the commutation relations between those operators lead to quantisation of the field.

Our examination of this subject will be somewhat superficial but in enough depth to get a feeling for the principles involved. Our model field will be an assembly of harmonic oscillators.

To begin, let's summarise our knowledge of classical fields.

16.2 CLASSICAL FIELDS

We've seen how the Lagrangian for a scalar field is formulated from a consideration of the principle of least action. There are expressions for the Lagrangian for other types of fields, such as the Dirac field for a particle with spin ½ (e.g. electrons) and the electromagnetic field, which is a vector field for particles with spin 1 (e.g. photons).

The Lagrangian is determined by the nature of the field. Often, a trial and error approach is taken and the resulting equations of motion compared against those already known to validate the choice.

As mentioned in Chapter 13, the potential function in the Lagrangians is in general:

$$V(x) = V_0 + C_1V(x) + C_2V(x)^2 + C_3V(x)^3 + \ldots \tag{16.1}$$

We usually deal with potentials of the first few orders of x. The significance of these higher orders will become apparent when we discuss the interactions that occur between quantum fields.

16.2.1 Scalar Field (Spin 0)

For a scalar field, Section 13.3 showed that the Lagrangian is expressed:

$$\begin{aligned}\mathcal{L} &= \frac{1}{2}(\nabla\varphi)^2 - V(\varphi)\\ &= \frac{1}{2}\left(\frac{\partial\varphi}{\partial t}\right)^2 - \frac{1}{2}\left(\vec{\nabla}\varphi\right)^2 - V(\varphi)\end{aligned} \tag{16.2}$$

This leads to the Euler–Lagrange equation:

$$\begin{aligned}-\frac{\partial^2\varphi}{\partial t^2} + \frac{\partial\varphi}{\partial x_\mu} - \frac{\partial V(\varphi)}{\partial\varphi} &= -\frac{\partial^2\varphi}{\partial t^2} + \vec{\nabla}^2\varphi - \frac{\partial V(\varphi)}{\partial\varphi} = 0\\ \nabla^2\varphi + \nabla_\varphi V(\varphi) &= 0\end{aligned} \tag{16.3}$$

which can be written in terms of the Lagrangian as:

$$\frac{\partial}{\partial x_\mu}\frac{\partial\mathcal{L}}{\partial(\nabla\varphi)} - \frac{\partial\mathcal{L}}{\partial\varphi} = 0 \tag{16.4}$$

This is the Klein–Gordon equation. The Klein–Gordon equation is the equation of motion for a scalar field.

In a later section, we will have cause to use the kinetic term of this, which, in the case of a mass m in the scalar field, becomes:

$$\left(\partial^\mu\partial_\mu + m^2\right)\varphi = 0 \tag{16.5}$$

The superscript μ on the differential reminds us of the negative sign on the space terms associated with the d'Alembertian.

The field function is of the general form:

$$\varphi(x_u) = \varphi(t, x, y, z)) \tag{16.6}$$

With a suitable choice of potential term consistent with the relativistic energy equation, the Euler–Lagrange equation is in the form of the Klein–Gordon equation.

The factor ½ is inserted by convention. The significance of it is that it leads to conventional representations of kinetic energy as $\frac{1}{2}mv^2$ if we were to be applying these equations to a mechanical system such as a particle of mass m.

16.2.2 Dirac Field (Spin ½)

For an electron, we treat the wave equation $\psi(x_\mu)$ as if it were the equation for a field. The complex wave function becomes a complex field function. In the case of the scalar field, the dynamics of a particle were treated as the dynamics of a field. The dynamics of the electron field are similar to those for the scalar field discussed above. However, here the physical interpretation of the wave function, now the "field" function, is one of probabilities.

The Lagrangian is:

$$\mathcal{L} = \bar{\Psi}\left(i\gamma_\mu\partial_\mu - m\right)\Psi \tag{16.7}$$

and the Euler–Lagrange equation of motion gives the Dirac equation:

$$\left(i\gamma_\mu\partial_\mu - m\right)\Psi = 0 \tag{16.8}$$

The associated wave functions are:

$$\Psi\left(x_u\right) = u\left(p\right)e^{-ip_\mu \cdot x_\mu} \tag{16.9}$$

Note that this Lagrangian is first order in $\partial_\mu\Psi$.

16.2.3 Vector Field (Spin 1)

For the scalar and Dirac field, the procedure was fairly straightforward because all we did was substitute a field ϕ or ψ in place of some variable (e.g. x, in the case of the mechanical system of Section 9.6). There was no direction attached to these fields; they were scalars. This worked because the differential operator acts on a scalar to give a vector, the gradient. However, we cannot do this for a vector field. The Lagrangian has to be expressed as a scalar.

For the electromagnetic field, the vector field A_μ was combined with the velocity to produce the required scalar that ultimately led to the equations of motion that we recognise as Maxwell's equations.

The Lagrangian for the electromagnetic field, with charges and currents, is thus expressed:

$$\mathcal{L} = -\frac{1}{4}F_{\mu\nu}F^{\mu\nu} - \frac{1}{2}\left(\partial_\mu A_\mu\right)^2 \tag{16.10}$$

The Euler–Lagrange equation gives the equation of motion as:

$$-\partial_\mu F^{\mu\nu} = J_\mu \tag{16.11}$$

Analysing this equation in the context of the four-potential gives us Maxwell's equations.

The electromagnetic field is represented by the electromagnetic field tensor $F_{\mu\nu}$ which is succinctly defined as:

$$F_{\mu\nu} = \partial_\mu A_\nu - \partial_\nu A_\mu \tag{16.12}$$

16.3 THE HARMONIC OSCILLATOR

16.3.1 Commutator

The mechanics of the harmonic oscillator are an essential part of quantum field theory. A typical example of the harmonic oscillator for a mechanical system is a mass suspended by a spring as shown in Figure 2.1. In this system, a mass, initially at an equilibrium position $x = 0$, is given a small perturbation δx and oscillates around the equilibrium position with a frequency ω, being acted upon by a restoring force from the spring such that:

$$F = -m\omega^2 x \tag{16.13}$$

In the most fundamental mode of oscillation, the product of $m\omega^2$ is a constant and is the spring constant, k.

The displacement of the mass from the equilibrium position at time t is:

$$x(t) = Ae^{i\omega t} \tag{16.14}$$

The total energy possessed by the mass is the sum of the kinetic and potential energies.

$$E = \frac{1}{2}mv^2 + \frac{1}{2}m\omega^2 x^2 \tag{16.15}$$

In terms of momentum p and mass m, this is stated:

$$E = \frac{p^2}{2m} + \frac{1}{2}m\omega^2 x^2 = H \tag{16.16}$$

This total energy is called the Hamiltonian H.

In a mechanical oscillator, both x and p are functions of time. As the displacement x increases, the momentum p of the mass decreases. At the extreme ends of travel, the momentum is zero and the displacement is a maximum. At the equilibrium position, the displacement is zero and the momentum is a maximum. The values of both x and p oscillate and, in a classical system, can be independently measured at any time.

Note that:

$$\begin{aligned} \frac{\partial H}{\partial x} &= m\omega^2 x = \frac{\partial p}{\partial t} = F \\ \frac{\partial H}{\partial p} &= \frac{p}{m} = \frac{\partial x}{\partial t} \end{aligned} \tag{16.17}$$

These equations of motion are called Hamilton's equations and are equivalent to the Euler–Lagrange equations.

In a *quantum* mechanical system, we replace momentum and position with quantum mechanical operators:

$$\hat{p} = -i\hbar\frac{\partial}{\partial x}; \ \hat{x} = x \tag{16.18}$$

The Hamiltonian operator, for a quantum harmonic oscillator, is thus:

$$\hat{H} = \frac{\hat{p}^2}{2m} + \frac{1}{2}m\omega^2\hat{x}^2 \tag{16.19}$$

Now, consider the two operators $\hat{p}\hat{x}$ acting on some function $\psi(x)$:

$$\hat{p}\hat{x}\psi(x) = \hat{p}x\psi(x) = -i\hbar\frac{\partial(x\psi(x))}{\partial x} = -i\hbar\left(x\frac{\partial\psi(x)}{\partial x} + \psi(x)\right) \tag{16.20}$$

Compare this with $\hat{x}\hat{p}$ acting on $\psi(x)$:

$$\hat{x}\hat{p}\psi(x) = -i\hbar x\frac{\partial\psi(x)}{\partial x} \tag{16.21}$$

Taking the difference of the two:

$$\begin{aligned}(\hat{x}\hat{p} - \hat{p}\hat{x})\psi(x) &= -i\hbar x\frac{\partial\psi(x)}{\partial x} + i\hbar\left(x\frac{\partial\psi(x)}{\partial x} + \psi(x)\right)\\ &= i\hbar\left(-x\frac{\partial\psi(x)}{\partial x} + x\frac{\partial\psi(x)}{\partial x} + \psi(x)\right)\\ &= i\hbar\psi(x)\end{aligned} \tag{16.22}$$

That is, for a quantum mechanical harmonic oscillator, using quantum mechanical operators:

$$(\hat{x}\hat{p} - \hat{p}\hat{x}) = i\hbar \tag{16.23}$$

The difference of the product of two operators in the form above is called the commutator. When two operators are said to commute, the commutator is equal to zero. In a classical system, with algebraic operators $\hat{p}$ and $\hat{x}$, this difference would be zero. For a quantum mechanical oscillator, the commutator is not equal to zero. The operators do not commute. This is a manifestation of the Heisenberg uncertainty principle. The above equation is called the commutation relation between the momentum and position operators.

The commutator between two operators is given the shorthand form:

$$(\hat{x}\hat{p} - \hat{p}\hat{x}) = [\hat{x}, \hat{p}] \tag{16.24}$$

or, more generally,

$$(\hat{A}\hat{B} - \hat{B}\hat{A}) = [\hat{A}, \hat{B}] \tag{16.25}$$

Just like operators, commutators follow certain algebraic identities, the ones of interest to us being:

$$\begin{aligned}[\hat{A}\hat{B}, \hat{C}] &= \hat{A}[\hat{B}, \hat{C}] + [\hat{A}, \hat{C}]\hat{B}\\ [\hat{A}, \hat{B}\hat{C}] &= \hat{B}[\hat{A}, \hat{C}] + [\hat{A}, \hat{B}]\hat{C}\end{aligned} \tag{16.26}$$

As distinct from an algebraic operator, with quantum mechanical operators, it is the operator that represents the physical quantity.

16.3.2 Energy Levels

16.3.2.1 Energy Levels

Classical theory predicts that the total energy of a harmonic oscillator can take on any value. That is, we just simply alter the amplitude of the vibration to change the energy. We might ask what happens in a quantum system? We've actually already seen the result. When Planck quantised black body electromagnetic radiation, he found that the energy of the oscillation of the field could only take on discrete values such that:

$$E = n\hbar\omega \tag{16.27}$$

where $n = 0, 1, 2, 3\ldots$

These are the eigenvalues of the Hamiltonian. The eigenfunctions satisfy the wave equation, and the eigenvalues are allowed energies that follow from the boundary conditions of the eigenfunctions.

In Planck's black body radiation, the simple harmonic motion of the molecules in the walls of the chamber resulted in electromagnetic waves, standing waves, within the cavity. Planck's radiation law tells us that the energy levels (the amplitudes), for any one frequency, are evenly spaced and are discrete levels – but also, the spacing between these levels depends on the frequency.

16.3.2.2 Power Series Method

The solution to the Schrödinger equation for a particle in a quantum harmonic oscillator potential is arrived at using a power series method. The solution will not be given in detail here, but the results for the first few eigenfunctions obtained are:

$$\begin{aligned}
\psi_o &= A_o e^{-u^2/2} \\
\psi_1 &= A_1 u e^{-u^2/2} \\
\psi_2 &= A_2\left(2u^2 - 1\right)e^{-u^2/2} \\
\psi_3 &= A_3\left(2u^3 - 3u\right)e^{-u^2/2} \\
\psi_4 &= A_4\left(4u^4 - 12u^2 + 3\right)e^{-u^2/2}
\end{aligned} \tag{16.28}$$

where u is:

$$u = \sqrt{\frac{m\omega}{\hbar}}x \tag{16.29}$$

The values of A_n can be found by normalisation. The polynomial in u takes on the form of what is called a Hermite polynomial.

For the ground state, the wave function is:

$$\psi_0 = A_0 e^{-\frac{m\omega}{2\hbar}x^2} = A_0 e^{-\frac{\sqrt{km}}{2\hbar}x^2} \tag{16.30}$$

Figure 16.1(a) shows the shape of the eigenfunctions as a function of position from the equilibrium position for the first three energy levels. The square of the eigenfunction gives the positional probability of the particle between the extremes of travel x_1 and x_2. Also shown in the figure is a plot of the potential energy $V(x)$.

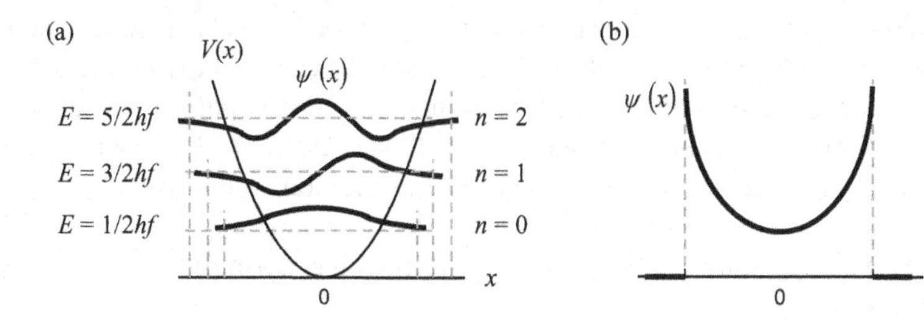

FIGURE 16.1 Eigenfunctions for the first three energy levels of a simple harmonic oscillator shown against the potential function and spaced according to the respective energy level.

The kinetic energy of the mass would be the inverse of this. The vertical position of the plots for the eigenfunctions have been drawn to illustrate the even spacing of those levels.

Interpreted as a mass on a spring or a swinging pendulum, the particle undergoes oscillations about an equilibrium position $x = 0$. In a classical oscillator, the position of the mass is bound between the extremities of the travel, where the potential energy is the greatest and the kinetic energy is zero. If the amplitude of the oscillation changes, then the extremities move further apart or closer together. To aid our understanding, we will assume that the frequency of the oscillation is a constant and depends only on the mass m and the spring constant k. The only degree of freedom is the amplitude. In practice, different modes of vibration can be excited in such a system whereby the vibration of the mass is the result of a superposition of frequencies or harmonics, each with their own amplitude.

Consider the shape of the probability distribution for a classical system as shown in Figure 16.1(b). The probability is highest at the ends of the limits of travel, where the mass spends most time, and is lowest at the equilibrium position where, for a certain Δx, it spends the least time (it has the greatest velocity). Outside the limit of bounds of displacement, the probability of finding the mass there is zero. In the classical limit of our day-to-day world, this is what is usually measured. Increases in energy are accommodated by increases in amplitude of the motion.

In a quantum mechanical system, things appear quite differently. At the lowest energy, say $n = 0$, the mass does not gently bob upwards and downwards (or backwards and forwards) over a wide swing but instead spends most of its time near the equilibrium position and hardly any time near the classical prediction of the extremities of the motion. It is not really possible to say exactly what the "amplitude" of the oscillation is, since the outer bounds of the motion are no longer fixed but taper off exponentially compared to the classical limits. If more energy is fed into the system, the shape of the probability density function changes. At $n = 1$, the mass appears to spend no time at all at the equilibrium position, nor near the extremes of travel, but in two places somewhere in between either side of the equilibrium position. It is as though the mass slows down to a stop halfway through its journey, then speeds up again near the classical extreme position, before coming back the other way. In this mode of oscillation, there will be a new probability distribution within which the increased energy is stored as potential energy and released as kinetic energy as the mass oscillates. That is, the range of x over which we have a reasonable probability of finding an electron has increased compared to the $n = 0$ case. We might call this the effective amplitude. Since this is happening at the same frequency as before, the time taken for a complete round trip from any point in x and back again remains unchanged. At $n = 2$, the mass spends more time at the equilibrium position again but also spends some time near the classical extremes of travel. At this mode, there will be yet a larger effective amplitude.

So, although there is still one degree of freedom (the amplitude), the shape of the positional probability of the mass around the equilibrium position changes as energy in the system is increased or decreased. At $n = 15$ or so, $\psi^*\psi$ begins to resemble the classical shape. In the limit $n \longrightarrow \infty$ the classical shape is achieved.

Remember though, that for a harmonic oscillator potential, the energy levels are evenly spaced, and the spacing between them is linearly dependent on the frequency of the oscillation. The frequency of oscillation, in turn, depends on the spring constant k and the mass m. Note also that in a quantum oscillator, there is an amplitude (probability) for the particle to be outside the classical extremities of the oscillation.

These modes of oscillation are not temporal frequency harmonics that we might associate with the vibrations of a stretched string. The modes of oscillation of $\psi(x)$ are *spatial* frequencies and relate to the probability of position x of the mass at some time t. The oscillation of the mass in time is the angular frequency ω.

Consider the lowest energy state ψ_0. When this eigenfunction is normalised, the expression becomes:

$$\psi_o = \left(\frac{m\omega}{\pi\hbar}\right)^{\frac{1}{4}} e^{-\frac{m\omega}{2\hbar}x^2} \tag{16.31}$$

That is, the amplitude of the space variations in probability contains a time frequency term. We saw the dependence of the energy stored in a harmonic oscillator in Section 2.3. In the classical limit, it is expressed:

$$E_T = \frac{1}{2}m\omega^2 A^2 \tag{16.32}$$

In our model system, ω is a constant because the mass and the spring stiffness are both constants. Any increase in energy goes into the amplitude.

As n increases to a very large number, for a particular frequency of oscillation, the spacings between energy levels remain evenly spaced, but, compared to the overall total energy being stored in the system, those spacings are small. This is the classical limit. The energy spacings approach a continuum. Any changes in amplitude tend to look like smooth changes, and the probability distribution approaches the classical shape.

16.3.2.3 Operator Method

Dirac approached the quantum harmonic oscillator in terms of the Hamiltonian operator as a product of two factors. In one dimension, this is:

$$\hat{H} = \frac{1}{2}\left(\frac{\hat{p}_x^2}{m} + m\omega^2\hat{x}^2\right) = \frac{1}{2m}\left(-i\hat{p}_x + m\omega\hat{x}\right)\left(i\hat{p}_x + m\omega\hat{x}\right) \tag{16.33}$$

Now, $\hat{p}_x$ and $\hat{x}$ are quantum mechanical operators. They are non-commuting operators in the sense that the product $\hat{x}\hat{p}_x \neq \hat{p}_x\hat{x}$. The difference is Planck's constant:

$$\hat{x}\hat{p}_x - \hat{p}_x\hat{x} = i\hbar \tag{16.34}$$

Also:

$$\hat{p}_x\hat{x} - \hat{x}\hat{p}_x = -i\hbar \tag{16.35}$$

Thus:

$$\begin{aligned}
\frac{1}{2m}\left(-i\hat{p}_x + m\omega\hat{x}\right)\left(i\hat{p}_x + m\omega\hat{x}\right) &= \frac{1}{2m}\left(\hat{p}_x^2 - im\omega\hat{p}_x x + im\omega x\hat{p}_x + m^2\omega^2\hat{x}^2\right) \\
&= \frac{1}{2m}\left(\hat{p}_x^2 + m^2\omega^2\hat{x}^2\right) + \frac{im\omega}{2m}\left(\hat{x}\hat{p}_x - \hat{p}_x\hat{x}\right) \\
&= \frac{1}{2m}\left(\hat{p}_x^2 + m^2\omega^2\hat{x}^2\right) - \frac{1}{2}\hbar\omega
\end{aligned} \tag{16.36}$$

That is, in Dirac's scheme, the Hamiltonian operator is a product of two factors plus an amount $1/2\,\hbar\omega$:

$$\hat{H} = \frac{1}{2m}\left(-i\hat{p}_x + m\omega\hat{x}\right)\left(i\hat{p}_x + m\omega\hat{x}\right) + \frac{1}{2}\hbar\omega \tag{16.37}$$

Note that when both momentum and position are equal to zero, there remains a ground state energy $1/2\,\hbar\omega$. This is the $n=0$ energy state in Section 16.3.2.2. It is a manifestation of the Heisenberg uncertainty principle. Both momentum and position cannot be zero at the same time in a bound system. The system is bound by the simple harmonic oscillator potential.

Since the zero-point energy is an additive constant, for nearly all of our analyses to follow this can be ignored since our interest will be in the differences in energies rather than their absolute values. The zero-point energy can be added in again when required.

For later convenience, a factor $\hbar\omega$ is applied to the numerator and denominator of the factors. Ignoring the zero-point energy for the moment, the Hamiltonian is thus:

$$\hat{H} = \frac{\left(-i\hat{p}_x + m\omega\hat{x}\right)}{\sqrt{2m\hbar\omega}} \frac{\left(i\hat{p}_x + m\omega\hat{x}\right)}{\sqrt{2m\hbar\omega}} \hbar\omega \tag{16.38}$$

The two factors in the Hamiltonian are given the operator symbols $\hat{a}^+$ and $\hat{a}^-$:

$$\begin{aligned} \hat{a}^+ &= \frac{\left(-i\hat{p}_x + m\omega\hat{x}\right)}{\sqrt{2m\hbar\omega}} \\ \hat{a}^- &= \frac{\left(i\hat{p}_x + m\omega\hat{x}\right)}{\sqrt{2m\hbar\omega}} \end{aligned} \tag{16.39}$$

They are themselves operators, although most times the hat symbol is not written. The Hamiltonian is thus:

$$\hat{H} = a^+a^-\omega\hbar \tag{16.40}$$

Or to be more precise:

$$\hat{H} = a^+a^-\hbar\omega + \frac{1}{2}\hbar\omega \tag{16.41}$$

Note out of interest that:

$$\hat{H} = a^-a^+\hbar\omega - \frac{1}{2}\hbar\omega \tag{16.42}$$

Our aim is to determine the eigenvalues, that is, the quantum mechanical energy levels $\hat{E}$.

To proceed, let us examine the commutator relation between a^- and a^+.

$$\left[a^-, a^+\right] = a^-a^+ - a^+a^- \tag{16.43}$$

$$\left[\frac{\left(i\hat{p}_x + m\omega\hat{x}\right)}{\sqrt{2m\hbar\omega}}, \frac{\left(-i\hat{p}_x + m\omega\hat{x}\right)}{\sqrt{2m\hbar\omega}}\right] = \frac{1}{2m\hbar\omega}\left[i\hat{p}_x + m\omega\hat{x}, -i\hat{p}_x + m\omega\hat{x}\right]$$

$$= \frac{1}{2m\hbar\omega}\left((i\hat{p}_x + m\omega\hat{x})(-i\hat{p}_x + m\omega\hat{x})\right.$$
$$\left. -(-i\hat{p}_x + m\omega\hat{x})(i\hat{p}_x + m\omega\hat{x})\right)$$
$$= \frac{1}{2m\hbar\omega}\left(\hat{p}_x^2 + im\omega\hat{p}_x\hat{x} - im\omega\hat{x}\hat{p}_x + m^2\omega^2\hat{x}^2\right.$$
$$\left. -\hat{p}_x^2 + im\omega\hat{p}_x\hat{x} - im\omega\hat{x}\hat{p}_x - m^2\omega^2\hat{x}^2\right)$$
$$= \frac{2im\omega}{2m\hbar\omega}\left(\hat{p}_x\hat{x} - \hat{x}\hat{p}_x\right)$$
$$= \frac{i}{\hbar}(-i\hbar)$$
$$= 1$$

Likewise:

$$\left[a^+, a^-\right] = -1 \tag{16.44}$$

The factors a^+ and a^- are operators. They are called raising and lowering operators. To see why, we can see that the commutator $\left[H, a^+\right]$ is, by the operator identities in the previous section:

$$\left[\hat{H}, a^+\right] = \hat{H}a^+ - a^+\hat{H} = \left[a^+a^-\hbar\omega, a^+\right] = \hbar\omega a^+ \tag{16.45}$$

Similar results occur for other commutators, and the combinations are:

$$\left[\hat{H}, a^-\right] = \hat{H}a^- - a^-\hat{H} = \left[a^+a^-\hbar\omega, a^+\right] = -\hbar\omega a^-$$
$$\left[a^+, \hat{H}\right] = a^+\hat{H} - \hat{H}a^+ = \left[a^+, a^+a^-\hbar\omega\right] = -\hbar\omega a^+ \tag{16.46}$$
$$\left[a^-, \hat{H}\right] = a^-\hat{H} - \hat{H}a^- = \left[a^+, a^+a^-\hbar\omega\right] = \hbar\omega a^-$$

Now, let us find:

$$\hat{H}a^+ = a^+\hat{H} + \left[\hat{H}, a^+\right]$$
$$= a^+\hat{H} + \hbar\omega a^+ \tag{16.47}$$
$$= a^+\left(\hat{H} + \hbar\omega\right)$$

Also:

$$\hat{H}a^- = a^-\hat{H} - \left[a^-, \hat{H}\right]$$
$$= a^-\hat{H} - \hbar\omega a^- \tag{16.48}$$
$$= a^-\left(\hat{H} - \hbar\omega\right)$$

The Hamiltonian operator, which is equivalent to the energy E, when operating on the raising or lowering operators results in an increase or a decrease in the energy by one quantum, $\hbar\omega$.*

This operator procedure avoided the complexities of dealing with the eigenfunctions directly. To see the connection with the eigenfunctions, consider the ground state from the polynomial solution:

$$\psi_0 = A_0 e^{-\frac{m\omega}{2\hbar}x^2} \tag{16.49}$$

Now apply the raising operator:

$$\begin{aligned}\psi_1 &= a^+\psi_0 \\ &= \frac{\left(-i\hat{p}_x + m\omega\hat{x}\right)}{\sqrt{2m\omega\hbar}} A_0 e^{-\frac{m\omega}{2\hbar}x^2}\end{aligned} \tag{16.50}$$

and allowing the momentum (and displacement) operator to act, we obtain:

$$\begin{aligned}\psi_1 &= \frac{A_0}{\sqrt{2m\omega\hbar}}\left(-i\left(-i\hbar\frac{\partial}{\partial x}\right)e^{-\frac{m\omega}{2\hbar}x^2} + m\omega x e^{-\frac{m\omega}{2\hbar}x^2}\right) \\ &= \frac{A_0}{\sqrt{2m\omega\hbar}}\left(m\omega x e^{-\frac{m\omega}{2\hbar}x^2} + m\omega x e^{-\frac{m\omega}{2\hbar}x^2}\right) \\ &= \frac{A_0 m\omega x}{\sqrt{2m\omega\hbar}}(2)e^{-\frac{m\omega}{2\hbar}x^2} \\ &= A_0\sqrt{2}\left(\sqrt{\frac{m\omega}{\hbar}}x\right)e^{-\frac{m\omega}{2\hbar}x^2}\end{aligned} \tag{16.51}$$

which is in agreement with the result for ψ_1 from the polynomial solution.

Returning to our mechanical system of a mass and a spring, the application of the raising operator (energy fed into the system) "raises" the amplitude of the oscillation, but what's more, it changes the shape of the probability distribution of location for the mass. The frequency of the oscillation is the same as before and is a constant. An energy change will be referred to as a change in amplitude where this means a change in the shape of the probability distribution. Application of the lowering operator lowers the amplitude, but we can only go as low as the ground state.

The quantisation of the harmonic oscillator worked in this way because the physical quantities of momentum and position were replaced with quantum mechanical operators which did not commute. Further, the relationships shown above were a result of the energy function being a quadratic in p and x. The harmonic oscillator has one degree of freedom – the amplitude of the oscillation.

16.3.3 The Harmonic Oscillator Field

A scalar field $\phi(x)$, independent of time, might be represented by an array of an infinite number of independent quantum harmonic oscillators. This is called a free field in the sense that each oscillator is independent of the others.

---

* Note, whether or not operators a^+ and a^- perform a raising or lowering of energy levels depends on whether they operate on H or H operates on them.

For our array of harmonic oscillators, any individual oscillator may change its state (by changing its amplitude as a result of energy leaving or entering the system). For a single oscillator, there is a spectrum of amplitudes, or energy levels, available. These correspond to the eigenfunctions $\psi_0, \psi_1, \psi_2, \psi_3 \ldots \psi_n$. An oscillator with eigenfunction ψ_n would have an energy E_n. The energy is built up in evenly spaced steps $\hbar\omega$, and so a wave function ψ_n would have an energy $E_n = n\hbar\omega$.

We are now in the position in which Boltzmann found himself with moving gas molecules. It is impossible to track the energy of each individual molecule, but it is possible to say something about the number of molecules which might exist in a particular energy or at least, an energy interval. Planck extended this into the quantum analysis of the electromagnetic standing waves in a cavity. In the case of black body radiation, there was a range of frequencies present, each with a spectrum of amplitudes, the spacing of which depended upon the frequency of that standing wave.

So, instead of dealing with an infinite number of oscillators, we divide them up according to their energy content. For simplicity, the zero-point energy will be set to zero. In the array of oscillators, there will be N_0 oscillators with energy $0E$, N_1 oscillators with energy $1E$, N_2 oscillators with energy $2E$, etc. We might call N_n the number of oscillators with an energy E_n where n is going from zero to infinity. That is, N_n is the occupation number for each energy level.

The total energy E_T of the array is thus:

$$E_T = \sum_{n=0}^{\infty} N_n E_n \tag{16.52}$$

Each energy level differs from the one above or below it by an amount $\hbar\omega$. At the nth energy level, the energy stored in one oscillator is $n\hbar\omega$. The total energy of the system is thus:

$$E_T = \sum_{n=0}^{\infty} N_n n\hbar\omega \tag{16.53}$$

In practice, the higher energy level states will not be filled, and so there is a practical limit to the total energy. N_n, the occupation number, gives the number of oscillators in the array which have an energy E_n.

Expressed in terms of the raising and lowering operators, the energy in each oscillator is the Hamiltonian $H = a^+a^-\hbar\omega$, and so we might say that the product a^+a^- gives us the number of oscillators N_n at some energy E_n. *

$$N_n = a_n{}^+a_n \tag{16.54}$$

N_n is sometimes written as an operator, the number operator which acts on the index n.

The total energy is thus:

$$E_T = \sum_{n=0}^{\infty} a_n^+ a_n^- n\hbar\omega \tag{16.55}$$

Now, we re-interpret the role of the raising operator and call it the creation operator. When the creation operator acts upon an oscillator which has an energy E_n, it essentially increases the number of oscillators that exist in the $E_{n+1} = E_n + \hbar\omega$ energy state. N_n decreases by 1 and N_{n+1} increases by 1. When a lowering operator, now called an annihilation operator, acts upon an oscillator with energy E_n, it decreases the number of oscillators that have that energy and increases the number of oscillators that have an energy $E_{n-1} = E_n - \hbar\omega$. N_n decreases by 1 and N_{n-1} increases by 1.

* When moving our point of view from a single oscillator to a field, the notation usually changes. The creation operator is labelled $a^\dagger$ and the annihilation operator just a. For our introduction to quantum field theory, we will stay with our original notation.

If we keep applying the lowering operator to an oscillator, there is a point where the oscillator cannot go to a lower energy state – it has stopped oscillating.

Note the shift in focus here. Previously, for a particle, the raising and lowering operators acted on a particle and its energy was raised or lowered as a consequence. Here, the creation operator creates particles in the next higher energy state, while the annihilation operator removes particles from the current energy state. We are primarily now talking about energy states and the number of particles which enter or leave them.

We are now going to shift our view again slightly and say that instead of talking about an array of harmonic oscillators, as if there were something like an infinite mattress of masses and springs stretched out before us, we now say that the oscillators are represented by their actual *oscillations*. If we see an oscillation in the array of oscillators, we know there is an oscillator there. The array of oscillators is now called a field.

When operating on the vacuum, the raising operator creates an oscillation and puts it into the $n=1$ energy state. Successive applications of the raising operator lift that oscillator to a higher energy state. When the annihilation operates on the $n=1$ state and lowers it to the ground state, that oscillation disappears – the oscillation has been annihilated; the occupation number, the number of oscillations, for the ground state remains at zero.

Perhaps you can see where this is going. The oscillations themselves become particles. Occupation numbers in each energy shift around as the creation and annihilation operators work. In the ground state, the particles, the oscillations in the field, have become annihilated.* The changes of energy in the field take place in discrete steps, and these are reflected in the changing occupation numbers. We've gone from a single particle system to a multi-particle system.

The ground state is the lowest energy state. In the simplest view, with no zero-point energy, the ground state of an oscillator is a state of zero energy. That is, no oscillations of that oscillator.

The ground state has a different meaning to the concept of the vacuum. The vacuum is a series of energy states, E_n, in which the occupation number for all of the states is zero.

As seen above, there are oscillations in the ground state. For the quantum harmonic oscillator, the ground state is shown in Figure 16.1(a) for $n = 0$. An oscillator in the ground state is very slightly vibrating up and down spending most of its time near the equilibrium position. These oscillators, these oscillations, these particles, are called virtual particles because they are not directly observable. They do not have enough energy to emit a photon. A photon has a minimum energy of $\hbar\omega$. For the harmonic oscillator potential, the energy in the oscillators in the vacuum is of course $1/2\,\hbar\omega$.

16.3.4 Particles

The energy of one oscillator:

$$\hat{H} = a^+ a^- \hbar\omega \tag{16.56}$$

When a change of state occurs, the H operator acts on the creation operator a^+ so that the new energy of that oscillator is:

$$\hat{H}_{\text{new}} = H + \hbar\omega \tag{16.57}$$

The occupation number for that new energy level has increased by 1.

* Schrödinger's equation is usually applied to the wave function for an electron. Here, we are applying it to a particle in a harmonic oscillator potential. That is, a particle not restricted by any kind of selection rule. For example, in an atom, we cannot have two electrons (ignoring spin) occupying the same state, whereas here, we can have many more than one harmonic oscillator in the same energy state. Our particle is a spin zero boson, not a fermion.

The question is, what makes this change of energy occur? Why would an oscillator decide to suddenly increase or decrease its energy?

Physical systems tend to assume the lowest state of energy. It is a consequence of the second law of thermodynamics. For an oscillator, there is a certain tendency to decay from a higher state to a lower state, and the energy lost is gone as heat (i.e. it could be heat from friction in the atoms of the spring, or radiated heat, etc.). We might say that in one second, in an array of oscillators, there is a certain number of them that drop down one energy level. Energy is emitted as a result of this transition. It is called spontaneous emission.* Let us call the probability that an oscillator will drop down one energy level spontaneously as P_s.

But something else also happens. Because we all exist at a temperature above absolute zero, energy also gets absorbed by a body. At radiative equilibrium, the rate of absorption of energy equals the rate of emission. And so, for our oscillators, while some are shedding energy, others are accepting energy. Some oscillators are experiencing a lowering operation, while other are experiencing a raising operation. At equilibrium, the rate of absorption equals the rate of emission.

But things are not so quite symmetrically arranged. Einstein proposed that an electron in an atom has a greater chance of absorbing or emitting electromagnetic radiation if there is radiation of a certain frequency already present. This frequency is that given by the energy difference between the two states of the electron before and after the energy is absorbed or emitted – that is, $\hbar\omega$. This is called induced emission or absorption. Einstein found that these probabilities were equal. In fact, he found that the probability that an electron in an atom would emit or absorb a photon is increased by an amount $n + 1$ if there are already n photons present which had an energy equal to that of the final state.† That is, compared to if there are no photons present, the probability is increased for every photon of the higher energy state that is present. In practice, this means that an electron in an atom has a greater chance of emitting or absorbing a photon if one shines a light on that atom; the greater the intensity of the light, the greater the number of photons present and the greater chance that an electron in an atom will undergo a change of state. Of course, as seen in the photoelectric effect, the energy of those incident photons has to be greater than or equal to the energy of the final state of the electron.

In the case of a mechanical harmonic oscillator, we could say that a particular oscillator is more likely to change its amplitude if there are other oscillations present – a little like a forced oscillation at the resonant frequency of the mass and spring.

Let us call this probability P_i for the "incident" effect.

If there are N_u oscillators at the higher energy level, then the number per second which make the transition to the lower energy level would be $(P_i + P_s)N_u$. If there are N_l oscillators at the lower level, then the number per second which make the transition to the higher energy level would be P_iN_l.

At equilibrium, the same number of oscillators are changing their energies each second, thus:

$$\left(P_i + P_s\right)N_u = P_iN_l \tag{16.58}$$

Returning for a moment to Maxwell's gas molecules, the ratio of the number of molecules at one energy level to another is:

$$\frac{N_1}{N_2} = \frac{e^{-E_1/kT}}{e^{-E_2/kT}} = e^{-\Delta E/kT} \tag{16.59}$$

For the special case of a one-dimensional harmonic oscillator, Planck let $\Delta E = \hbar\omega$, and so, with two energy levels u and l, we have:

---

* In an atom, this does not mean all the electrons will fall down into the inner most energy level because of the Pauli exclusion principle. Electrons, however, will settle into the lowest energy configuration possible as the available energy levels are filled from the bottom up. No such impediment applies to our array of harmonic oscillators. They all, unless something else happens, settle down into the lowest amplitude of vibration for their respective natural frequencies.

† Note that in terms of an amplitude, the increase is $\sqrt{n+1}$.

$$\frac{N_u}{N_l} = \frac{P_i + P_s}{P_i} = \frac{e^{-E_u/kT}}{e^{-E_l/kT}} = e^{-\hbar\omega/kT}$$

$$P_i + P_s = P_i e^{-\hbar\omega/kT} \tag{16.60}$$

$$\frac{P_i}{P_s} = \frac{1}{e^{-\hbar\omega/kT} - 1}$$

According to Einstein, the probability P_i increases as the intensity of the incident radiation increases (i.e. the probability increases by $N + 1$ for every other photon present), and so P_i is, in the simplest case, proportional to the intensity, which in turn is directly related to the energy density.

$$I = \frac{P_s}{P_i} \frac{1}{e^{-\hbar\omega/kT} - 1} \tag{16.61}$$

This is very similar to the expressions shown in Section 5.6.5, the obvious conclusion being that a collection of harmonic oscillators can be modelled in terms of a collection of particle oscillators each of which carries an energy $\hbar\omega$. It might seem we have just repeated what Planck did in the first place, but he did this for electromagnetic waves modelled as standing waves of harmonic oscillators with one degree of freedom – the amplitude. Einstein worked it out for photons. In the above sections, we have worked it out for masses on springs. The asymmetry associated with the probabilities of spontaneous emission and induced emission and absorption in a sense are equivalent to the non-commutation relation between the quantum momentum and displacement operators.

The factor $N + 1$ comes up again in operator notation. We saw earlier that the Hamiltonian operator is written:

$$\hat{H} = a^+ a^- \hbar\omega + \frac{1}{2}\hbar\omega \tag{16.62}$$

Is the product a^+a^- the same as a^-a^+?

To answer this question, examine the commutator for a^- and a^+:

$$\left[a^-, a^+\right] = a^- a^+ - a^+ a^- \tag{16.63}$$

Thus:

$$\begin{aligned} a^- a^+ &= \left[a^-, a^+\right] + a^+ a^- \\ &= a^+ a^- + 1 \end{aligned} \tag{16.64}$$

But:

$$N = a^+ a^- \tag{16.65}$$

And so:

$$a^- a^+ = N + 1 \tag{16.66}$$

There is an asymmetry in the products of the operators, and it is this that leads to the discrete energy levels that pervade quantum theory.

The conclusion is that in an array of harmonic oscillators, the oscillations can be modelled in terms of particles. Each energy state has an occupation number which gives the number of oscillators which have a certain energy. When the harmonic oscillators represent the electric field, those particles are called photons.

There are various approaches used for quantising fields. In one method, the raising and lowering field operators are used in conjunction with the Hamiltonian to find the spectrum of energies. This procedure is called canonical quantisation. It is a difficult procedure and even more so when extended to the Dirac (spin ½) and electromagnetic (spin 1) fields, not to mention including relativity and normalisation.

In another method, quantisation is arrived at through the action principle and the Lagrangian. We will not examine this method here since, for our simple harmonic oscillator field, it is not necessary for our understanding. For more complicated fields, it provides a much easier pathway to quantisation than the canonical method.

16.3.5 The Quantum Field

The quantum harmonic oscillator field described in the previous section has been a free field in the sense that each oscillator is independent of the others.

In a solid, molecules are connected to one another by molecular bonds. Sound waves travel through the solid as displacements of molecules around their equilibrium positions, and, since all the molecules are connected with each other, elastic waves can travel through the solid. In our harmonic oscillator field, we might imagine that, as well as the masses being supported by springs, that also, the masses are connected sideways with each other by springs that cause the up and down movement of one oscillator to be transmitted to the one next to it. This is shown in Figure 16.2(a).

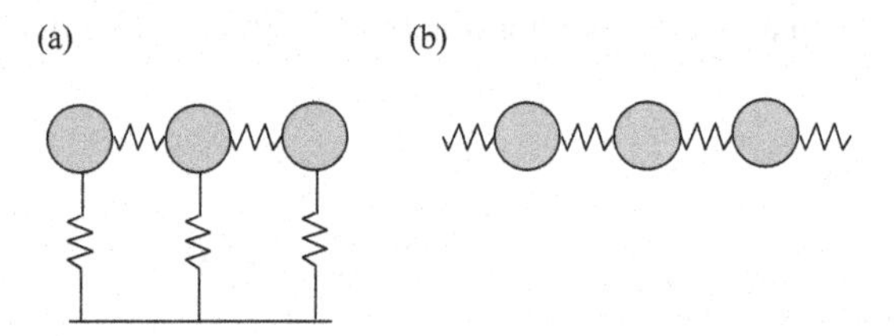

FIGURE 16.2 (a) Interconnected masses and springs for a simple harmonic oscillator model. (b) Horizontal springs only showing connection between masses to simulate a stretched string.

In fact, we don't even need to worry about the spring that supports the mass from below but instead, only those that act from the side as shown in Figure 16.2(b). The system of harmonic oscillators now resembles that of a stretched string, where the tension in the string takes the place of the spring constant as described in Section 2.7. Although it looks like the masses interact, they only do so for the purpose of transmitting the elastic restoring force and so are still considered free, but bound by a potential.

Waves of different frequencies may be superimposed on the string. The relative amplitude of these harmonics depend on how the string is excited.* Each particle in the string moves up and down, but waves can now propagate along the string with different frequencies. If the string is bounded at the ends, then standing waves may appear.

We can now describe the energy stored within the system in terms of the frequencies of the waves in the lateral direction rather than the displacements in the vertical direction.

In Section 2.3, it was shown that in a wave, both frequency and amplitude contribute to the energy in the same way:

$$E = \frac{1}{2} m\omega^2 A^2 \tag{16.67}$$

* For example, in a guitar, an initial deflection and release of the thumb produces standing waves at near the fundamental frequency, while a plectrum excites higher harmonics and the sound is brighter. The sharp edge of the plectrum introduces more localised distortions of the string thus exciting waves of shorter wavelength.

So, if the amplitude is kept constant and the frequency increased, the same amount of energy can be stored in the system compared to if the frequency were kept constant and the amplitude changed. Instead of focussing on amplitudes only, we can focus on frequencies as a measure of energy in the oscillator. In the more general case, the energy depends on both the component frequencies and the amplitude of each component. This is very much the situation encountered in Fourier analysis.

As far as the quantum field is concerned, a description of the energy state of the field is now given in terms of the distance x, which is now taken to be that along the string, and the sum of the deflections in the vertical direction for each component frequency. In this sense, we can determine the strength of the field at some coordinate x.

Just as we mapped the coordinate x to a scalar field $\phi(x)$ in Chapter 13, the vertical displacement of a point in the string is now given in terms of a field function $\phi(x,t)$. x is now the coordinate along the length of the string.

In canonical quantisation, dynamical variables are replaced by operators. Then, commutation relations are applied, and the commutation relations lead to the desired quantum mechanical expressions of interest – such as the eigenfunctions and eigenvalues, which in turn provide the expectation value for an observable such as the energy of the system.

In single particle quantum mechanics, momentum and position are replaced by momentum and position operators. In the case of the position operator $\hat{x}$, the operator is the same as the variable x.

In many-particle quantum theory, such as that described in the previous section for the harmonic oscillator field, raising and lowering operators are used, together with their commutation relations, to describe how energy in a system is quantised and how particles in the free field might be created or annihilated. In that formulation, we had a free field expressed in terms of energies.

In quantum field theory, the classical field $\phi(x,t)$ itself now becomes an operator $\hat{\varphi}(x)$. The conjugate momentum, $\pi(x)$, becomes the operator $\hat{\pi}(x)$.* As with single particle quantum mechanics, the operators $\hat{\pi}(x)$ and $\hat{\varphi}(x)$ do not commute.

Note carefully the definitions of these operators. They are operators that depend on the coordinate x. In that sense, at every point in space, there is a different operator. As with single particle quantum mechanics for the free field, we can avoid a Hamiltonian approach to quantisation and proceed to the equivalent creation and annihilation operator approach but this time applied to the field operators.

In the free field, creation and annihilation particle operators create particles in the energy eigenstates. For example, the creation operator operates on an eigenfunction $\psi_n(x)$ to create a new eigenfunction whose eigenvalue is $E_n + \hbar\omega$. The occupation number for that eigenstate, that energy level, has increased by one.

In the case of the quantum field, we want an expression that describes the energy of the field at some lateral position x. This is the sum of the energies of each component waveform at that position, which in turn depend on the frequency and amplitude of the component waveforms. We might call the component waveforms the eigenfunctions of the wave equation and use the same notation $\psi(x)$. Each eigenfunction $\psi_i(x)$ has an associated frequency ω_i. At each position x, then the field operator $\hat{\varphi}(x)$ is the sum represented by:

$$\hat{\varphi}^-(x) = \sum_{i=1}^{\infty} \hat{a}^- \psi_i(x) \tag{16.68}$$

or the Hermitian conjugate:

$$\hat{\varphi}^+(x) = \sum_{i=1}^{\infty} \hat{a}^+ \psi_i^*(x) \tag{16.69}$$

* A field possesses momentum by virtue of its energy density. In an electromagnetic wave, where the magnitudes of the E and B fields are changing sinusoidally with time, the Poynting vector describes the flow of energy through an area perpendicular to the direction of motion of the wave. The oscillating fields take the place of an oscillating mass in a mechanical system. The transfer of energy in a wave carries with it a transfer of momentum. Looking ahead, we might say it is the momentum carried by the photons.

The creation and annihilation operators $\hat{a}^+$ and $\hat{a}^-$ have within them the required quantisation or anti-commutation properties, and so the field described by these operators is a quantum field.

In the free field, the creation and annihilation operators $\hat{a}^+$ and $\hat{a}^-$ create or annihilate a single particle with energy E_n. The quantum field operator creates or annihilates particles at some position x.

To get a feeling for the quantum field operators, consider the action of the field creation operator on the vacuum. Just as the single particle creation operator created a free particle from the vacuum, the quantum field creation operator creates from the vacuum, at position x, a superposition of an infinite number of particles each with a different energy. That is, we can have more than one frequency of vibration existing on a string, each vibrating with its own frequency and each having its own energy. At some position x, the result is the superposition of these oscillations or, in other words, particles.

Can we have more than one particle existing at the same point? Yes. Photons of different frequency can exist in the same space – we might call such an occurrence white light. This happens mathematically because the creation and annihilation operators commute with themselves. For each type of boson, say a photon, a gluon or a Higgs boson, there is a different kind of field, but the basic mechanics of the field are those which are described above.

In practice, there is not an infinite number of particles created, because the step size of each energy level depends upon the frequency of the mode of vibration of the state, and so higher energy states cannot be filled unless that energy is supplied, via the creation operator, in the first place. An ultra-violet catastrophe of particles is avoided.

We might think of the quantum field as some kind of entity that pervades all space and in which ripples, or oscillations, appear and those oscillations act like particles and travel as if they were waves. As far as photons go, the concept has somewhat of an ironic similarity to the notion of the luminiferous aether of Chapter 8. The modern-day ethernet is well-named.

In this section, we are dealing with a spin-0 scalar field. For photons, we need to quantise the electromagnetic field, but this is beyond the scope of this book. For now, it is enough that we just accept that photons arise from the quantisation of a field. Although we are usually comfortable with the notion of an electron as a particle, since they were discovered as particles initially and behave for the most part like particles in relatively large-scale experiments, they too arise from quantum oscillations in the electron field, that is, the Dirac field. In Chapter 17, we are primarily interested in the interactions of particles, and so we won't take the above analyses any further at this point other than to write some expressions that deal with the interaction and propagation of particles.

16.4 PROPAGATORS

The movement of the particle within a quantum field is described in terms of what is called a propagator. The oscillation, the particle, carries four-momentum. For a mass m in a scalar field, the equation of motion comes from the Klein–Gordon equation. It is the kinetic term that is of interest here:

$$\left(\partial^{\mu}\partial_{\mu} + m^2\right)\varphi = 0 \tag{16.70}$$

When expressed in terms of momentum, this becomes:

$$\left(p_{\mu}^2 - m^2\right)\varphi = 0 \tag{16.71}$$

Our interest is in the propagation of the virtual particles of the vacuum, in particular, the photon. Virtual particles are unobservable, but they are known to exist because of the effect that have on observable particles. They are an inevitable consequence of quantum electrodynamics. The nature of them is predicted by the theory, and their effects have been demonstrated experimentally to a high precision.

In relativistic quantum mechanics, we always start with the relativistic energy equation, which in natural units is:

$$E^2 = p^2 + m^2 \tag{16.72}$$

It was previously shown that energy and momentum can be expressed in terms of a four-momentum. The time dimension is associated with energy E, and so the four-momentum consists of the energy E and the three spatial components of momentum.

$$p_\mu = (E, \vec{p}) \tag{16.73}$$

Using this terminology, we have:

$$p_\mu^2 = m^2 \tag{16.74}$$

Particles which obey this relationship are regarded as being "on the mass-shell", the shell being the shape of the surface when the relativistic energy equation is plotted in three-dimensional space. Observable, or real, particles that we measure in the laboratory are on the mass-shell. The Dirac current is used to determine the probability flow and hence the propagation amplitude for real particles.

The virtual particles of the vacuum are not on the mass-shell. Unlike real particles, there is a difference between the square of the four-momentum and the rest mass. A measure of the momentum carried off by the virtual particle when it interacts with a real particle is given by q^2 where:

$$q^2 = p_\mu^2 - m^2 \tag{16.75}$$

Or:

$$q^2 = E^2 - p^2 \neq m^2 \tag{16.76}$$

q^2 may be positive or negative depending on the nature of the interaction. In general, the propagation factor is defined as:

$$i\frac{1}{q^2 - m^2} \tag{16.77}$$

For a photon, the rest mass $m = 0$, and so the propagator becomes:

$$i\frac{1}{q^2} \tag{16.78}$$

We have so far only mentioned two directions of polarisation for photons, the transverse directions. In general, there are two more polarisations, one which might occur along the direction of travel (longitudinal polarisation) and one which is called the scalar, or time-like, polarisation. When these polarisations for photons over all directions are summed, it is found that:

$$\sum \varepsilon_\mu (\varepsilon_\nu)^* = -g_{\mu\nu} \tag{16.79}$$

where $g_{\mu\nu}$ is the metric tensor. Note, no one has ever seen a longitudinal or scalar polarised photon, but virtual photons can carry these polarisations.

When spin (i.e. polarisation for a photon) is included, the photon propagator is:

$$i\frac{\varepsilon_\mu(\varepsilon_\nu)^*}{q^2} = i\frac{g_{\mu\nu}}{q^2} \tag{16.80}$$

For a massive fermion, the propagator is written:

$$i\frac{\gamma_\mu q_\mu + m}{q^2 - m^2} \tag{16.81}$$

This brief account is enough for us to later describe the propagation of a virtual photon which occurs as a result of an interaction with a real particle.

16.5 INTERACTIONS

Particles such as electrons and photons are quanta of their respective fields. Particles are created and destroyed by the action of operators. A particle may be raised from the vacuum, given a range of energies and then annihilated. Do particles of one type interact with particles of another? That is, how do different types of fields interact? We can only give a brief account here, but it will be enough for our purpose in a later section when we determine the probability of particles interacting.

What happens when the particle interacts with another particle within a field? The field acts upon the particle, and the presence of the particle affects the local value of the field. By the principle of least action, the Lagrangian for each, the field and the particle, must combine so that the field and the particle move in a way which minimises the action. We examined a superficial case of a particle in a scalar field in Section 13.6. There, we found that the presence of the field adds to the Lagrangian for the particle as if it were an extra mass.

In terms of masses and springs, the potential function that governs the motion of the mass now contains additional high order terms:

$$V(x) = V_0 + C_1V(x) + C_2V(x)^2 + C_3V(x)^3 + \ldots \tag{16.82}$$

Of more interest to us right now is the interaction between a photon and an electron. We can say that the photon is represented by the electromagnetic field $F_{\mu\nu}F^{\mu\nu}$ and the electron is represented by the Dirac field. We went through this in a long-handed way in Section 15.8, but in a more shortened form, we can just combine the Lagrangians for the field A_μ and the electron ψ and include a term that quantifies the strength of the coupling. This extra term represents a conserved current.* The combined Lagrangian is written:

$$\mathfrak{L} = -\frac{1}{4}F_{\mu\nu}F^{\mu\nu} + \bar{\psi}(i\gamma_\mu\partial_\mu - m)\psi - e\bar{\psi}\gamma_\mu A_\mu\psi \tag{16.83}$$

e is the electronic charge, the coupling constant.

We don't need to take this analysis any further in this book. The message is simply that the Lagrangians for both interacting particles are added, and the strength of the interaction depends upon the magnitude of the quantised electromagnetic field (the photon) and the charge on the electron. In a later section, this will be used in connection with the propagator to characterise the interaction between an electron and a photon.

* The conserved current is a consequence of conservation of charge. It is an example of what is known as Noether's theorem.

Feynman Diagrams 17

17.1 INTRODUCTION

The interactions between particles can be described in several ways. In Newton's time, gravitation was considered a somewhat mysterious "action-at-a-distance" phenomenon. Newton himself could not explain exactly what gravity was. "I make no hypothesis" he says, but his universal law of gravitation certainly described how gravity acts and could be used to make verifiable predictions about the motion of particles over a very wide range of length scales.

Maxwell invoked Faraday's concept of a field when describing, in a mathematical way, the action of electric and magnetic forces on a charged particle. An electric or magnetic field does not require any action at a distance, except perhaps that the field exists in the first place due to the presence of a remote particle. The local value of the field acts upon the particle in question, and the force is applied according to Maxwell's equations. It is a classical theory and did not account for the observed manifestations of quantum behaviour of particles on a very small scale such as the photoelectric effect or the black body radiation spectrum. Quantum mechanics of the first part of the 20th century accounted for this behaviour in terms of wave-like probabilities.

We are going to now examine another method of describing the quantum interaction between particles – in particular, the interaction between light (photons) and matter. This novel treatment was developed by American physicist Richard Feynman in the 1940s.

17.2 QUANTUM ELECTRODYNAMICS

17.2.1 Path Taken by a Photon

Feynman asked us to consider a new point of view in relation to the passage of photons and their interaction with electrons in matter. It's all about ways of "counting the beans", in Feynman's words. We want to know an efficient way of calculating what happens without calculating the motion of every particle involved in the process. To do this, we begin with particles.

One of the fundamental principles of the theory is to recognise that photons and electrons are particles that travel from place to place over all possible paths. Each path contributes an amplitude to a cumulative sum of amplitudes, the square of which gives the probability that a particle travels from its starting point to a destination.

These amplitudes are probability amplitudes and are complex numbers. The actual probability is found from the square of the amplitude, much like how the square of the amplitude of the Schrödinger wave function is used to determine the probability of finding an electron at a certain place in the stationary state of an atom.

According to Feynman, the amplitude of each path, for a monochromatic source, is represented by an arrow which rotates at a fixed angular speed. For example, as shown in Figure 17.1, when a photon leaves a source at S and proceeds in a straight line to a detector D, the arrow may be thought to begin at the 12 o'clock position much like an imaginary stopwatch. As the photon proceeds, the arrow turns, and when the photon reaches its destination, the arrow stops turning.

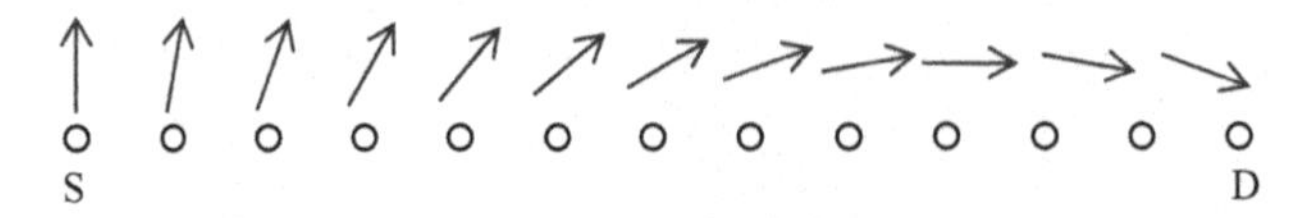

FIGURE 17.1 Photon travelling from a source S to a detector D showing rotation of an arrow along each increment of the path.

The rate of rotation of the Feynman arrow is connected with the periodic nature of the probability that a monochromatic source will emit a photon. For red light, the rate of rotation of the arrow is, in SI units, about 4.25×10^5 GHz (or a wavelength of about 700 nm if we were talking about a wave).

The length of the arrow is the amplitude. The square of the amplitude gives the probability that the particle will take this path. To see the significance of this, the direction of the final arrow is important, and this comes to light when we consider alternate paths from the source to the detector.

17.2.2 Alternate Paths

To get a feeling for the theory, let us examine the reflection of light from a mirror using Feynman's own examples. According to Feynman, photons from the source can take all possible paths from the source to the detector, even those which we might at first believe have no significance whatsoever to the final outcome.

As shown in Figure 17.2, each possible path is given a label z. We say that each of these paths is equally likely to occur – meaning that the length of the arrows, the amplitudes, are the same for each path under consideration.

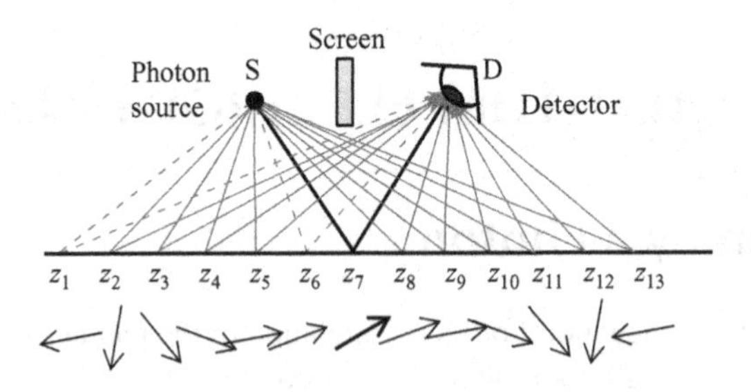

FIGURE 17.2 Possible straight-line paths from S to D taken by a photon reflecting from a surface. Each path is labelled. The final position of the arrow is shown below each path.

To our classical minds, it might be tempting to think that the path shown as a heavy line at z_7 is the only one of significance, but, as we shall see, every path makes a contribution to the final observed outcome.

It is evident that the photon taking path z_2 will take longer to reach D than the path at say, z_6, because there is a longer distance to travel. Thus, the arrow representing the amplitude of path z_2 will turn more compared to that of z_6 during the trip from S to D. Figure 17.3 shows a graph of the time taken by a photon for each path.

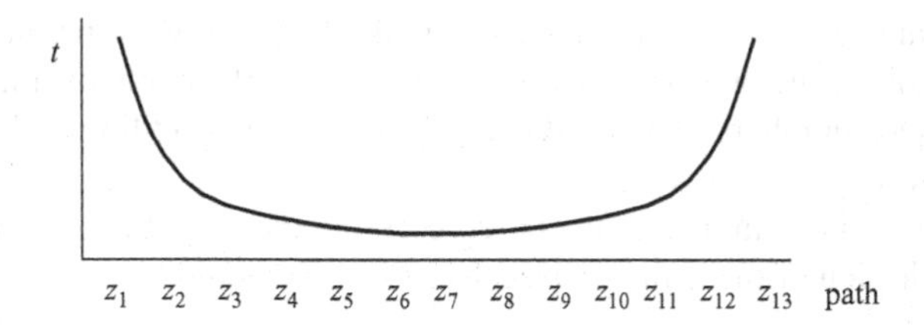

FIGURE 17.3 Representation of the time taken along each path z_1 to z_{13} for the photons shown in Figure 17.2.

The significance of this graph is that there is a greater change in the time taken from path to path for those near the edges (z_1, z_2, z_3, z_{11}, z_{12}, z_{13}) compared to those near the centre (z_5, z_6, z_7, z_8, z_9). That is, photons taking paths near to the centre take almost the same time to make the journey while those at the edges take quite different times depending how close to the edge they might strike the mirror. Thus, as shown in Figure 17.3, the arrows representing the amplitudes of paths near the edges rotate more from path to path. The arrows for those paths in the middle take much the same time.

We imagine that the arrow begins pointing vertically upward. The turning of the arrow stops when the photon reaches the detector. The final angle of the arrows indicates the time taken for a photon to take each of the possible paths. Note that the arrows change direction from path to path more towards the ends compared to those in the centre. Those near the centre of the mirror tend to line up, while those at the ends are pointing in quite different directions.

Adding the arrows head-to-tail, as shown in Figure 17.4, we find that the resultant amplitude A has quite a substantial length.

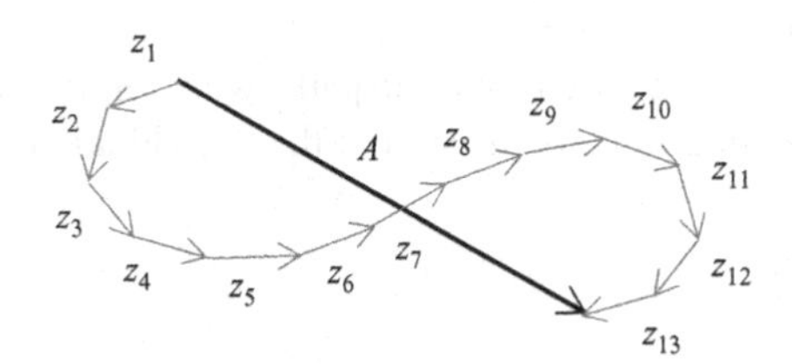

FIGURE 17.4 Vector sum of the arrows for each path z_1 to z_{13} and the resultant amplitude.

The physical significance of this is that the theory predicts a strong probability (A^2) that the photon will be reflected from the surface and arrive at the detector. That is, light will be reflected from the mirror in accordance with our everyday experience.

$$P(S \to D) = A^2 \tag{17.1}$$

Also, notice that the paths nearer to the centre of the mirror make more of a contribution to this final amplitude compared to those paths whose arrows twist and turn near the edges. Put another way, the paths that take the least time, the ones being reflected in the region near the centre of the mirror, are the ones that contribute most to the total probability of the process of reflection of photons from a mirror.

It is this amplification (from paths nearer to the centre) and cancellation (from paths nearer to the edges) of probabilities that is the central idea of quantum electrodynamics. The improbable paths do exist and play their part. Note also that nearby paths z_5, z_6 and z_8, z_9 contribute substantially to the final probability, not just the shortest path z_7.

The addition of arrows in this way is a diagrammatic method of what we now call the Feynman path integral, or sum over histories.

When an event can happen in a number of alternate ways, we *add* the amplitudes of each of those ways as vectors and then determine the final amplitude – the square of which gives the probability of that event occurring. That is, we *add* the amplitudes first as if they were complex numbers and then square the amplitude of the sum to get the probability. The length of the arrow is the magnitude of the complex amplitude. The turning of the arrow is the "phase" component of the complex amplitude.

We recall that complex numbers may be added and subtracted by adding and subtracting the real and imaginary parts separately.

Say for paths z_1 and z_2, the complex numbers representing the amplitudes could be $z_1 = a_1 + b_1 i$ and $z_2 = a_2 + b_2 i$. The vector addition of these first two paths $z = z_1 + z_2$ would be:

$$z_1 + z_2 = (a_1 + b_1 i) + (a_2 + b_2 i) = (a_1 + a_2) + (b_1 + b_2)i \tag{17.2}$$

When we add up all the paths, we obtain the final arrow:

$$\begin{aligned} z &= (a_1 + a_2 + a_3 + \ldots + a_{13}) + (b_1 + b_2 + b_3 + \ldots + b_{13})i \\ &= A + Bi \end{aligned} \tag{17.3}$$

The probability of the photon to travel from S to D is the square of the amplitude of the final arrow:

$$P(S \to D) = |z|^2 = z^* z = A^2 + B^2 \tag{17.4}$$

Each possible path does not have an individual probability even though we might think that some paths are very improbable. Each arrow has the same length – the same amplitude and therefore the same probability of occurring. The final probability of the event is calculated from the *vector* sum of the individual amplitudes. It is the phase component (the time taken) of the journey which determines which paths contribute to the final amplitude and probability.

Looked at another way, say there are two alternate paths with amplitudes A_1 and A_2. These amplitudes are complex numbers. The total amplitude is the sum of the individual amplitudes, and the probability is the absolute square of the result:

$$\begin{aligned} A &= A_1 + A_2 \\ P(A \to D) &= |A_1 + A_2|^2 \end{aligned} \tag{17.5}$$

17.2.3 Successive Steps

When an event occurs as a series of steps, there is a different procedure to follow to determine the final probability of the event occurring.

To see how this works, consider the path from S to D via the point z_7 as shown in Figure 17.5. We divide the journey up into two halves, from S to s_7 and then from s_7 to D.

The journey the photon takes from S to s_7 will take a certain time, as indicated by the imaginary stopwatch. Similarly, the journey from s_7 to D will take a certain time. The total time for the complete trip

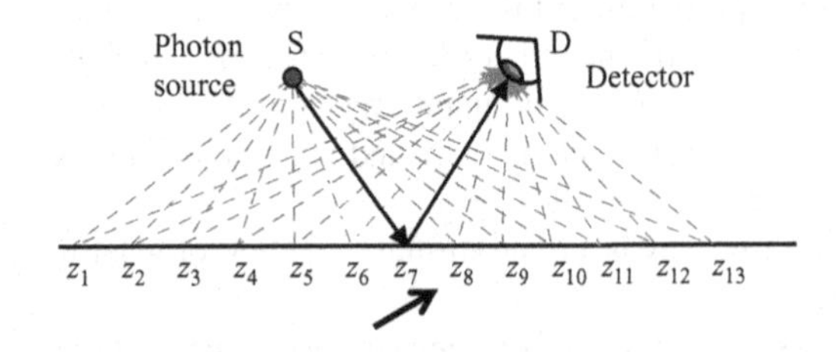

FIGURE 17.5 Representative path z_7 broken up into two halves to demonstrate how amplitudes combine when there is a series of events.

will be the sum of the times for each leg, and so the final direction of the resultant arrow will be the angle of turn for the first leg, plus the angle of turn for the second leg.

But, as is the case with any serial accumulation of probabilities, the total probability is the product of the two component steps, and so we *multiply* the amplitude of the first leg by the amplitude of the second leg. For the first leg, there is no reflection or absorption, and so the amplitude of this leg is 100%. That is, a photon leaving the source on its way to s_7 will get there. For a mirror, assuming 100% reflection, then the photon leaving s_7 will with certainty arrive at the detector, and so this amplitude is also 100%. The combined amplitude is in this case $1 \times 1 = 1$.

The angles associated with each step are added successively.

And so, we have arrows arranged such as in Figure 17.6.

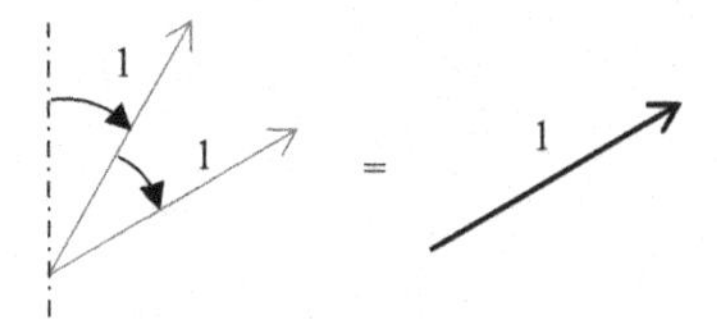

FIGURE 17.6 Summing the amplitudes for a series of two events.

This is a simple example, but the procedure holds for more complicated cases when, for example, there is partial reflection from the surface and the amplitude of the second leg is less than one.

When an event occurs as a result of two or more steps, we determine the amplitude for each step and then multiply the amplitudes to find the resultant amplitude, the square of which gives the probability of the combined event occurring.

So, say we have two complex numbers $z_1 = a_1 + ib_1$ and $z_2 = a_2 + ib_2$ whose amplitudes are to be multiplied. The amplitudes of the two numbers are A_1 and A_2 respectively.

$$\begin{aligned} A_1 &= \sqrt{a_1^2 + b_1^2} \\ A_2 &= \sqrt{a_2^2 + b_2^2} \end{aligned} \tag{17.6}$$

We can multiply these terms out directly, but it gets a little unwieldy. It is better to express the terms in trigonometric form. Thus, the product $z_1 z_2$ can be written:

$$\begin{aligned} z_1 &= A_1(\cos\theta_1 + i\sin\theta_1) \\ z_2 &= A_2(\cos\theta_2 + i\sin\theta_2) \\ z_1 z_2 &= A_1 A_2(\cos(\theta_1 + \theta_2) + i\sin(\theta_1 + \theta_2)) \end{aligned} \tag{17.7}$$

Note in the above how the angles are added and the amplitudes are multiplied. Letting $\theta = \theta_1 + \theta_2$ and $A = A_1 A_2$, we can write:

$$z = z_1 z_2 = A(\cos(\theta) + i\sin(\theta)) \tag{17.8}$$

The resultant complex number is z. The amplitude of z squared, that is, the resultant probability $P(z)$, is:

$$P(A \to D) = |z|^2 = z^* z = A^2 \tag{17.9}$$

In exponential format:

$$
\begin{aligned}
z_1 &= A_1 e^{i\theta_1} \\
z_2 &= A_2 e^{i\theta_2} \\
z_1 z_2 &= A_1 A_2 e^{i(\theta_1+\theta_2)}
\end{aligned}
\tag{17.10}
$$

Letting $\theta = \theta_1 + \theta_2$ and $A = A_1A_2$, we can write:

$$z_1 z_2 = Ae^{i\theta} \tag{17.11}$$

That is, the resultant amplitude A is A_1A_2.

Looked at another way:

$$
\begin{aligned}
P_1 &= A_1^2 \\
P_2 &= A_2^2 \\
P(A \to D) &= P_1 P_2 = A_1^2 A_2^2 \\
&= |A_1 A_2|^2
\end{aligned}
\tag{17.12}
$$

The final probability of the event when the event consists of two successive steps is calculated from the product of the individual probabilities.

17.3 THE BEHAVIOUR OF LIGHT

17.3.1 Straight-Line Path

Let us examine more closely the path of photons from a source to a detector. According to the theory, all possible paths that a photon might take must be considered and the amplitudes added, remembering that it is the phase of the amplitudes that is the important thing that determines that path's significance in the overall sum. For clarity, as shown in Figure 17.7, the possible paths near the direct path are drawn as a series of straight-line segments and those which are far away from a direct path as a wavy line.

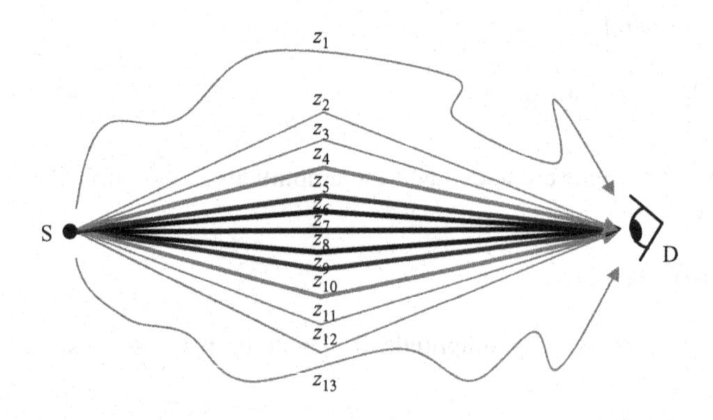

FIGURE 17.7 Possible paths taken by a photon from a source to a detector.

As shown in Figure 17.8, it is found that there is a larger difference in time from one path to the next for the "outer" paths compared to those which are close to the centre. Thus, the paths closer to the straight-line path z_7 make the largest contribution to the total amplitude for photons travelling from the source to the detector.

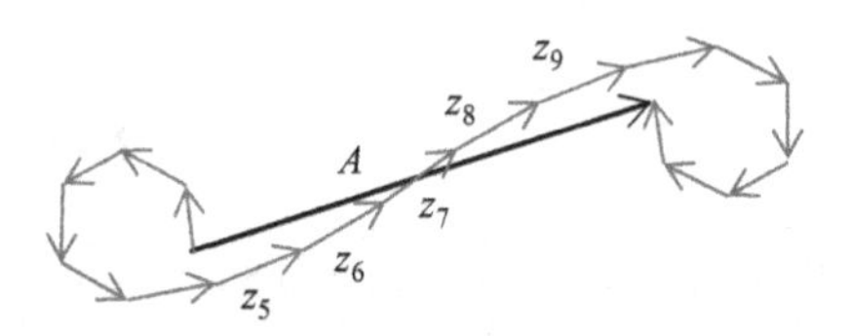

FIGURE 17.8 Vector sum of amplitudes to determine the resultant amplitude for the travel of a photon from S to D.

That is, if we were to sample a small corridor of space in the vicinity of z_7, we would find some photons taking the *slightly* longer paths: z_5 and z_6, z_8 and z_9, as well as the straight-line path z_7 make the most contribution to the total amplitude. In a sense, the photon takes all possible paths simultaneously, but the amplitudes for the outer ones cancel, while those near the straight-line path add together to produce an appreciable amplitude A, and therefore probability $P = A^2$, that photons travel, in what looks to us like a straight line, from S to D.

17.3.2 Single Slit Diffraction

Consider the passage of photons from S to D as they pass through a slit shown in Figure 17.9. Detectors are placed beyond the slit that register the arrival of a photon.

If the slit is fairly wide, then all we have done is to knock out the paths near the edges, and we are left with a significant amplitude for photons from S reaching the detector D_1 in more or less what we normally would call a straight-line path. There is not much difference in time taken by photons travelling near the "straight-line" route to D, and so the arrows for those paths tend to point in the same direction. However,

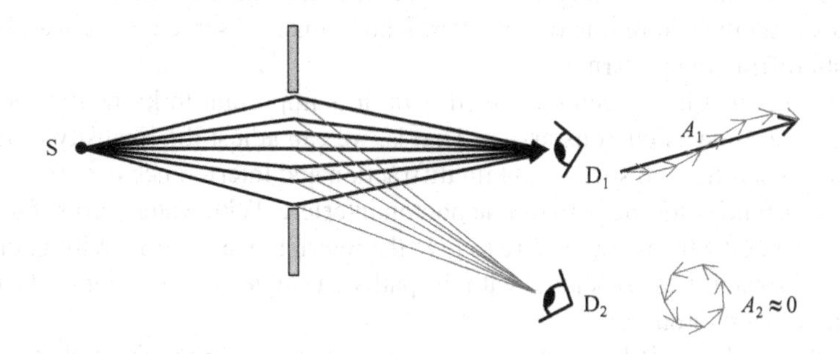

FIGURE 17.9 Possible paths taken by a photon travelling from S to D through a wide single slit.

the situation for photons taking a possible bent path to D_2 is different. These photons are unlikely to not reach a second detector at some arbitrary side position D_2 because the arrows representing the amplitudes for those paths would tend to cancel, there being a significant difference in time, *from path to path*, for those possibilities.

But consider what happens when the slit is made very small as shown in Figure 17.10.

Note, the variation in time from path to path is still higher for those going to D_2, but there is not such a significant variation from path to path for the arrows associated with those paths to form a circle and cancel the resultant amplitude.

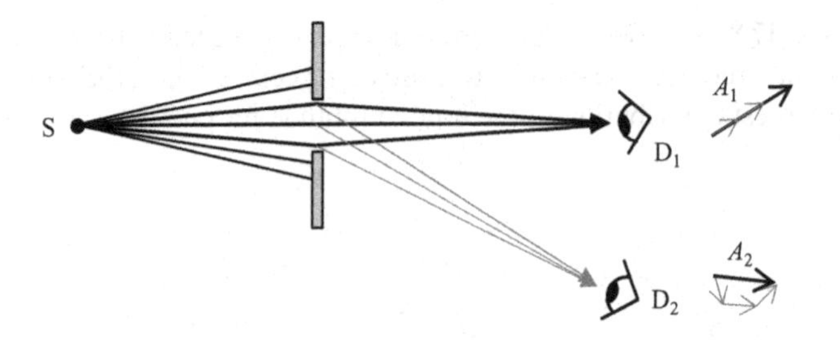

FIGURE 17.10 Possible paths taken by a photon travelling from S to D through a narrow single slit.

The lengths of the paths to D_1 are much the same, and so the amplitudes add up, and there is a net amplitude (somewhat reduced than before) for photons arriving at D_1. But what is more important is that the paths to D_2, although longer than those to D_1, are no longer accompanied by paths whose arrows would tend to cancel out, and so we are left with an amplitude that photons may indeed arrive at D_2 as well.

Going back to the example of a straight-line path from S to D, we can see that we need more than the single straight-line path z_7 to describe what happens; we need the adjacent paths which contribute to the final amplitude as well. If we put in a barrier in the form of a slit to narrow down the path to the detector, we find the light spreads out. The narrower the slit, the more spread out the light becomes. A photon needs the adjacent paths to find its way to D_1. If they are not there, by narrowing the slit, it is as if the photon becomes lost and goes through at wider angles looking for the target D_1 – thus giving us the phenomenon of diffraction.

17.3.3 Double Slit Interference

Consider now the well-known double slit experiment. Photons are emitted from the source and end up at a place on the right-hand side of the slits where we can place several detectors. It is found that depending on the spacing between the slits, there is an amplification and cancellation of amplitudes for detectors placed at certain distances from the centre line of slits. Physically, this is observed as light and dark bands – the classical double-slit diffraction pattern.

In all the processes that have been described here, it is important to know that the response of the detector is built up over time as photons arrive. That is, we can adjust the intensity of the source so that photons are sent one at a time, and still we obtain diffraction and interference over time. The significance of this is that the probabilities for the path of one photon interfere. With water waves, for example, it is the waves emanating from both slits that interfere to give the interference pattern. With photons, the interference arises from the probabilities associated with the path of a single photon. That is, the photon interferes with itself, not with other photons.

A particular photon does not know which path was taken by the previous photon. Photons do not spread out and go through both slits. All we can say is that there is an amplitude for a photon to go through one slit and an amplitude to go through the other. In fact, there is an amplitude for the photon to take an infinite number of paths from the source to the detector. All that matters is that we observe a photon leave the source, and we wish to calculate the probability that we will observe that photon arrive at the detector.

The immeasurable part is what happens on the way. That is, we cannot look at the particles (whether they be photons or electrons) as they pass through the slit and expect to see a diffraction pattern beyond the slits because the act of viewing them at the slits is taking a measurement, or an observation – which is the end of an event. One then has to analyse a new event from the slit to the screen. That is, the process is a "successive steps" affair. When this is done, the interference pattern disappears, and we obtain a single slit pattern for each slit.

17.4 ACTION

We might well ask, what is this "stopwatch" with the imaginary hand that turns?

A wave function in general is expressed:

$$y(x,t) = Ae^{i(kx-\omega t)} \tag{17.13}$$

Or, by de Broglie, we can express the wave function for the electric field for a photon as:

$$E(x,t) = E_o e^{i\left(\frac{px}{\hbar}x - \frac{E}{\hbar}t\right)} \tag{17.14}$$

Whether it be a matter wave or an electromagnetic wave, we have the product of an amplitude and a phase. We separate the spatial and time components so as to write:

$$E(x,t) = E_o e^{i\frac{px}{\hbar}x} e^{-i\frac{E}{\hbar}t} \tag{17.15}$$

In this equation, the first exponential, multiplied by E_0, is the "amplitude" which is then multiplied by the second, time-like, exponential. We did the same thing earlier with the matter wave for an electron.

The scenario for the single and double slit experiments could also be done with electrons. The probability amplitude for a particle, such as an electron or a photon, has phase information. The amplitude is a complex number. For a photon, we write the amplitude as:

$$E_o e^{i\frac{px}{\hbar}x} \tag{17.16}$$

For an electron, we write:

$$\psi(x) = u(\vec{p})e^{i\frac{\vec{p}}{\hbar}x} \tag{17.17}$$

The amplitude is a periodic function in x. It is a complex number. The amplitude squared gives the probability. The space exponential provides the phase or the "turning" of the stopwatch. For a particle with a definite momentum, there is a direct connection between the time and the distance that a particle might travel, and so we describe the turning of the stopwatch in terms of time rather than the distance travelled by the particle.

The phase component becomes important when we need to add or multiply the arrows. When an event can happen in several different ways, we add the amplitudes as vectors and then square the result to obtain the probability. When an event occurs in a succession of steps, we multiply the amplitudes and add the angles algebraically.

The probability amplitude for an arrow, say z, is:

$$z = a + bi \tag{17.18}$$

If A is the amplitude of z, then in complex number form, z is expressed:

$$z = Ae^{i\theta} \tag{17.19}$$

This complex number, representing one of the paths from S to D, contains an amplitude term, A, and the phase term, $e^{i\theta}$. The phase, the turning of the stopwatch, is in essence the action S:

$$\theta = \frac{S}{\hbar} \tag{17.20}$$

The complex number representation of the amplitude becomes:

$$z = Ae^{iS/\hbar} \tag{17.21}$$

If we express the path, let's say in one dimension, as a function of time, $x(t)$, the particle's action for this path is given by:

$$S = \int_{t_1}^{t_2} x(t)\,dt \tag{17.22}$$

The integral is calculated for each path under consideration. Those paths which have the *least action* contribute most to the resultant probability. This is because when changes in S are small in comparison to Planck's constant, there is not so much turning of the stopwatch compared to the paths where changes in S are large. The least action is calculated by a sum of the Lagrangian computed along the path.

When we are considering paths near to the one that gives the least action, the first order changes in action are zero, and these arrows tend to align. As the first order terms become more important compared to the second order terms, the arrows become more misaligned and the vector sum of the wave function for those paths cancels.

This unique method of expressing the concepts of quantum mechanics was formulated by Feynman in 1942 and vividly illustrated in his book "The Strange Theory of Light and Matter" from which the above examples are taken.

17.5 FEYNMAN DIAGRAMS

17.5.1 Feynman Diagrams

Feynman described the motion of particles such as electrons and photons in space-time using certain rules of propagation (travel from one place to another) and interaction. To see how this is done, we will consider a two-dimensional version of events to make things manageable.

For the interaction between electrons and photons, we recognise three fundamental events that can happen:

- An electron can move from place to place.
- A photon can move from place to place.
- An electron can absorb or emit a photon.

Let's examine the consequences of these.

Figure 17.11(a) shows an electron at rest in a local coordinate system at position A. Its world line is vertical, indicating that its position x does not change with time. If the electron were moving to the left, travelling from say A to B, then its world line would be tilted over to the left as shown in Figure 17.11(b). If the electron were moving to the right, the world line would tilt to the right as shown in Figure 17.11(c).

Now, with an electron moving to the right, say for some reason it emits a photon as shown in Figure 17.12(a). The world line of a photon is at 45° on these axes. Let's say the photon moves off to the

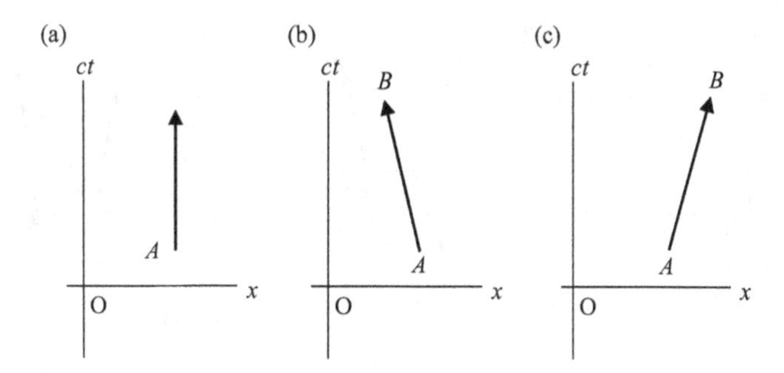

FIGURE 17.11 (a) World line for an electron at rest. (b) World line for an electron travelling to the left. (c) World line for an electron travelling to the right.

right. The photon is shown as a wiggly line. We might ask, what happens to the electron's world line after the photon has been emitted?

Because the photon has momentum, the electron recoils to the left. This is shown in Figure 17.12(b).

Now let's say there is another electron nearby and this second electron absorbs the photon emitted by the first electron. As seen in Figure 17.12(c), that electron is deflected to the right.

What are we seeing here? Like charges repel. It is as though the presence of the second electron stimulates the emission of a photon from the first. The act of emission and absorption of a photon causes the two electrons to move away from one another. There is an interaction between the electron and the

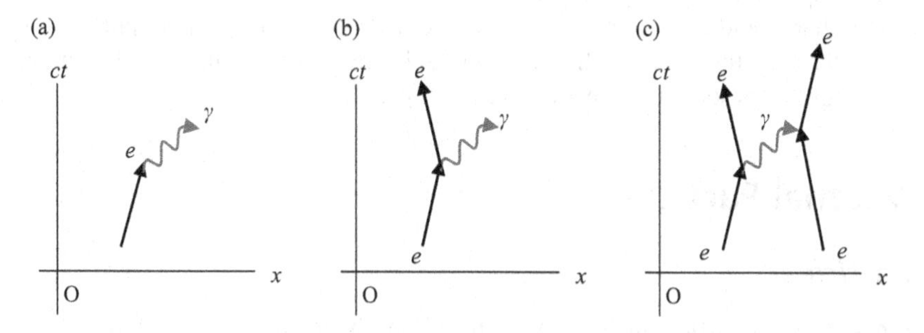

FIGURE 17.12 (a) An electron emits a photon. (b) World line of electron after emitting the photon. (c) Two electrons approach each other, there is an exchange of a photon, and the two electrons move away from each other.

photon when the photon is emitted and another interaction between the photon and the second electron where the photon is absorbed. These interactions occur where the lines meet on the space-time diagram – the vertices.

In the example above, we considered the emission and absorption of a photon by an electron.

Consider the example of Figure 17.12(c) which is drawn again in Figure 17.13(a). At first, it was said that the photon is emitted by the electron on the left and then later absorbed by the electron on the right. However, the interaction can be also viewed as the photon being emitted from the second electron, travelling instantaneously towards the first electron, Figure 17.13(b) or even travelling backwards in time, Figure 17.13(c) and then being absorbed by the first electron. When we look at the final outcome, two electrons approaching each other and then moving away from each other, we cannot tell what actually happened – we can't really tell which electron emitted or absorbed the photon or when these events happened. All we can see are the external particles coming together and moving apart. Another possibility for our event above is that two photons are involved.

The photon in Figure 17.13 is called an exchange particle. Since we don't know which way the photons travelled, we say they are exchanged. Particles which appear and disappear within the

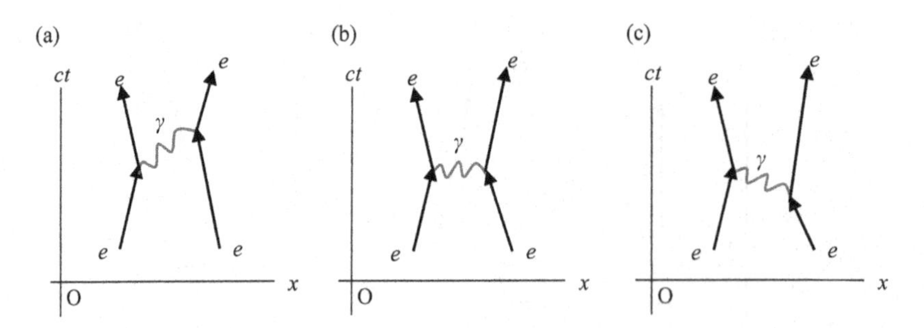

FIGURE 17.13 An interaction between two electrons showing (a) photon emitted from the first electron and then later being absorbed by the second, (b) photon instantaneously being emitted and absorbed and (c) photon being emitted by second electron, travelling backwards in time and being absorbed by the first electron.

interaction zone are called virtual particles. Virtual particles transmit forces. For example, a virtual photon transmits the electromagnetic force. Force-carrying particles are called bosons. There are other bosons besides photons; their function is the transmission of nuclear forces within the nucleus. Bosons have integer spin. Fermions are particles, such as electrons, protons and neutrons, with half-integer spin. Fermions can also appear as virtual particles within the interaction zone. Particles which enter and leave the interaction zone are the measurable particles. They are referred to as external particles.

Much like in the double slit experiment, we cannot know what happened in between the initial and final states. We can only know what goes in and what comes out. Inside the interaction zone, everything that could happen that would lead to the same outcome has a probability, an amplitude, to happen. This includes travel faster than the speed of light, travel backwards in time and travel in more complicated paths than the straight lines shown above – all are possible.

17.5.2 External Particles

17.5.2.1 Particles

Plane wave functions for a particle such as an electron have the form:

$$\Psi = u(x,t)e^{i(p_x x - Et)} \tag{17.23}$$

Or in more general terms:

$$\Psi = u(p_\mu)e^{-ip_\mu \cdot x_\mu} \tag{17.24}$$

Here, $\mu = 0,1,2,3$ stands for t, x, y, z. The amplitude term is the Dirac spinor. The p_μ term is the four-momentum. The four-momentum consists of the energy E and the three spatial components of momentum p.

$$p_\mu = (E, \vec{p}) \tag{17.25}$$

So, in the wave equation, for the spatial terms, $\mu = 1, 2, 3$, $p_\mu = p_x, p_y, p_z$ and $x_\mu = x, y, z$. For time-like term, $\mu = 0$, $p_\mu = E$ and $x_\mu = t$.

For external particles entering the interaction zone, we use the amplitude of the wave function to represent that particle, that is, the Dirac spinor $u(p_\mu)$. For external particles leaving the interaction zone, we use the amplitude of the adjoint wave function to represent that particle, the adjoint Dirac spinor $\bar{u}(p_\mu)$.

The reason for this is the connection with the concept of probability flow. If the interaction zone is our volume V, then we have a probability current across it. Probability flow is described by the four-vector probability current:

$$J_\mu = \bar{u}\gamma_\mu u \tag{17.26}$$

where γ_μ are the gamma matrices and u and $\bar{u}$ are the Dirac spinor and adjoint spinors respectively.

17.5.2.2 *Antiparticles*

For antiparticles entering the interaction zone, we use the charge conjugate wave function ψ', and for those leaving, we use its adjoint. Thus, the probability flow across the interaction zone is represented by:

$$J_\mu = \bar{v}\gamma_\mu v \tag{17.27}$$

17.5.2.3 *Photons*

In the case of a photon entering or leaving the interaction zone, we have the four-potential wave function:

$$\mathbf{A}_\mu = \boldsymbol{\varepsilon}_\mu\left(p_\mu\right)e^{-ip_\mu x_\mu} \tag{17.28}$$

The probability for an external particle, such as an electron, to go from place to place, that is, to *propagate* from one place to another, is determined from the amplitude of the particle's wave function, which in turn comprises the spinor:

$$\boldsymbol{\varepsilon}_\mu\left(p_\mu\right) \tag{17.29}$$

For a photon, it is the amplitude of the four-potential that is used.

17.5.3 Interactions

17.5.3.1 *Vertex Factor*

An electron has a charge $q_e = -1.602\times 10^{-19}$ Coulombs. We might ask ourselves, what might this charge be in natural units? We remember that in natural units, all quantities are expressed in terms of energy. In natural units, the electronic charge is expressed in terms of a dimensionless "fine structure constant" α.

The fine structure constant has several physical interpretations but was originally introduced by Sommerfeld in 1916 to account for the fine structure of line spectra of the hydrogen atom. Sommerfeld proposed that this fine structure arose from a relativistic correction to the electron mass as the velocity of an orbiting electron changed as it traversed a very slightly elliptical path. Later experiments by Goudsmidt and Ulenbeck in 1925 showed that the fine structure of line spectra was a result of electron spin. Spin represented the two possible directions of angular momentum of the electron as it interacted with the internal magnetic field of an atom.

The fine structure constant depends upon the magnitude of the charge and for an electron, is given by:

$$\alpha = \frac{q_e^2}{4\pi\hbar c} \approx \frac{1}{137} \tag{17.30}$$

In natural units, the electron charge is given the symbol e, and so α is expressed:

$$\alpha = \frac{e^2}{4\pi}$$
$$e = \sqrt{\alpha 4\pi} \tag{17.31}$$

There are four fundamental forces in nature. In particle physics, forces between particles of matter (fermions) occur due to the exchange of massless particles (bosons). For example, the strong nuclear force is thought to be a result of the exchange of gluons between quarks. The electromagnetic force manifests itself as the exchange of photons between charged particles (e.g. electrons). The strength of the interaction is described by what is called a coupling constant g. We met the coupling constant briefly in Section 16.5. For electromagnetic interactions, the coupling constant g is the charge e. However, often it is better to express the strength of the interaction in terms of α because α is dimensionless and therefore gauge invariant. The value of α is different for different types of interaction; for example, the interaction between quarks takes place via an exchange of what are called gluons, and the strength of that interaction is such that $\alpha \sim 1$.

When an electron emits or absorbs a photon, this happens at a vertex in the Feynman diagram. It is an interaction. In fact, in quantum electrodynamics, the interaction between light and matter, it is the only interaction.

Now, we need to quantify this interaction. So far, we have only dealt with an electron and a photon. The electron, being negatively charged, interacts with the photon. If we take a step back to classical physics, we know that the electrostatic Coulomb force (qE), and the magnetic force (qvB), depend upon the magnitude of the charge. It is not surprising therefore that the strength of the electron–photon interaction depends upon the magnitude of the charge. The strength of the interaction is given by e, and this certainly affects the probability of the interaction occurring, but we need more. For example, what about spin? Does the spin of the electron affect the interaction? Yes.

The strength of the interaction depends on spin since different spins allow two electrons to occupy the same energy level. Spin is included in our measure of the strength of the interaction by virtue of the Dirac gamma matrices.

The strength of the electron–photon interaction is therefore given by the vertex factor:

$$ie\gamma_\mu \tag{17.32}$$

where e is given as above in natural units. i is included as a phase term.

As we found earlier, the interaction between a particle and a field, which is essentially the interaction between one particle and another, adds terms to the Lagrangian. The Lagrangian, when summed over a path, determines the action, which in turn gives phase information about the probability for an event – say, a particle travelling from one place to another – to occur. It is no surprise therefore to find that the vertex factor is a quantitative measure for the amplitude that the event, that is in this case, the interaction, will take place. The stronger the interaction, the more likely the event will occur.

17.5.3.2 Photon Propagator

In a particle interaction, such as that shown in Figure 17.13, there is an electron interacting with another electron through the exchange of a photon. It is called a virtual photon. Virtual photons (or more correctly, virtual particles) do not appear in the tally of incoming and outgoing particles and are the immeasurable components of the interaction. Both bosons (e.g. photons) and fermions (e.g. electrons) may appear as virtual particles.

Virtual particles "propagate" from one vertex to another. Virtual particles do not obey momentum–energy conservation laws. They can borrow from the "vacuum" as long as the debt is repaid after the interaction.

Although we have mentioned that virtual particles such as a virtual photon have a force-carrying role, what we really mean is that there is momentum and energy transfer involved when such a particle moves from one place to another.

Virtual particles travel from one place to another in the interaction zone, that is, from one vertex to another. At each vertex, energy, momentum and charge are conserved. The strength of the interaction between an electron and a photon is given by the vertex factor. Where there are two vertices in a diagram, they are labelled μ and ν.

As we saw in Section 16.4, the propagation factor is defined as:

$$i\frac{1}{q^2 - m^2} \tag{17.33}$$

For a photon, the rest mass $m = 0$, and so the propagator becomes:

$$i\frac{1}{q^2} \tag{17.34}$$

And when spin (i.e. polarisation for a photon) is included, the photon propagator is:

$$i\frac{\varepsilon_\mu (\varepsilon_\nu)^*}{q^2} = i\frac{g_{\mu\nu}}{q^2} \tag{17.35}$$

A factor i is included as a phase term, the details of which need not concern us for now.

17.5.4 Electron–Photon Interactions

Electron–photon interactions can take six different forms. If we represent a photon by the symbol γ, a positron by p and an electron by e, the interactions possible are given in Table 17.1.

Note that in each interaction, charge is conserved but particle number is not. These interactions are the basic building blocks of quantum electrodynamics. Each interaction occurs at a vertex in the diagram.

TABLE 17.1 Interactions between an electron and a photon

INTERACTION	*EVENT*
$e \to e + \gamma$	Electron emits a photon
$e + \gamma \to e$	Electron absorbs a photon
$\gamma \to e + p$	Photon creates an electron positron pair
$p + \gamma \to p$	Positron absorbs a photon
$p \to p + \gamma$	Positron emits a photon
$e + p \to \gamma$	Electron and photon annihilate and produce a photon

17.6 COMPONENTS OF A FEYNMAN DIAGRAM

Table 17.2 shows the significance of the elements of a Feynman diagram in mathematical terms.

TABLE 17.2 Elements of a Feynman diagram

ITEM	*SYMBOL*	*TYPE*	*MEANING*	*FUNCTION*
Incoming particle	$u(p_\mu)$	4-spinor	p_μ = four-momentum	Amplitude of the wave function
Outgoing particle	$\bar{u}(p_\mu)$	adjoint 4-spinor	p_μ = four-momentum	Amplitude of the adjoint wave function
Incoming antiparticle	$v(p_\mu)$	4-spinor	p_μ = four-momentum	Amplitude of the charge-conjugated wave function
Outgoing antiparticle	$\bar{v}(p_\mu)$	adjoint 4-spinor	p_μ = four-momentum	Amplitude of the adjoint charge conjugated wave function
Incoming photon	$\varepsilon_\mu(p_\mu)$	2-spinor	μ = scalar and vector potentialsϕ, A_x, A_y, A_z p = four-momentum	Amplitude of the four-potential wave function
Outgoing photon	$\varepsilon_\mu(p_\mu)^*$	2-spinor	μ = scalar and vector potentials ϕ, A_x, A_y, A_z p = four-momentum	Conjugate of the amplitude of the four-potential wave function
Vertex (involving an electron of charge e)	$ie\gamma_\mu$	Vertex factor	γ_μ = Dirac matrices γ_0, γ_1, γ_2, γ_3 e = electronic charge expressed as $\sqrt{4\pi\alpha}$ where α is the fine structure constant	The amplitude that an electron will emit or absorb a photon. Also represents the strength of the interaction between the charged particle and the photon The γ matrix includes the spin of the electron into the interaction i is a phase term
Virtual photon (boson, integer spin)	$i\dfrac{1}{q^2}$ with no spin, or $i\dfrac{\varepsilon_\mu(\varepsilon_\nu)^*}{q^2} = -\dfrac{ig_{\mu\nu}}{q^2}$ with spin included	Propagator with no rest mass	q = four-momentum of the virtual particle ε_μ = polarisation at the μ vertex ε_ν = polarisation at the ν vertex $g_{\mu\nu}$ = metric tensor	Connects vertices μ and ν together Note, the sum of the product of the spin matrices is equal to the negative of the metric tensor $g_{\mu\nu}$ i is included as a phase factor
Virtual fermion (1/2 integer spin)	$\dfrac{\gamma_\mu q_\mu + m}{q_\mu^2 - m^2}$	Propagator with rest mass	m = rest mass (natural units) γ_μ = Dirac matrices γ_0, γ_1, γ_2, γ_3 q_μ = four-momentum	Connects vertices μ and ν together

17.7 A FEYNMAN DIAGRAM

We are now ready to examine the case of two electrons approaching each other and then scattering away from each other.

As shown in Figure 17.14(a), the Feynman diagram for this consists of the world lines for the incoming particles, the world lines for the outgoing particles, the world lines for the virtual particles and the interactions at the vertices.

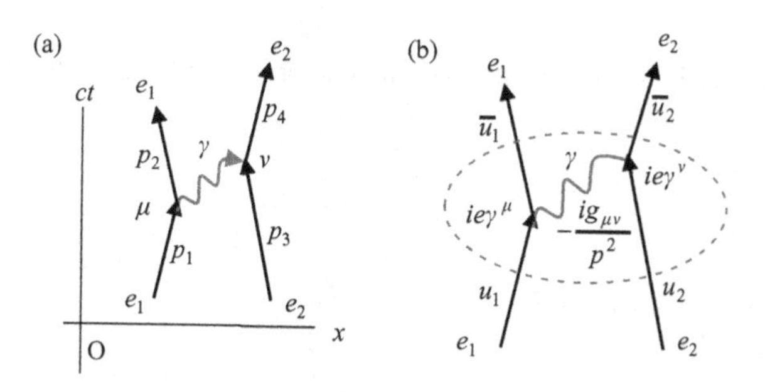

FIGURE 17.14 (a) Feynman diagram showing the interaction between two electrons. (b) Feynman diagram showing particles, propagators and the interaction terms for the interaction.

Each line in the Feynman diagram represents an amplitude for something to occur. The first thing to occur might be the amplitude for the incoming electron e_1 to travel from place to place. That is, from its initial starting position at the bottom of the diagram to the vertex μ in the diagram. Another thing that has an amplitude to occur is that the outgoing electron e_1 might travel from the vertex μ to its final place. The same amplitudes happen for the second electron e_2 and its travel to and from the vertex ν.

Each vertex in the diagram also represents an amplitude for something to occur. In Figure 17.14(a), the vertex factor at μ represents the amplitude that the electron e_1 would emit a photon, and the vertex factor at ν is the amplitude for the electron e_2 to absorb a photon.

The virtual photon propagates from μ to ν, and so there is an amplitude for that as well.

We identify each amplitude in the Feynman diagram according to Table 17.2.

The observable events are the initial electrons approaching and the deflected electrons moving away. What happens in the interaction zone is unobservable and immeasurable.

Now, when something happens as a series of successive steps, we *multiply* the amplitudes of each step together and then square the result to find the probability of the combined event. In this case, for the two electrons to travel from their initial starting points to their final destinations, there is a series of successive steps.

Since the terms in the Feynman diagram are matrices, the multiplications will be matrix multiplications, and the final result will be a matrix and is given the symbol M. The amplitude (squared) of the resulting matrix gives the probability for the event as a whole to occur. That is, in Figure 17.14(b), the amplitude (squared) of M is the probability that the electron will scatter off another electron.

The matrix M is thus constructed from:

$$\mathcal{M} = \left(\bar{u}_1(p_2)ie\gamma^\mu u_1(p_1)\right)\left(\frac{-ig_{\mu\nu}}{q^2}\right)\left(\bar{u}_2(p_4)ie\gamma^\nu u_2(p_3)\right) \tag{17.36}$$

This expression can be rearranged so that:

$$\mathcal{M} = \frac{ig_{\mu\nu}}{q^2} e^2 \left(\bar{u}_1 \gamma^\mu u_1\right)\left(\bar{u}_2 \gamma^\nu u_2\right) \tag{17.37}$$

Note the significance of the probability currents $\bar{u}\gamma^\mu u$ involved. Because these currents are vectors, the dot product is appropriate and so:

$$\mathcal{M} = \frac{ig_{\mu\nu}}{q^2} e^2 J_1 \cdot J_2 \tag{17.38}$$

The matrix M is the amplitude for the event to occur. The probability *P* is determined from the square of the amplitude:

$$\begin{aligned} P &= |\mathcal{M}|^2 = \mathcal{M}\mathcal{M}^* \\ &= \left(\frac{ig_{\mu\nu}}{q^2} e^2 J_1 \cdot J_2\right)\left(\frac{-ig_{\mu\nu}}{q^2} e^2 J_1 \cdot J_2\right) \\ &= \left(\frac{g_{\mu\nu}}{q^2} e^2 J_1 \cdot J_2\right)^2 \end{aligned} \tag{17.39}$$

In summary, a Feynman diagram is an equation involving amplitudes. There are some significant practical difficulties. For example, as mentioned previously, events within the interaction zone can take many forms and may consist of interactions with other virtual particles, loops, creation and annihilation of particles – these correspond to the many different paths taken by a photon in the earlier examples with the stopwatch. Inclusion of these other possible paths leads to great complexity in the final expression for $\mathcal{M}$.

17.8 RENORMALISATION

Well, we might say, this is all very nice, but what does it mean? It means we have "counted the beans", but only for one of an infinite number of possibilities for the way in which this event can happen. Much like the mirror, we have attended to only one of the possible paths. It turns out we have worked it out for the path which has the greatest amplitude, or probability, to happen because this is the one with the minimum number of vertices. But, as we know, the multitude of other paths contribute to the physics of the process. Those other paths are needed to ensure that the action along the path of the events modelled is the one of least action.

The true picture – whether it be reflection from the centre of the mirror, or diffraction from single and double slits or scattering of two electrons – emerges only after the sum of all the possible amplitudes is taken. This sum is the Feynman path integral, or the Feynman sum over histories.

Feynman diagrams are typically drawn for other possible events that connect the incoming and outgoing particles and the matrix element determined for each. When an event can occur in several different ways, the total amplitude is given by the vector sum of the individual amplitudes. The total probability is determined from the square of the total amplitude:

$$P = |\mathcal{M}|^2 = \left(\mathcal{M}_1 + \mathcal{M}_2 + \mathcal{M}_3 \ldots\right)\left(\mathcal{M}_1^* + \mathcal{M}_2^* + \mathcal{M}_3^* \ldots\right) \tag{17.40}$$

But when do we stop? If there is an infinity of possible paths, how do we know which ones to include in the sum?

For quantum electrodynamics, the interactions between electrons and photons, the strength of the interaction is:

$$\alpha \approx \frac{1}{137} \tag{17.41}$$

This is fairly small compared to say that between quarks where α is close to 1, and so for electromagnetic interactions, higher order diagrams, that is, higher order terms in the Hamiltonian, can be ignored.

In the account given above, we have been careful to avoid the numerous infinities that crop up in canonical quantisation of the field and also in the probabilities associated with Feynman diagrams. These infinities were eventually removed by using known physical quantities of the masses and charges of the particle involved as boundary conditions for the computations. This process is known as renormalisation.

17.9 CONNECTION TO EXPERIMENTAL RESULTS

Particles are the quanta of their respective fields. Forces between particles arise from an exchange of particles. Although the theory is treated ultimately via the Lagrangian, the experimental observables in a particle accelerator are what are called the scattering cross-section and the transition rate.

When we talk about particles, we talk about their energy state. When an electron, say in an atom, is in a particular energy state, the solutions ψ to the wave equation, whether it be the Schrödinger or Dirac equation, give the probability of where the electron might be or the energy it might have or the momentum it might possess, etc. The eigenvalues of the Schrödinger equation are energy states that an electron might have before or after an interaction with a photon. In a Feynman diagram, this probability is expressed in terms of the probability current flowing in and out of the interaction zone.

When an electron emits or absorbs a photon, its energy state changes. Usually, this means the change of energy when the electron emits or absorbs a photon while it is in a bound state, such as in an atom. Whether or not an electron will emit or absorb a photon is a probabilistic event. The probability of the electron making a transition from one state to another is expressed in terms of a transition rate, that is, the number of transitions per unit time from an initial state to a final state.

In particle physics, the transition rate Γ is found from perturbation theory. In perturbation theory, the perturbed Hamiltonian is expanded and solved for the allowed energy states.

For example, if the potential term $V(x)$ in a non-relativistic quantum system is perturbed by a small amount λ, then the Hamiltonian for the system becomes:

$$\hat{H} = \hat{H}_0 + \lambda\hat{H} \tag{17.42}$$

and the perturbed energy levels are:

$$\langle E_n \rangle = \langle \psi_n | \hat{H} | \psi_n \rangle + \sum_{p \neq n} \frac{\left| \langle \psi_p | \hat{H} | \psi_n \rangle \right|^2}{E_n^0 - E_p^0} \tag{17.43}$$

n is the final state and p is the initial state. $\langle E_n \rangle$ is called the transition matrix element T. Additional loops and events within the interaction zone of a Feynman diagram are represented by higher order terms in the Hamiltonian.

The transition rate Γ is given by the transition matrix element T, multiplied by the density of energy states. That is, the density of states of the final state $\rho(E_f)$.

$$\Gamma = 2\pi|T|^2\rho(E_f) \tag{17.44}$$

This is known as Fermi's Golden Rule.

When particles collide, the collision is called a scattering. The cross-section is a quantitative measure of the scattering.

The cross-section is the effective cross-sectional area of the target particles for the interaction to occur with an incoming particle. This is expressed as:

$$\sigma = \frac{\text{Number of interactions per unit time}}{\text{Number of incoming particles per unit area per unit time}}$$

It is the differential cross-section that is usually used.

$$\frac{\partial\sigma}{\partial\Omega} = \frac{\text{Number of interactions per unit time into a solid angle}}{\text{Number of incoming particles per unit area per unit time}}$$

where Ω is a solid angle, an angle swept out in three dimensions. Because particles do not in general have a clearly defined shape or outline, the cross-section is not generally related to the geometric size of a particle but more to its ability to interact with another particle.

In quantum field theory, the equivalent treatment is given in terms of what is called the S matrix. Elements of the S matrix can be found via experiment results using Fermi's Golden Rule but can also be calculated in terms of the Lagrangian. These experiments are typically done in a particle accelerator when particles collide with high velocity.

In quantum field theory, the essential elements of the S matrix are used to construct the Feynman invariant amplitude $\mathcal{M}$ matrix. This provides theoretical information about the transition rate and the differential cross-section which can then be compared to experimental results. The $\mathcal{M}$ matrix is also the product of a series of Feynman diagrams. The Feynman diagrams allow rapid calculation of what might be a very tedious computation in quantum field theory.

Conclusion

18

In 1900, Planck applied the concepts of statistical mechanics to the phenomenon of black body radiation to resolve the ultraviolet catastrophe, and thus opened the door to the development of quantum physics.

In the 1920s, we've seen how Schrödinger's equation was used to describe the probability associated with the position and energy of a particle. Dirac extended this to particles moving with a relativistic velocity, accounted for spin and predicted the existence of antimatter. He further extended this to the study of quantum oscillations of underlying universal fields.

By 1940, Feynman described how a photon or an electron, when travelling from place to place, do so over all possible paths simultaneously. The particle, when it arrives at its destination, does not have a definite history. It is the sum over all the possible histories that determines the probability of where the particle will go according to the sum of the probability amplitudes for each of the possible paths taken.

Feynman introduced his diagrams, then called "Feynman graphs", in a paper entitled "Space-Time Approach to Quantum Electrodynamics" published in *Physical Review*, Volume 76, in 1949. At the time, the diagrams were not well-received. Scientists of the day found them unfamiliar and complicated. At the time, problems with infinities with the traditional Hamiltonian approach had not yet been resolved, and these resurfaced again in the Feynman diagrams approach. Some 20 years later, when Feynman, Schwinger and Tomonaga established the process of renormalisation, they reached their full potential. It was a younger generation of physicists, largely led by Freeman Dyson, that led to their eventual popularity.

The Feynman theory of quantum electrodynamics is a particle-based theory. There are no explicit field interaction calculations, only the amplitudes involved with the motion and interaction of particles. However, we've seen that the form of the amplitudes, the vertex factors and the propagators have a connection to quantum field theory. The Feynman rules, the definitions and arrangements of vertices and propagators, can be derived from the Lagrangians of quantum field theory. The Feynman diagrams and associated rules are a short-hand version of quantum field theory.

In terms of fields, we can say that the vertices of a Feynman diagram show the interaction between the electric field of the electron and the electromagnetic field of the photon. The focus has gone from electromagnetic fields of Faraday and Maxwell, to the particles of Planck and Einstein and then to quantum fields where particles are, in effect, localised amplitudes in their corresponding field.

Although originally applied to electrons and photons, and then to the quantum fields of other fundamental particles, Feynman's sum over histories is now being used to model the evolution of the universe. In the normal computation of the action, we observe the initial state and the final state, and then compute the sum of the probability amplitudes for each of the infinite number of ways the initial state can reach the final state. As we know, the initial and final states are supposed to be fixed. However, for the case of the universe, we don't have an observable fixed "initial state" to start from. We do have a final state of sorts – the present. However, we can keep guessing different initial states and compute the integrals going backwards until the minimum action is arrived at. We might even take the process forward into the future and determine what might happen next.

We should realise that all these approaches to a description of reality are models. We use models to make accurate predictions as to what we might observe, and even then, what we observe is to some extent subjective. Models have been used for centuries to explain phenomena, some well-known models being the geocentric scheme of the solar system of Ptolemy and the heliocentric model of Copernicus.

Models are models and nothing more. Do the planets, for example, consult Kepler's laws before making their way around the sun? Electrons, for example, do not bother solving Schrödinger's equation when they wish to go from place to place, nor do they compute the Feynman sum over histories. It all happens "automatically", and the model we wish to use, whether it be Maxwell's equations, Schrödinger's equation, the Dirac equation or Feynman's sum are conveniences for us who wish to make predictions or explain phenomena. There may well be other interpretations or models as-yet unformulated which might be just as effective. After the success of Schrödinger and Dirac, who would have thought that yet another, and often more productive, point of view, the Feynman diagrams, would then arise? The success of a model is measured by its comparison to experimental data, and in that context, the theory of quantum electrodynamics is regarded as one of the most successful ever formulated.

We should always remind ourselves of Plato's theory of forms. We are but observers who never see the real nature of things, but only their shadow. Despite this, a model provides understanding and the ability to predict. It also provides us with a satisfying point of reference from which all things have come.

In this book, the journey from classical physics to Feynman diagrams has been presented in a series of steps that assume basic knowledge of classical physics as a starting point. We began our journey in about 1900. It might surprise the reader to know that we leave the story at about 1950. The rapid development in those 50 years belies the richness and imagination that underpins the very brief treatment given here. Now in the 21st century, with the application of quantum field theory and sum over histories to other fundamental particles and interactions, for the first time in history, we are on the brink of understanding how everything came to be, an extraordinary circumstance in the evolution of living creatures. We should not miss out. Most scientists, even those with a background in physics, may feel this new knowledge has become so arcane that it is forever out of reach. I hope that this book provides a gentle slope towards an understanding so that you, the reader, may be inspired to go deeper and further, and appreciate what has been accomplished so far.

Appendix

APPENDIX FOR CHAPTER 2

Euler's Formula

The power series expansion of e^x:

$$e^{ix} = 1 + ix + \frac{(ix)^2}{2!} + \frac{(ix)^3}{3!} \dots$$

$$= 1 + ix + i^2 \frac{x^2}{2!} + i^3 \frac{x^3}{3!} \dots$$

$$= 1 + ix - \frac{x^2}{2!} - i\frac{x^3}{3!} + \frac{x^4}{4!} + i\frac{x^5}{5!} + \dots$$

$$= \left(1 - \frac{x^2}{2!} + \frac{x^4}{4!} - \dots\right) + i\left(x - \frac{x^3}{3!} + \frac{x^5}{5!} - \dots\right)$$

But: $\sin x = x - \frac{x^3}{3!} + \frac{x^5}{5!} - \frac{x^7}{7!} + \dots$ and $\cos x = 1 - \frac{x^2}{2!} + \frac{x^4}{4!} - \frac{x^6}{6!} + \dots$

So:

$$e^{ix} = \cos x + i \sin x$$

Euler's formula.

APPENDIX FOR CHAPTER 8

Integration by Parts

$$\int \frac{x}{\sqrt{c^2 - x^2}} dx$$

$$u = x;\ du = dx$$

$$dv = \frac{1}{\sqrt{c^2 - x^2}} dx;\ v = \sin^{-1} \frac{x}{c}$$

$$= x\sin^{-1}\frac{x}{c} - \int \sin^{-1}\frac{x}{c}dx$$

$$= x\sin^{-1}\frac{x}{c} - c\int \frac{1}{c}\sin^{-1}\frac{x}{c}dx$$

$$= x\sin^{-1}\frac{x}{c} - c\left[\frac{x}{c}\sin^{-1}x/c + \sqrt{1-x^2/c^2}\right]_0^x$$

$$= x\sin^{-1}\frac{x}{c} - c\left[\frac{x}{c}\sin^{-1}x/c + \sqrt{1-x^2/c^2} - 1\right]$$

$$= x\sin^{-1}\frac{x}{c} - x\sin^{-1}x/c - c\sqrt{1-x^2/c^2} + c$$

$$= c\left(1-\sqrt{1-x^2/c^2}\right)$$

$$= \frac{c}{\sqrt{1-v^2/c^2}} - c$$

Binomial Theorem

The binomial theorem provides an expansion of a function of the form $(a+b)^n$ where n is a positive integer, as a finite sum.

$$(1+x)^n = 1 + nx + \frac{n(n-1)}{2!}x^2 + \ldots + \frac{n(n-1)\ldots(n-r+1)}{r!}x^r + \ldots + x^n$$

When n is not a positive integer, the expansion takes the form of an infinite series:

$$(1+x)^n = 1 + nx + \frac{n(n-1)}{2!}x^2 + \ldots + \frac{n(n-1)\ldots(n-r+1)}{r!}x^r + \ldots$$

When $x << 1$, only the first two terms are significant, and so: $(1-x)^{-1/2} \approx 1 + \frac{1}{2}x$

For the case of $n = ½$, then $(1-x)^{1/2} \approx 1 - 1/2\,x$

APPENDIX FOR CHAPTER 9

Commutators

$$\frac{d}{dt}\langle O\rangle = \int \left(\Psi^*\hat{O}\right)\frac{d}{dt}\Psi + \left(\hat{O}\Psi\right)\frac{d}{dt}\Psi^* dx$$

$$\hat{H}\Psi = i\hbar\frac{\partial}{\partial t}\Psi$$

$$\frac{d}{dt}\Psi = \frac{1}{i\hbar}\hat{H}\Psi; \ \frac{d}{dt}\Psi^* = -\frac{1}{i\hbar}\left(\hat{H}\Psi\right)^*$$

$$\frac{d}{dt}\langle O\rangle = \frac{1}{i\hbar}\int\left(\Psi^*\hat{O}\right)\hat{H}\Psi + -\hat{O}\Psi\left(\hat{H}\Psi\right)^* dx$$

$$= \frac{i}{\hbar}\int \hat{O}\Psi\left(\hat{H}\Psi\right)^* - \Psi^*\hat{O}\hat{H}\Psi dx$$

but $\int \hat{O}\Psi\left(\hat{H}\Psi\right)^* dx = \int \Psi^*\hat{H}\hat{O}\Psi dx$

therefore $\frac{d}{dt}\langle O\rangle = \frac{i}{\hbar}\int \Psi^*\hat{H}\hat{O}\Psi - \Psi^*\hat{O}\hat{H}\Psi dx$

$$= \frac{i}{\hbar}\int \Psi^*\left(\hat{H}\hat{O} - \hat{O}\hat{H}\right)\Psi dx$$

$$\frac{d}{dt}\langle O\rangle = \frac{d}{dt}\langle \Psi|\hat{O}|\Psi\rangle = \frac{i}{\hbar}\int \Psi^*\left[\hat{H},\hat{O}\right]\Psi dx = \frac{i}{\hbar}\left\langle \Psi\middle|\left[\hat{H},\hat{O}\right]\middle|\Psi\right\rangle$$

$$= \frac{i}{\hbar}\left\langle\left[\hat{H},\hat{O}\right]\right\rangle$$

APPENDIX FOR CHAPTER 10

The Gamma Matrices

Pauli–Dirac representation of the gamma matrices:

$$\gamma_0 = \begin{bmatrix} 1 & 0 & 0 & 0 \\ 0 & 1 & 0 & 0 \\ 0 & 0 & -1 & 0 \\ 0 & 0 & 0 & -1 \end{bmatrix} \quad \gamma_1 = \begin{bmatrix} 0 & 0 & 0 & 1 \\ 0 & 0 & 1 & 0 \\ 0 & -1 & 0 & 0 \\ -1 & 0 & 0 & 0 \end{bmatrix}$$

$$\gamma_2 = \begin{bmatrix} 0 & 0 & 0 & -i \\ 0 & 0 & i & 0 \\ 0 & i & 0 & 0 \\ -i & 0 & 0 & 0 \end{bmatrix} \quad \gamma_3 = \begin{bmatrix} 0 & 0 & 1 & 0 \\ 0 & 0 & 0 & -1 \\ -1 & 0 & 0 & 0 \\ 0 & 1 & 0 & 0 \end{bmatrix}$$

Abbreviated form:

$$\gamma_0 = \begin{bmatrix} I & 0 \\ 0 & -I \end{bmatrix} \quad \gamma_k = \begin{bmatrix} 0 & \sigma_k \\ -\sigma_k & 0 \end{bmatrix}$$

The σ_k matrices are the Pauli spin matrices:

$$\sigma_x = \begin{bmatrix} 0 & 1 \\ 1 & 0 \end{bmatrix} \quad \sigma_y = \begin{bmatrix} 0 & -i \\ i & 0 \end{bmatrix} \quad \sigma_z = \begin{bmatrix} 1 & 0 \\ 0 & -1 \end{bmatrix}$$

Properties of the gamma matrices:

$$(\gamma_0)^2 = 1$$

$$(\gamma_{1,2,3})^2 = -1$$

$$\gamma_o\gamma_k = -\gamma_k\gamma_o$$

$$\gamma_j\gamma_k = -\gamma_k\lambda_j \quad k \neq j$$

APPENDIX FOR CHAPTER 11

Probability Flow

$$\int_{x_1}^{x_2} \Psi \frac{\partial^2\Psi^*}{\partial x^2} dx + \int_{x_1}^{x_2} \Psi^* \frac{\partial^2\Psi}{\partial x^2} dx$$

$$u = \Psi; \; du = d\Psi$$

$$dv = \frac{\partial^2\Psi^*}{\partial x^2} dx; \; v = \frac{\partial\Psi^*}{\partial x}$$

$$\int udv = uv - \int vdu$$

$$\int_{x_1}^{x_2} \Psi \frac{\partial^2\Psi^*}{\partial x^2} dx = \Psi \frac{\partial\Psi^*}{\partial x} - \int_{x_1}^{x_2} \frac{\partial\Psi^*}{\partial x} d\Psi$$

$$= \Psi \frac{\partial\Psi^*}{\partial x} - \left[\Psi \frac{\partial\Psi^*}{\partial x}\right]_{x^1}^{x2}$$

and

$$u = \Psi^*; \; du = d\Psi^*$$

$$dv = \frac{\partial^2\Psi}{\partial x^2} dx; \; v = \frac{\partial\Psi}{\partial x}$$

$$\int udv = uv - \int v\,du$$

$$\int_{x_1}^{x_2} \Psi^* \frac{\partial^2\Psi}{\partial x^2} dx = \Psi^* \frac{\partial\Psi}{\partial x} - \int_{x_1}^{x_2} \frac{\partial\Psi}{\partial x} d\Psi^*$$

$$= \Psi^* \frac{\partial\Psi}{\partial x} - \left[\Psi^* \frac{\partial\Psi}{\partial x}\right]_{x^1}^{x2}$$

$$\int_{x_1}^{x_2} \Psi \frac{\partial^2 \Psi^*}{\partial x^2} + \Psi^* \frac{\partial^2 \Psi}{\partial x^2} dx = \Psi \frac{\partial \Psi^*}{\partial x} - \left[\Psi \frac{\partial \Psi^*}{\partial x} - \Psi^* \frac{\partial \Psi}{\partial x} \right]_{x^1}^{x_2} + \Psi^* \frac{\partial \Psi}{\partial x}$$

$$= \Psi \frac{\partial \Psi^*}{\partial x} + \Psi^* \frac{\partial \Psi}{\partial x} + \left[\Psi^* \frac{\partial \Psi}{\partial x} - \Psi \frac{\partial \Psi^*}{\partial x} \right]_{x^1}^{x_2}$$

$$= \frac{\partial \Psi^* \Psi}{\partial x} + \left[\Psi^* \frac{\partial \Psi}{\partial x} - \Psi \frac{\partial \Psi^*}{\partial x} \right]_{x^1}^{x_2}$$

$$= \left[\Psi^* \frac{\partial \Psi}{\partial x} - \Psi \frac{\partial \Psi^*}{\partial x} \right]_{x^1}^{x_2}$$

APPENDIX FOR CHAPTER 15

Maxwell's Equations

1. Gauss' law: (Electric charge)

$$q = \varepsilon_0 \oiint \mathbf{E} \cdot d\mathbf{S}$$

$$= \varepsilon_0 \iiint \nabla \cdot \mathbf{E} \, dV$$

$$q = \iiint \rho \, dV$$

$$\nabla \cdot \mathbf{E} = \frac{\rho}{\varepsilon_0}$$

2. Gauss' law: (Magnetism)

$$\oiint_S \mathbf{B} \cdot d\mathbf{S} = 0$$

$$\iiint \nabla \cdot \mathbf{B} \, dV = 0$$

$$\nabla \cdot \mathbf{B} = 0$$

3. Faraday's law:

$$\oint \mathbf{E} \cdot d\mathbf{l} = -\frac{\partial \Phi}{\partial t}$$

$$\iint (\nabla \times \mathbf{E}) \cdot d\mathbf{S} = -\frac{\partial}{\partial t} \iint \mathbf{B} \cdot d\mathbf{S}$$

$$\nabla \times \mathbf{E} = -\frac{\partial}{\partial t} \mathbf{B}$$

4. Ampere's law:

$$\oint \mathbf{B}\cdot d\mathbf{l} = \mu_0 I + \mu_0\varepsilon_0 \frac{\partial\varphi}{\partial t}$$

$$\frac{1}{\mu_o}\oint \mathbf{B}\cdot d\mathbf{l} = I + \varepsilon_0 \frac{\partial}{\partial t}\oiint \mathbf{E}\cdot d\mathbf{S}$$

$$= \frac{1}{\mu_o}\iint (\nabla\times\mathbf{B})\cdot d\mathbf{S}$$

$$\frac{1}{\mu_o}\iint (\nabla\times\mathbf{B})\cdot d\mathbf{S} = I + \varepsilon_0 \frac{\partial}{\partial t}\oiint \mathbf{E}\cdot d\mathbf{S}$$

$$= \oiint \mathbf{J}\cdot d\mathbf{S} + \varepsilon_0 \frac{\partial}{\partial t}\oiint \mathbf{E}\cdot d\mathbf{S}$$

$$\nabla\times\mathbf{B} = \mu_0\mathbf{J} + \mu_0\varepsilon_0 \frac{\partial}{\partial t}\mathbf{E}$$

Index

Printed in the United States
By Bookmasters